AF703392

Geographic Information Systems: Applications in Natural Resource Management

Geographic Information Systems: Applications in Natural Resource Management

Edited by Pedro Anderson

SYRAWOOD
PUBLISHING HOUSE
New York

Published by Syrawood Publishing House,
750 Third Avenue, 9th Floor,
New York, NY 10017, USA
www.syrawoodpublishinghouse.com

Geographic Information Systems: Applications in Natural Resource Management
Edited by Pedro Anderson

© 2023 Syrawood Publishing House

International Standard Book Number: 978-1-64740-401-7 (Hardback)

This book contains information obtained from authentic and highly regarded sources. Copyright for all individual chapters remain with the respective authors as indicated. All chapters are published with permission under the Creative Commons Attribution License or equivalent. A wide variety of references are listed. Permission and sources are indicated; for detailed attributions, please refer to the permissions page and list of contributors. Reasonable efforts have been made to publish reliable data and information, but the authors, editors and publisher cannot assume any responsibility for the validity of all materials or the consequences of their use.

Trademark Notice: Registered trademark of products or corporate names are used only for explanation and identification without intent to infringe.

Cataloging-in-publication Data

Geographic information systems : applications in natural resource management / edited by Pedro Anderson.
p. cm.
Includes bibliographical references and index.
ISBN 978-1-64740-401-7
1. Geographic information systems. 2. Natural resources--Data processing. 3. Information storage and retrieval systems--Geography. I. Anderson, Pedro.
G70.212 .G46 2023
910.285--dc23

TABLE OF CONTENTS

PREFACE

A geographic information system (GIS) refers to a type of computer based information system that displays and analyzes geographically referenced information. It makes use of information that is related to a particular geographical location. GIS can display a wide range of data on a single map such as plants, streets and buildings. GIS can be utilized for various purposes including resource management, environmental impact assessment, historical research, scientific research, cartography, marketing, asset management, urban planning, logistics, criminology, and sales. The primary application of GIS in natural resource management is to manage environmental problems and disasters such as soil erosion, earthquake, flood, drought, and landslide. It addresses major issues such as habitat loss, pollution, climate change, and population increase. The use of GIS in natural resource management also entails data modeling and analysis. This book outlines the applications of geographic information systems in natural resource management. It will serve as a reference to a broad spectrum of readers.

Various studies have approached the subject by analyzing it with a single perspective, but the present book provides diverse methodologies and techniques to address this field. This book contains theories and applications needed for understanding the subject from different perspectives. The aim is to keep the readers informed about the progresses in the field; therefore, the contributions were carefully examined to compile novel researches by specialists from across the globe.

Indeed, the job of the editor is the most crucial and challenging in compiling all chapters into a single book. In the end, I would extend my sincere thanks to the chapter authors for their profound work. I am also thankful for the support provided by my family and colleagues during the compilation of this book.

Editor

1

A Hybrid GIS and AHP Approach for Modelling Actual and Future Forest Fire Risk under Climate Change Accounting Water Resources Attenuation Role

Gianluigi Busico [1,*], Elisabetta Giuditta [1], Nerantzis Kazakis [2] and Nicolò Colombani [3]

[1] DiSTABiF—Department of Environmental, Biological and Pharmaceutical Sciences and Technologies, Campania University "Luigi Vanvitelli", Via Vivaldi 43, 81100 Caserta, Italy; elisabetta.giuditta@unicampania.it

[2] Department of Geology, Laboratory of Engineering Geology & Hydrogeology, Aristotle University of Thessaloniki, 54124 Thessaloniki, Greece; kazakis@geo.auth.gr

[3] Department of Materials, Environmental Sciences and Urban Planning, Polytechnic University of Marche, Via Brecce Bianche 12, 60131 Ancona, Italy; n.colombani@univpm.it

* Correspondence: gianluigi.busico@unicampania.it

Abstract: Forest wildfires usually occur due to natural processes such as lightning and volcanic eruptions, but at the same time they are also an effect of uncontrolled and illegal anthropogenic activities. Different factors can influence forest wildfires, like the type of vegetation, morphology, climate, and proximity to human activities. A precise evaluation of forest fire issues and of the countermeasures needed to limit their impact could be satisfactory especially when forest fire risk (FFR) mapping is available. Here, we proposed an FFR evaluation methodology based on Geographic Information System (GIS) and the analytic hierarchy process (AHP). The study area is the Campania region (Southern Italy) that, for the last 30 years, has been affected by numerous wildfires. The proposed methodology analyzed 12 factors, and AHP was used for weight assignment, offering a new approach to some parameters. The method divided the study area into five risk classes, from very low to very high. Validation with fire alerts showed a good correlation between observed and predicted fires (0.79 R^2). Analyzing the climate projections, a future FFR for 2040 was also assessed. The proposed methodology represents a reliable screening tool to identify areas under forest fire risk, and can help authorities to direct preventive actions.

Keywords: forest fire risk; GIS; climate change; water resources; groundwater

---

1. Introduction

Fires can be considered as an abiotic and natural ecological factor whose characteristics are different according to the various terrestrial ecosystems. Wildfires represent a globally important and a fundamental process of many ecosystems, such as the Mediterranean area, playing a key role in ecosystem dynamics, and in the retention of species that have evolved in response to fire [1]. Most wildfires are natural processes, triggered mainly by lightning, more rarely by volcanic eruptions and spontaneous combustion, and represent one of the primary ecological change agents across the forests [2]. However, forest fires are also triggered by anthropogenic factors and several causes of anthropic wildfires have been identified, such as: (i) culpably, namely related to carelessness or negligence; (ii) arson, associated with the will of individual to obtain personal benefits, and (iii) unknown, not easily linkable to a specific cause [3]. The human influence on forest fires is also well documented in studies of contemporary fire patterns, where fire management activities and land use changes have been implicated [4], as well as the role of humans in changing the pattern, season, and frequency of fires through human-caused ignitions [4,5]. Growing fire problems, especially in forests, has prompted investigation into alternative approaches for preventing and reducing forest fire risk (FFR), such as land use planning, the creation of areas of defensible space adjacent to structures in fire-prone areas [3,6], and prescribed burning [7–9]. However, it is important to highlight how these preventive techniques are meant to be applied in natural fire-prone areas, so it is necessary to screen accurately the forest area subject to risk of fire [10,11]. Therefore, to be able to accurately intervene in cases of FFR prevention, it is mandatory to know the correct location of areas subject to risk of fire. It is possible to provide an FFR map and thereby to minimize the fire frequency and eventually avert damages [11,12]. A precise evaluation of forest fire problems and of the countermeasures to be applied can be satisfactory when FFR mapping is available [11]. The knowledge of the spatial distribution of areas under fire risk has become necessary to improve fire prevention strategies and tactics [13]. In this context, geographical information system-based (GIS-based) FFR assessment could represent a powerful tool [14]. Different approaches have been proposed for the modelling and assessment of FFR and behavior. The most applied are physics-based method like FIRETEC [15] and LANDIS-II [16], and statistical methods. The latter are often combined with GIS tools [17,18] and include different methodologies, such as multiple linear and logistic regression [11,19,20], fuzzy logic, and weight of evidence [21] among others. Each one of these methodologies suffers from some problems, e.g., physical methods can simulate potential fire behavior through a set of mathematical equations, but are more suitable for small areas, because they require very detailed information. Statistical methods, instead, can be applied to large areas and are easier compared to the other methods, but in some cases, due to the lack of linearity of the fire regime, erroneous results could occur. To eliminate these drawbacks, it is essential to use different types of factors related to fire ignition and behavior with the same spatial resolutions. On the other hand, the use of spatial data with different ages, scales, and resolutions can bring uncertainties and imprecisions in the FFR assessment [22–24]. In many countries, such as China [25], Spain [26], France [27], Brazil [18], Portugal, Corsica, and Greece [28,29], these tools have been successfully used and incorporated in forest management plans. Also, for the Italian peninsula, some studies have been realized. For instance, Semerato et al. [30] assessed the vulnerability to fire during the summer period in the Apulian region (South Italy) using a fuzzy logic system, and Poldini et al. [31] analyzed the fire damage to different types of vegetation, integrating phytosociological maps of vegetation, geomorphology, and climate.

1.1. *Factors Influencing Fire Ignition and Behavior*

There are many factors influencing fire occurrence in forests in terms of fire ignition and fire behavior, including climate, vegetation (or land cover), topography, human activity [32], and soil texture [11,33]. Climate is surely one of the main agents responsible for fire ignition, while mean annual temperature and precipitation were commonly used as climatic variables for fire regimes as they are the main parameters that control fuel moisture content and the general characteristics of climate conditions [32,34]. Indeed, an important variation of air temperature can affect the severity and frequency of forest fire and, together with the modification of soil moisture induced by rainfall variation [35], can further change the fire behavior. Climate projections for future scenarios have shown a worldwide temperature increase and a higher frequency of extreme climate events, e.g., rainstorms, which could further aggravate the current FFR due to lighting. Indeed, for certain regions, extreme climate events have shown a great impact on forest fire activity [36]. The type of vegetation, as land cover type, has been shown to be especially relevant as a factor influencing fire ignition, together with climate variables [37]. Vegetation types differ in fuel loads and flammability as well as on fuel continuity [38,39]. Topography, in terms of elevation, slope, and aspect, directly influences vegetation composition and fuel structure and often determines where and why fires will occur and spread [32,40]. Topography is an important factor that influences both fire ignition and behavior. Indeed, slope is the primary factor for fire progression because it will define the fire velocity. The fire will travel faster if the slope is steeper [41]. Thus, elevation is the main parameter that influences precipitation, temperature, humidity, and evapotranspiration [42]. Regarding aspect, it influences soil moisture and wind speed, which are strongly linked to fire behavior [43]. Moreover, human activities can directly affect fires in terms of ignition or suppression [44,45], and can also indirectly induce changes in fire occurrence by modifying the spatial pattern of vegetation [46]. The latter is also correlated to the spatial distribution and density of fires too [47].

1.2. *Aim of the Research*

In this study, a hybrid FFR methodology was produced whereby factor weights were assigned with the analytic hierarchy process (AHP) to 12 chosen parameters and validated using the forest fire alarm of the moderate resolution imaging spectroradiometer (MODIS) database for 2018–2019. The methodology has the object to assess the global FFR, involving in the evaluation 12 parameters responsible for fire triggering and behavior. The use of AHP can overcome the problem of subjectivity in weight assignment. Contrary to the other FFR applications performed around the world, here, a new way to consider the risk associated with the distance to water courses was proposed for the FFR assessment and was then compared with previous literature assumptions. The distance from water courses has been considered an attenuating factor for FFR, reducing the probability of fire ignition and modifying fire behavior. Moreover, among the chosen parameters, climatic factors like temperature and precipitation are mainly subject to future changes. So, using climate projections for temperature and precipitation in the Mediterranean region, an FFR assessment was also provided for 2050.

2. Study Area

The study area is the Campania region (Figure 1) in the Southern Italian Peninsula, with a total area of 13.595 km^2. The main regional boundaries are the Central Apennine Chain to the West and the Tyrrhenian Sea to the East. Morphologically (Figure 1b), the Apennine Mountains domain the region, with a maximum height of 1880 m above sea level (a.s.l.), and by two large flat coastal plains, the Campana and Sele plains. These two plains also host the main rivers and canals and are characterized by large land reclamation works. This region has a typical Mediterranean climate with a cold winter and dry summer. Wet summers characterize the coastline while severe winters are specific to the mountainous area. Large differences of temperature occur between the mountainous and coastal zones. Indeed, the maximum and minimum temperatures can vary 6–8 °C between the mountain peaks and the coastal plains [48,49]. The mean annual rainfalls range from 800 to 2000 mm depending on the area and show clear differences in terms of elevation and proximity to the sea. The highest precipitation amounts occur in the Apennine chain, with values of 1200–1500 mm/year, up to 2000 mm/year. Lower values are registered in the coastal plains, with values around 800 mm/year [50]. From a geological point of view, the region is mainly made up of sedimentary and volcanic rocks. With respect to the land cover (Figure 1c), agricultural fields cover more than 50% of the territory and are mainly located in flat areas, followed by forests which occupy around 30%, covering 4800 km^2, and urban settlements. Regarding the forest species, the coastline and lower altitudes are characterized by Mediterranean shrub, where oak holms are the predominant species, as well as pine plantations. Following the increase in altitude, oak tree, mixed forest, and beeches can be found, in sequence. The Campania region has been subject to about 900 forest fires per year between 1990 and 2017, and most of these fires were recorded in the summer period from June to September [51]. A particularly dramatic year for the number of forest fires was 2017, with 1199 fires which burned 17,694 hectares of forests, the highest total yearly number of hectares of forest burned since 2007 [52]. Moreover, in the last two years (2018–2019), data from Moderate Resolution Imaging Spectroradiometer (MODIS) Active Fire Data [53] has shown over 1400 fire alerts only in the Campania region (Figure S1). For this reason, attention was again focused on forest fires in the Campania region in terms of actions to prevent FFR. In addition, the Campania region hosts 3400 km^2 of sites under special protection (Natura 2000) and 90% of these are occupied by forests. For example, three protected areas in the Campania region that are dominated by pine plantations and have been classified as highly vulnerable to fire are: Castel Volturno Nature Reserve, Cilento, Vallo di Diano e Alburni National Park, and Riserva Statale Tirone, Alto Vesuvio in the Vesuvio National Park. In these areas, several forest fire prevention measures have been carried out from the early 1990s to 2015 in order to reduce FFR, especially during the summer months.

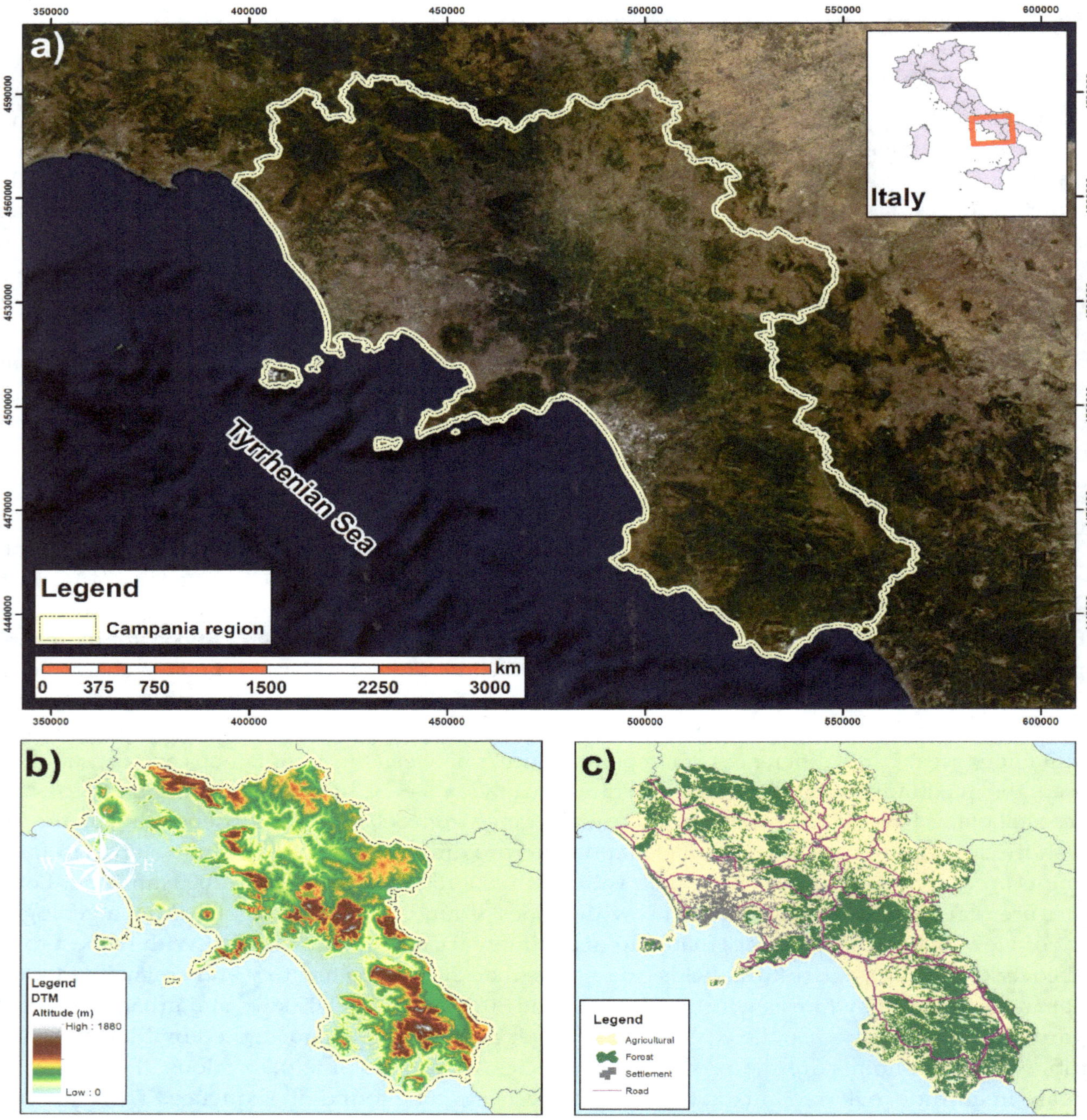

Figure 1. Overview of the study area: (**a**) area localization, (**b**) morphology and (**c**) land use classification.

3. Material and Methods

3.1. Data Collection

A complete geo-database has been built collecting data from local authorities to perform the FFR assessment. Data were collected in shapefile format and then converted into a raster image using "topo to raster" function tool in ArcGIS 10.2 with a cell resolution of 20 × 20 m. A complete list of the collected data and the respective sources is shown in Table 1. The morphological factors including slope degree, slope aspect, and elevation were all produced starting from a digital terrain model (DTM) with a resolution of 20 m. Climatic data for the realization of temperature and precipitation maps were obtained by analyzing 14 meteorological stations located in the Campania region, and these are available online from the Ministry of Agriculture website [54]. The mean value of precipitation and temperature, calculated for the two periods 1990–2018 and 2030–2050, were used for the actual and future assessments. In this case, the raster was produced interpolating the punctual data using ordinary kriging technique starting from point measurement and retaining always a resolution of 20×20 m. In this work, the distance from urban infrastructure, roads, springs, water courses, and agricultural fields was also considered. Euclidean distance was calculated in ArcGIS 10.2, producing a raster file reflecting buffer zones of 100, 200, 300, 400, and over 500 m, and then classified. Forest type and vegetation cover, representing the fuel material, were recorded using Corine Land Cover (CLC) classification of 2018, and classified according to six forest classes. The location of 118 springs were taken from the online dataset of ARPAC [55] together with the shape file of the main perennial water course (Campania's main rivers). Finally, soil textures were extracted analyzing the soils classification from "I Sistemi di Terre della Campania" [56].

Table 1. List of collected data for the generation of the Forest Fire Risk (FFR) parameters.

Data	Parameters	Source
DTM	Slope	Advance Spaceborne Thermal Emission and Reflection Radiometer (ASTER) [57]
	Aspect	
	Elevation	
CLC 2018	Distance from settlement	Copernicus, Land Monitoring Service [58]
	Distance from agricultural	
	Forest Type	
Meteorological data	Precipitation 2018	MIPAAF. Agrometeorological Online Database 2018 [54]
	Temperature 2018	
CORDEX Database	Temperature 2050	EURO-CORDEX 2019 [59]
	Precipitation 2050	
MODIS	Fire alerts	Fire Global Forest Watch [53]
Hydrographic network	Distance from river	ARPAC: Regional environmental protection agency [55]
Springs location	Distance from springs	ARPAC: Regional environmental protection agency [55]
Soil map	Soil texture	Di Gennaro. [56]

3.2. *Assignment of the Parameters' Weight*

With the help of a weighted hierarchical analysis (Analytical Hierarchy Process—AHP) method developed by Saaty [60], a comprehensive measurement of each factor was determined by giving priority to, and classifying the variables in a pairwise comparison scale, and then defining a linear hierarchy of importance among the factors (precipitation (PCP), temperature (TMP), altitude (ELV), slope (SLP), aspect (ASP), soil (SOL), forest type (FRS), distance from roads (ROD), distance from settlements (URB), distance from agricultural fields (AGR), distance from water courses (WTC), and distance from springs (SPR)). The parameters are ranked according to their relative importance, and hence their weights can be determined. In order to provide a variable hierarchy, a preliminary weight has been attributed to each factor based on their different impact on FFR. The importance values assigned to a factor strictly interfere with the outcome, so this is considered the most important step in the entire process. Initially, the pairwise comparison was obtained in a 12×12 matrix in which the diagonal elements are equal to one. The values of each row correspond to the importance between two parameters. The preliminary weight assignment to define the relative importance among the parameters were decided consulting the available literature. The main assumption was to give a higher weight to the bio-physical factors (forest type, slope, aspect, temperature, precipitation), according with Amalina et al. [14] and Pouthagy et al. [11] who analyzed the influence of variables on FFR using different data mining techniques, and stated how rainfall, temperature, and land cover were the most important parameters, followed by slope and anthropogenic factors. Wu et al. [32] similarly clarified that fire regimes at the regional scale are strongly linked, mainly to climate, vegetation, topography, and then human activities. Relative weights of each parameter have been assigned considering these findings. Hence, TMP is equally important as PCP, however, it is much more important than SOL and ELV. The lower impact of elevation is because its effect is considered yet inside the climatic and vegetation factors. After the pairwise comparison, the values were normalized, and the weights of each parameter were estimated. The consistency of the application was checked using the consistency ratio (CR), according to the following equation:

$$\mathrm{CR} = \mathrm{CI}/\mathrm{RI} \quad (1)$$

where RI is the random index and CI is the consistency index. The consistency ratio has been calculated CR = 0.09, which is lower than the threshold (0.1), and hence the consistency of the weights is affirmed. In Saaty [60], details for the analytical hierarchy process application can be found. The pairwise comparison of the criteria significance and the weights are shown in Table 2.

Table 2. Pairwise comparison of the preparatory factors.

Parameter	TMP	PCP	SLP	FRS	URB	AGR	SPR	ROD	WTC	ASP	ELV	SOL	Weight
TMP	1	1	2	2	3	3	4	4	4	4	5	5	1.88
PCP	1	1	2	2	3	3	4	4	4	4	5	5	1.88
SLP	0.5	0.5	1	1	2	2	3	3	3	3	4	4	1.22
FRS	0.5	0.5	1	1	2	2	3	3	3	3	4	4	1.22
URB	0.33	0.33	0.5	0.5	1	1	2	2	2	2	3	3	0.75
AGR	0.33	0.33	0.5	0.5	1	1	2	2	2	2	3	3	0.75
SPR	0.25	0.25	0.33	0.33	0.5	0.5	1	1	1	1	2	2	0.44
ROD	0.25	0.25	0.33	0.33	0.5	0.5	1	1	1	1	2	2	0.44
WTC	0.25	0.25	0.33	0.33	0.5	0.5	1	1	1	1	2	2	0.44
ASP	0.25	0.25	0.33	0.33	0.5	0.5	1	1	1	1	2	2	0.44
ELV	0.2	0.2	0.25	0.25	0.33	0.33	0.5	0.5	0.5	0.5	1	1	0.27
SOL	0.2	0.2	0.25	0.25	0.33	0.33	0.5	0.5	0.5	0.5	1	1	0.27

3.3. Sensitivity Analysis

The proposed FFR methodology was further validated using a single parameter sensitivity analysis (SA) [61]. SA supports the researcher to analyze the significance of subjectivity components and produces useful information on the importance of weighting values assigned to the corresponding parameters. SA has been widely used and proposed to validate index and rating methods in different fields of study, including FFR assessment [62–66]. The effective weighting was calculated using the following equation:

$$W = \left(\frac{PrPw}{V}\right) \times 100 \quad (2)$$

where W is the effective weighting of each parameter, Pr is the rating value (single parameters map), Pw is the weighting value assigned with AHP, and V is the overall value of the applied index (final FFR map). The SA was applied for 2018 and 2050 analysis to investigate the changing weight of the single parameters while retaining the same Pw.

3.4. Forest Fire Risk Assessment

In this study, different parameters have been considered as the main factors controlling FFR. Among these, morphological characteristics of the study area, are the main FFR's parameters controlling fire behavior and ignition [10,40,67,68]. The impacts of slope are mainly related to the fire behavior, while aspect and elevation mainly affect ignition. Climate, instead, is considered a key cause of fire triggering [34,69]. Temperature and precipitation are mainly responsible of fire ignition. Another parameter responsible for fire ignition is land cover, which represents the primary fuel material [70]. Together with the aforementioned parameters, human activities and anthropogenic infrastructures can have a big influence on FFR too. In the anthropogenic factor, all the human activities that can trigger fires are taken into account, like roads and settlements [43]. An increased risk was associated with a decreased distance from anthropogenic factors, because easily accessible areas are characterized by a high risk for both criminals and unintentionally triggered fires. Distance from water courses and springs are other common parameters utilized for the assessment of FFR [71]. They are usually classified in the anthropogenic factors because human activity, like the presence of camps near riversides and springs, can affect the occurrence of forest fires. So, usually, the risk connected to water courses and bodies is considered higher as the fire approaches the water system. In this work, however, the presence of water is seen as a mitigation factor, since surface waters can influence the soil's moisture, thereby reducing the risk of fire. Thus, in this work, the risk has been considered inversely proportional to the distance from the water courses (rivers) and springs. Finally, soil texture was also considered an indicator of fire behavior [35]. The FFR empirical model was developed according to Equation (3), and it is based on the statistical weights for each variable calculated with AHP.

$$\begin{aligned}FFR = 0.27 \times ELV + 1.22 \times SLP + 0.44 \times ASP + 1.88 \times PCP + 1.22 \times FRS + 1.88 \times TMP +\\ 0.44 \times ROD + 0.75 \times AGR + 0.75 \times URB + 0.44 \times WTC + S0.44 \times PR + 0.27 \times SOL\end{aligned} \quad (3)$$

where ELV is elevation, SLP is slope, ASP is aspect, PCP is precipitation, FRS is forest type, TMP is temperature, ROD is distance from main road, AGR is distance from agricultural field, URB is distance from settlement, WTC is distance from main water course, SPR is distance from springs, and SOL is soil texture. Each parameter has been classified into five risk classes, namely very high, high, medium, low, and very low. Classes and ratings are shown in Table 3 and a general conceptual model is shown in Figure 2. Elevation and slope were classified using equal interval according to site characteristics, for example, considering elevation, there was a general tendency of higher fire ignition risk at lower elevations [72]. Aspect, instead, was classified according to current literature [18,25,73] considering common patterns of a lower ignition risk in northern aspects. The land cover was classified attributing a higher risk to grassland and conifers, together with pine plantation, medium risk to broad leafed forest, and lower risk to mixed forest according with You et al. [25]. The climatic factor was classified using equal intervals according to site characteristics, following the general assumption that a lower risk is connected to lower temperatures and higher precipitation. Meanwhile, a higher risk was related to higher temperatures and lower precipitation.

Table 3. System indicators with classes and ratings for the assessment and mapping of FFR.

Precipitation (mm)		Temperature (°C)		Forest Type	
Classes	**Risk**	**Classes**	**Risk**	**Classes**	**Risk**
<800	Very High	<13	Very Low	Grassland, sclerophyll	Very High
800–900	High	13–14	Low	Conifer	High
900–1000	Medium	14–15.5	Medium	Broad leafed forest	Medium
1000–1100	Low	15.5–17	High	Transition forest	Low
>1100	Very Low	>17	Very High	Mixed forest	Very Low
Distance from Springs (m)		**Distance from Water (m)**		**Soil Texture**	
Classes	**Risk**	**Classes**	**Risk**	**Classes**	**Risk**
<100	Very Low	<100	Very Low	Clay	Very Low
100–200	Low	100–200	Low	Silty, Silty clay	Low
200–300	Medium	200–300	Medium	Loamy	Medium
300–400	High	300–400	High	Loamy sand	High
>400	Very High	>400	Very High	Sand	Very High
Elevation (m)		**Slope (°)**		**Aspect**	
Classes	**Risk**	**Classes**	**Risk**	**Classes**	**Risk**
<300	Very High	<5	Very Low	0–70	Very Low
300–600	High	5–25	Low	70–140	Low
600–900	Medium	25–35	Medium	140–210	Very high
900–1200	Low	35–45	High	210–280	High
>1200	Very Low	>45	Very High	>280	Medium
Distance from Agricultural (m)		**Distance from Urban Settlements (m)**		**Distance from the Main Road (m)**	
Classes	**Risk**	**Classes**	**Risk**	**Classes**	**Risk**
<100	Very High	<100	Very High	<100	Very High
100–200	High	100–200	High	100–200	High
200–300	Medium	200–300	Medium	200–300	Medium
300–400	Low	300–400	Low	300–400	Low
>400	Very Low	>400	Very Low	>400	Very Low

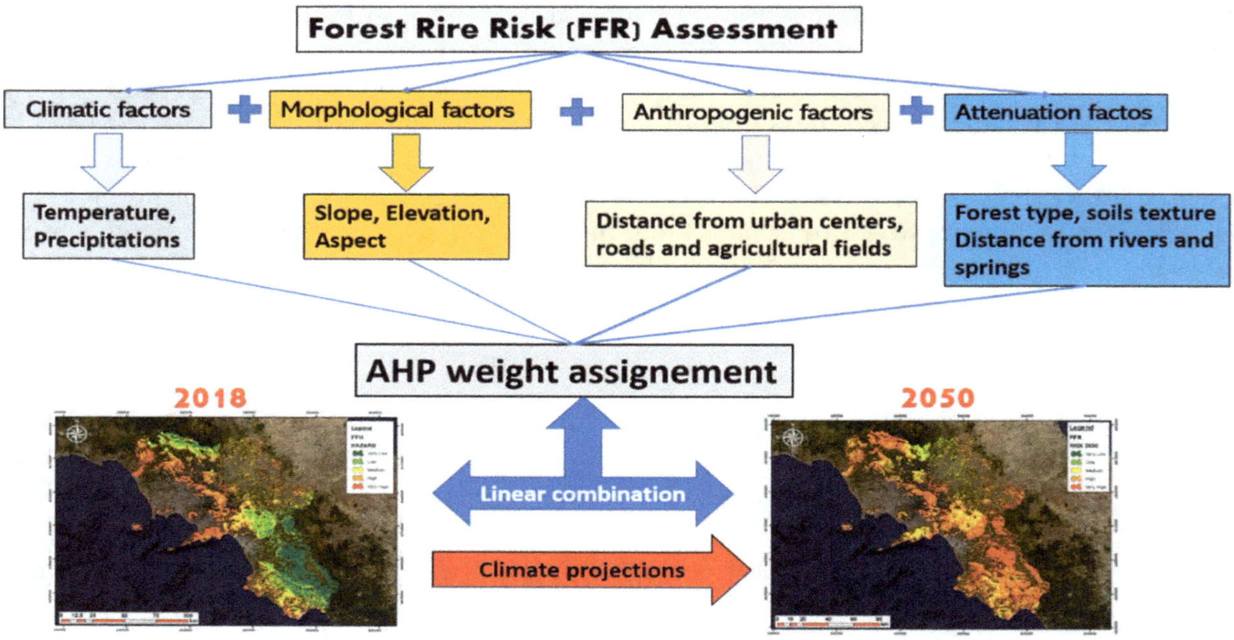

Figure 2. Conceptual model of FFR application.

3.5. Climate Predictions

In Italy, climate parameters like temperature and precipitation have been and will remain subject to severe changes in the coming years [74]. To verify the impact of future climate change on FFR, temperatures and precipitation, predictions from the EURO-CORDEX [59] database have been taken into consideration. The data was downloaded using ICHEC-EC_EARTH system model under EUR 11 domain. The reference period was 2030–2050 under RCP4.5 scenario, which is the most likely scenario used in predictive models. Twenty grid points, chosen inside the region (Figure S1), were used to obtain the climate projection of temperature and precipitation. The data were obtained in monthly average values for both precipitation and temperature for the chosen period. Generally, for the Mediterranean basin, a general average increase of 1.5–2.0 °C of temperature has been predicted, with markedly high temperatures during the summer period [75,76]. Regarding precipitation, a general decrease is also forecast with a markedly seasonal regime, i.e., a large decrease of precipitations during the summer period and a slight increase during the winter period [77]. Moreover, an increase of extreme events, like intense storms or extended drought periods will be likely to happen and these factors could negatively influence FFR [78]. In Figure 3, the actual (1990–2018) and future (2030–2050) climate projections for the Campania region are presented. In Figure 3a,b, it is possible to appreciate how the amount of mean precipitation will decrease from a minimum of 700 mm/year to 530 mm/year, and from a maximum of 1200 to 930 mm/year. The spatial distribution suggests an unchanged situation for Campania's plains, wherein the amount of precipitation will remain stable at around 800–900 mm/year. Menawhile, in the mountains, the precipitation will become less abundant. Regarding the temperature, minor changes will happen, recording a variation of ±1–2 °C. Nevertheless, the temperature distribution will experience large changes. In fact, the mountainous zone will record higher mean temperatures >16 °C, where actual temperatures are between 13 and 15 °C. Higher temperatures and less abundant precipitation can generate favorable conditions for fire triggering, and in areas normally characterized by a low fire risk.

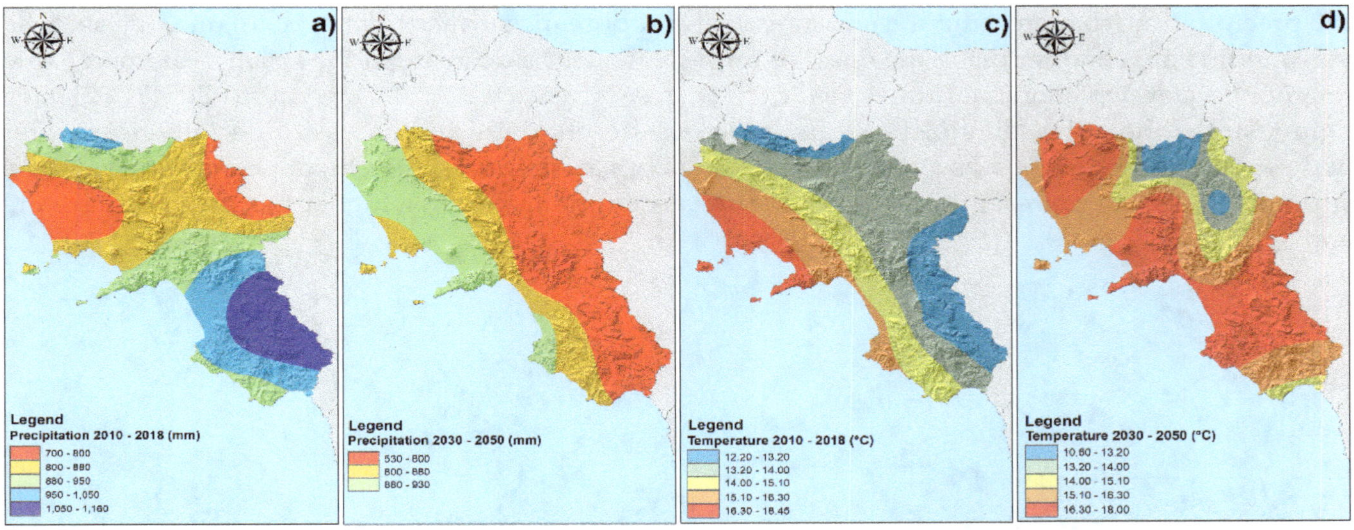

Figure 3. Mean precipitations and temperatures distribution maps for the Campania region, from EURO-CORDEX database (2019). Panel (**a**) shows the 2010–2018 precipitations, panel (**c**) the 2010–2018 temperatures, panel (**b**) the 2030–2050 projection for precipitations and panel (**d**) the 2030–2050 projection for temperatures.

4. Results

4.1. FFR Assessment for 2018

The FFR map for 2018 was produced following Equation (3) using the map algebra function in ArcMap 10.2 and applied only on the forest territory. The initial importance ranking of all parameter, useful for weight assignment was done taking in consideration the current bibliography. In agreement with You et al. [25], Eskandari [79] and Eugenio et al. [18] higher impacts have been attributed to morphological and climatic factor followed by forest type and anthropogenic factors. The maps for each parameter are shown in Figure 4. The final map (Figure 5) was classified by dividing the final index into five classes of risk, as follows: very low (<22), low (23–26), medium (26–28), high (28–31), and very high (>31) using geometrical intervals. The study area results were equally divided between the five classes. The very high class occupies 21% of the area and is mainly located near the Tyrrhenian Sea towards West (Figure 5a,b) and in the Apennine mountains in the center of the region (Figure 5c). Those areas are characterized by high temperatures and less abundant precipitation, which causes both vegetation and soil litter to dry, creating the conditions for a higher susceptibility to fire. Looking at the temperature and precipitation maps, those areas are marked by a very high risk for both parameters (Figure 4). Furthermore, here, the vegetation is represented by Mediterranean shrubs and maritime pines plantations, which are species highly susceptible to fire, and the presence of a sandy-loamy soil textures can further increase fire propagation since this kind of soil easily lose its moisture [80]. These parts of the Campania region are also the ones with a higher population density, agricultural fields, and urban centers, where human activities connected to roads, agricultural practices, or illegal burning can further increase the FFR. Towards the East, following the higher altitudes, the FFR decreases as temperatures decrease and the extent of precipitation increases. Here, altitude also influences the vegetation types that are less prone to fire triggering. The presence of perennial rivers and springs also contributes to decreasing FFR in these areas. Contrary to Hong et al. [21,73], You et al. [25], and Amalina et al. [14], according to whom a higher risk was associated with a minor distance to water courses and springs, in this work, the opposite scenario has been taken in consideration. Usually, the distance from rivers and springs belongs to the anthropogenic factor and it is thought to negatively influence the occurrence of fire, since the existence of camping sites and holoiday spots where human presence in their proximity could increase the FFR. In this application, the presence of water courses and springs is considered an attenuating factor. Such a factor can directly contribute, together with the precipitation, to increasing the soil moisture and phreatic surface (water table), making wet the surrounding area and reducing the onset of fires. Obviously, a decrease in the springs' discharge due to groundwater depletion is a further factor which may increase the FFR. Different studies [81,82] have demonstrated how the soil moisture can reduce the depth of fire and its risk, especially near water bodies. The methodology was applied considering both assumptions: (i) a higher risk near rivers and springs (ATR), that represent the main literature assumption, and (ii) a higher risk away from rivers and springs (AFR).

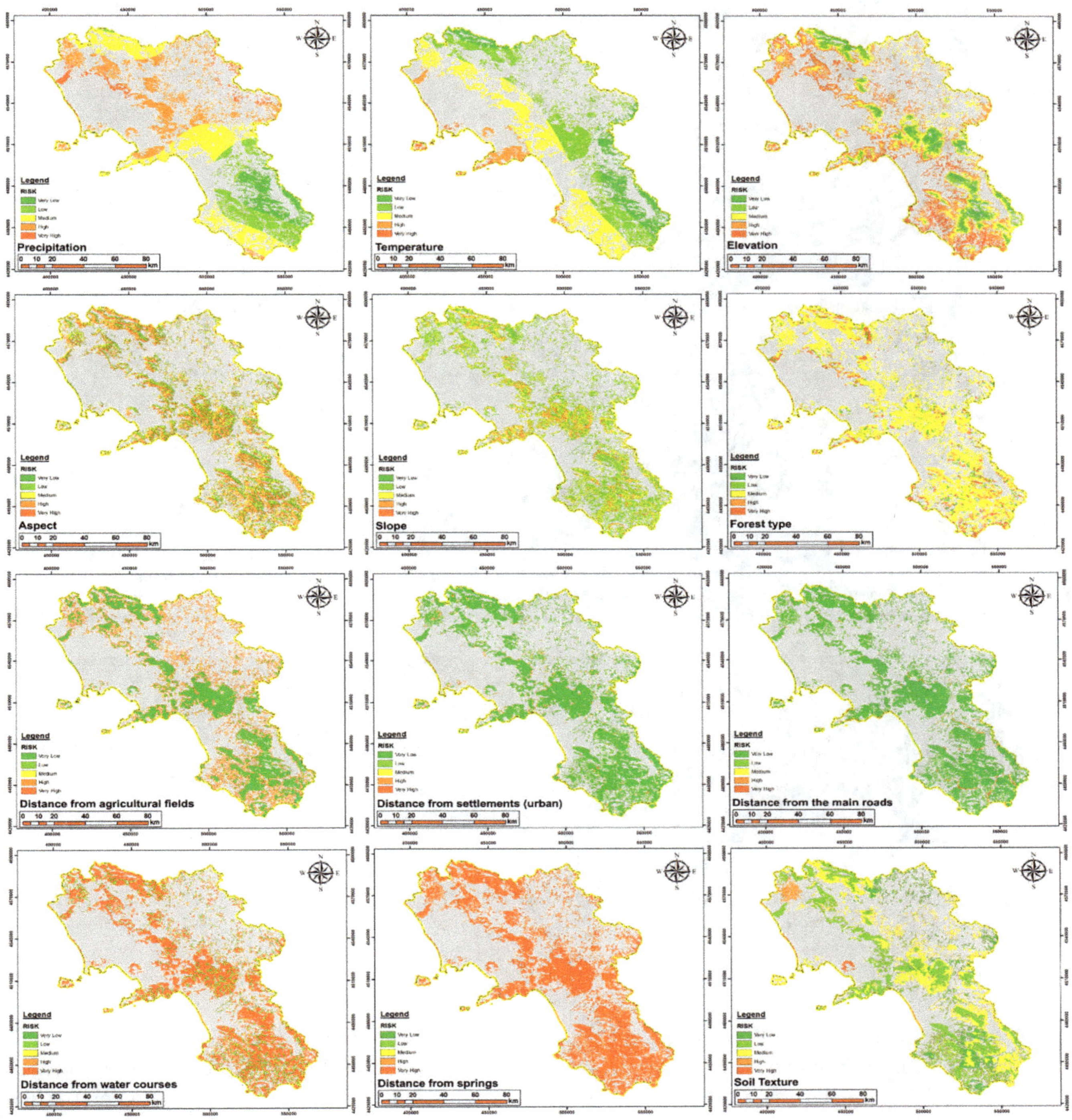

Figure 4. Risk map for all the 12 parameters involved in FFR assessment.

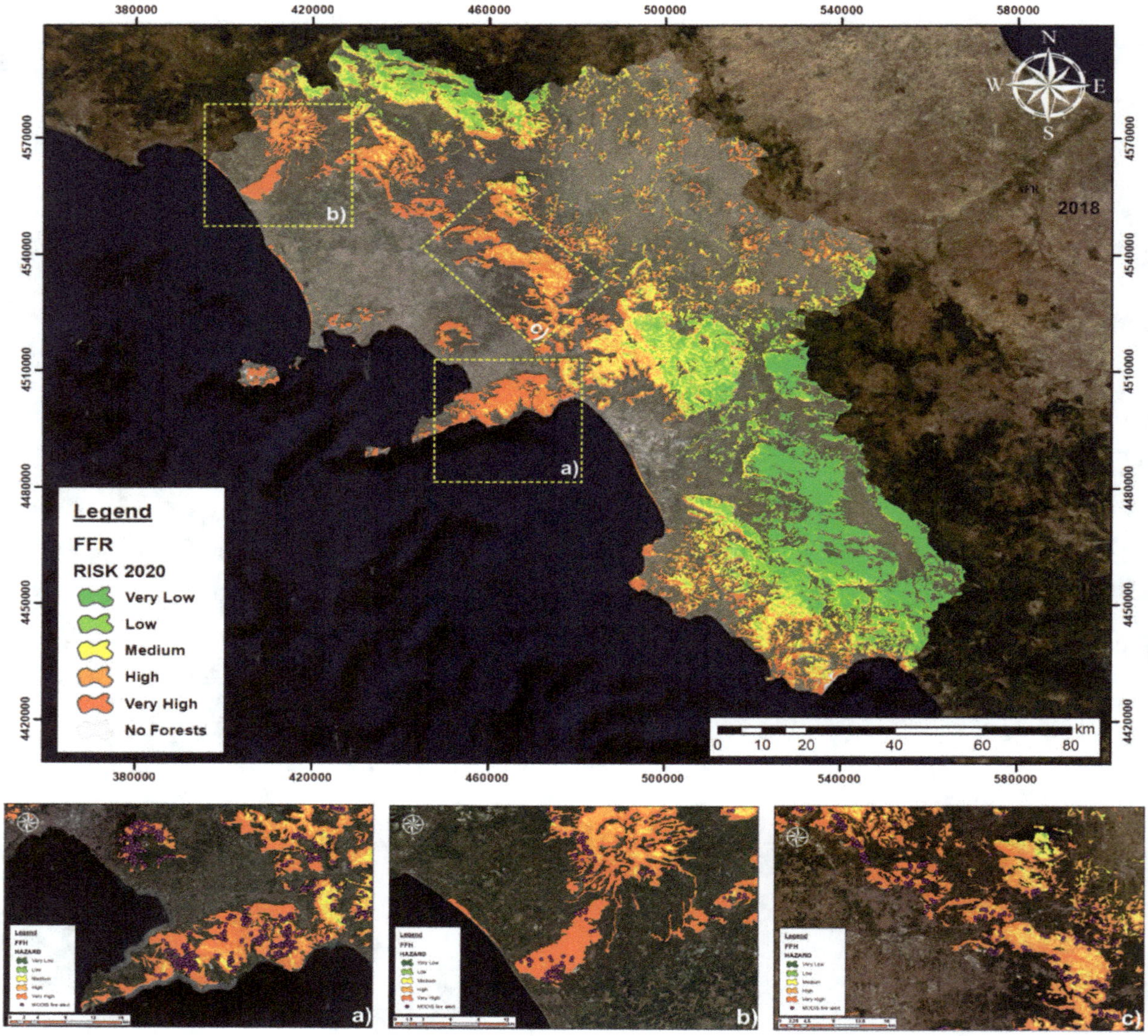

Figure 5. FFR map of the Campania region for the year 2018. Figure a, b, and c represents a zoom on those areas characterized by higher FFR.

4.2. FFR Validation

After the assessment of the proposed FFR, we proceed with the validation of the methodology, which was possible thanks to the fire alerts available in the MODIS database. As mentioned in the introduction, 1400 fire alerts were registered in the Campania region for the period between 01/2018 and 06/2019. A shapefile of fire alerts was thus projected onto the FFR map to visualize any possible correlation between fire alerts and FFR. Using the function export values in ARCGIS 10.2, it was possible to extract the corresponding risk classification according to each fire alert location. Then, each risk classification of the original FFR rank was divided into three sub-categories for a total of 15 classes of risk, organized in ascending order from 0 (very low) to 15 (very high), and rescaled to the number of fire alarms to fit the equation X = Y, merely to help the eye. Then, the number of fire alerts for each class was calculated to gain an R^2 correlation. The results are shown in Figure 6, where the R^2 values for the FFR map considering ATR and AFR are presented. The FFR map with AFR assumption obtained a higher R^2 (0.79) with respect to the FFR map with ATR (0.51). Figure 5a–c also offers a graphical output of the correlation, whereby it is possible to appreciate how the fire alerts (purple dots) overlap with the areas characterized by a higher fire risk, making this screening methodology reliable.

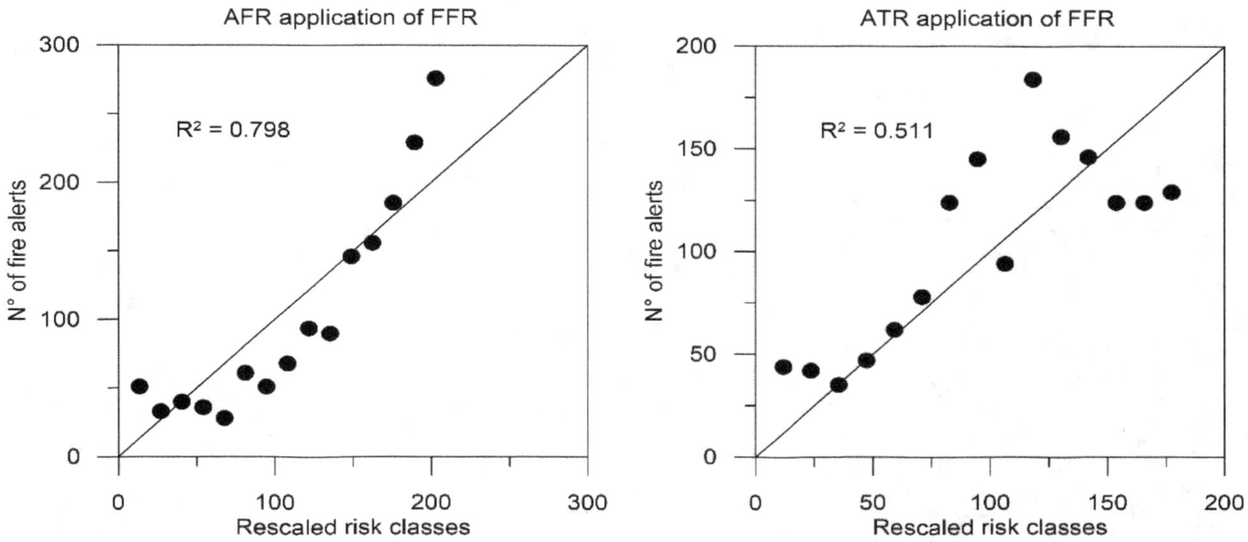

Figure 6. Representation of statistical correspondence between FFR classification and fire alerts.

4.3. FFR Assessment for 2050

The validation with actual data and the good results obtained allow for the use of the same methodology for the assessment of FFR for future scenarios. In this case, the period 2030–2050 was considered due to the availability of climatic data. All parameters and classifications were maintained, except for temperature and precipitation, which were replaced with the corresponding climatic projections for 2050. As discussed before, climate projections for the Campania region provided a less abundant increase of both precipitation and temperature. This phenomenon will be more striking at higher altitudes, where precipitation and temperature will register a large reduction and increase, respectively (Figure 3). For the future FFR assessment, the parameters were classified using the value reported in Table 1. The final FFR map for 2050 was obtained using Equation (2). The same final ranking of 2018 map was applied also for 2050 assessment with the classification in 5 classes of risk to make the 2 maps comparable. The map was classified as follows: very low (<22), low (23–26), medium (26–28), high (28–31) and very high (>31). The map for 2050 is shown in Figure 7a and highlights two main patterns: (i) the very low and low risk will shrink over the region occupying only 1% of the territory, compared to 20% of 2018 (Figure 7c) and (ii) the spreading of the high and very high risk over more than 80% of the territory (Figure 6c), also in zones previously characterized by medium-low risk. Figure 7b shows the patterns of FFR change between 2018 and 2050. The areas that appear to be more prone to deterioration are those located in the South, where a sharp transition from very low and low risk to very high risk takes place. The same happens, but to a smaller area, in the Northern part of the Campania region. This is mainly due to the changed regimes of temperature and precipitation. In those areas, a mean temperature increase of 2–3 °C and a reduction in precipitation of 100 mm/year are predicted. Figure S2 shows the classification risk for temperatures and precipitations in 2050, compared with the classification risk in 2018. The location of higher risk zones will worsen from 2018 to 2050 in the Northern and Southern portion of the Campania region.

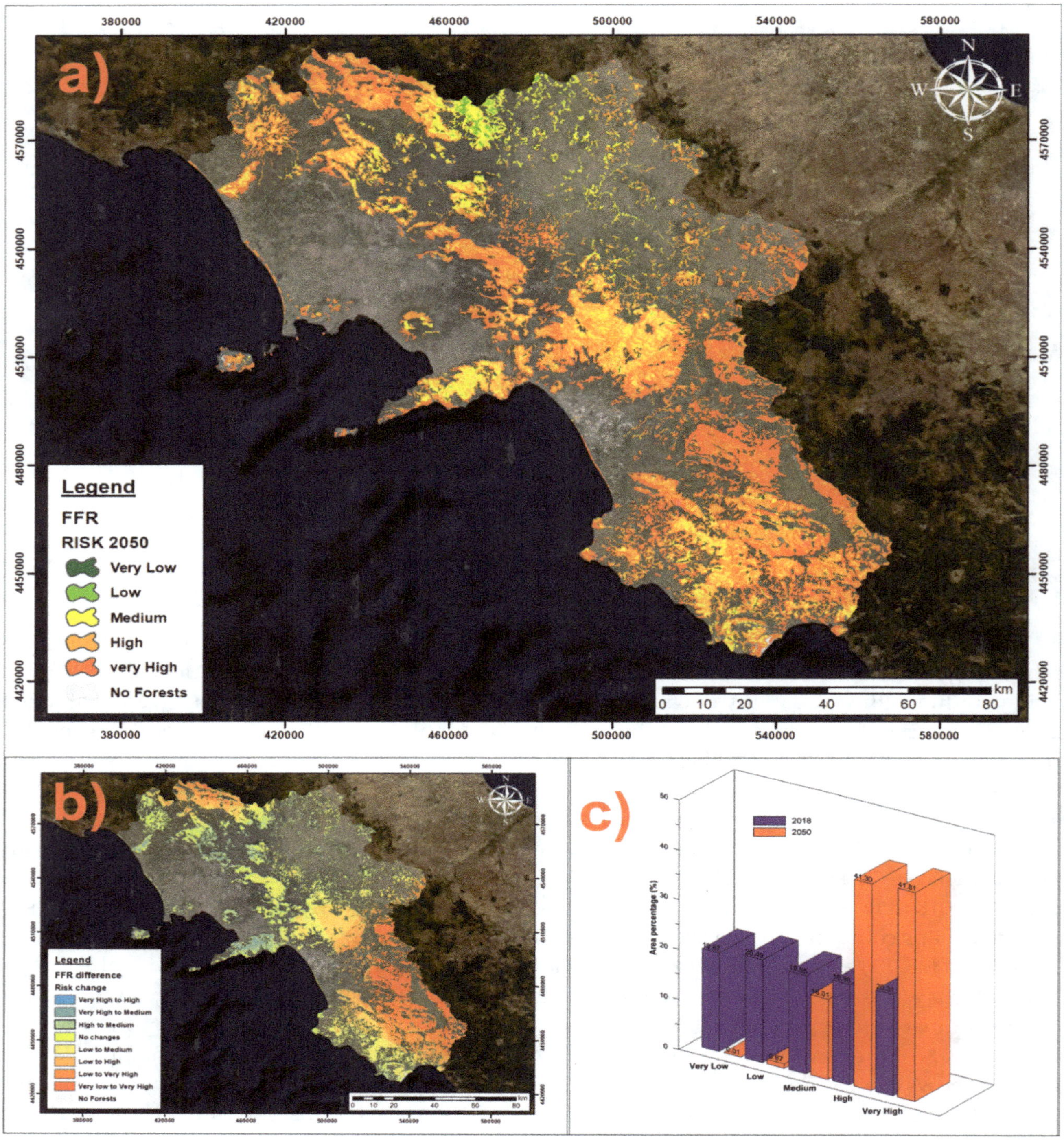

Figure 7. (**a**) FFR map of the Campania region for the year 2050, (**b**) risk changes on the region from 2018 to 2050, (**c**) risk classes distribution for 2018 and 2050.

4.4. SA Application

Finally, SA was applied for both applications of the FFR model. The results are shown in Table 3. Regarding the 2018 application, among the 12 parameters considered, the two with the higher impact results were the climatic factors followed by the forest type. These three parameters influence almost 50% of the results. Of the morphological factors, only slope seems to have a moderate impact on FFR, followed by aspect and elevation. Accordingly, with AHP weight assignment, the proximity to water courses, to springs and soil properties, also shows a moderate weight, influencing almost 20% of the results, emphasizing the importance of these three parameters in mitigating the FFR. Among the anthropogenic factors, the one with the higher weight is the proximity to agricultural fields, probably due to the large extension of the study area, followed by the distance to settlements and roads. These results agree with the preliminary weights assigned through AHP confirming the robustness and reliability of the methodology. Concerning the 2050 application, the SA application (Table 4) highlights how the effective weights of temperatures and precipitations become higher with respect to the 2018 FFR map despite having the same relative weight. Moreover, the role of the attenuating factors, like proximity to rivers, springs, and soil properties, lose half of their weight and effect.

Table 4. Sensitivity analysis results for FFH. Wherein ELV is elevation, SLP is slope, ASP is aspect, PCP is precipitation, FRS is forest type, TMP is temperature, ROD is distance from main road, AGR is distance from agricultural field, URB is distance from settlement, WTC is distance from main water course, SPR is distance from springs, and SOL is soil texture.

Parameters	Assigned Weight	Relative Weight 2018	Relative Weight 2050
TMP	1.88	15.75	22.22
PCP	1.88	20.2	26.32
SLP	1.22	11.26	9.03
FRS	1.22	15.44	11.56
URB	0.75	3.32	2.69
AGR	0.75	6.84	5.63
SPR	0.44	8.54	6.8
ROD	0.44	0.48	1.48
WTC	0.44	7.42	5.61
ASP	0.44	5.03	4.02
ELV	0.27	3.35	2.75
SOL	0.27	2.37	1.89

5. Discussion

The availability of the resulting maps can represent a crucial requirement in the realization of sound forest management plans, and in the execution of preventive actions intended for FFR reduction. The opportunity to know which area is susceptible to fire could allow to mitigate the problem and to better focus preventive actions. Some of these actions, like prescribed burning, have demonstrated to be very effective in fire vulnerable zones. Indeed, during the summer 2017, a major wildfire occurred in the Vesuvio National Park (Southern Italy), which destroyed about 1550 hectares of forest. Here, around 306 hectares were found to be very damaged, and 187 hectares slightly damaged. Among the slightly damaged areas, there were parcels treated with prescribed fires in 2014 and 2016 according to a program promoted in the Campania Region and the SMA Campania with the technical and scientific support of the GDL "Forest Fire Management" of SISEF [83]. This area is characterized by very high FFR in both applications of 2018 (Figure 5b) and 2050 (Figure 7). The application also highlights the importance to correctly consider the role of some parameters inside the evaluation. In fact, the application considering ATR results to be more reliable of the one AFR, remarking the importance of superficial and groundwater in preventing and mitigate FFR. The future scenario depicts a sharp deterioration of a worrying situation, whereby the temperatures increase at higher altitudes also, which will make thousands of hectares of forests more vulnerable to fires. The worsening for the 2050 resulted to be mainly driven by climate changes, as suggested by SA, in fact less precipitations and higher temperature are predicted together with extreme events that could negatively influence fire occurrence. This study highlighted a more severe scenario of the one reported in Michetti and Pinar [74] and Faggian [84], who predicted an increase of fire events and area burnt by 25% and 75%, respectively in central Southern Italy. The FFR mapping tool represents a user-friendly methodology for FFR assessment worldwide since the input data are usually easily accessible from any local authority. Furthermore, accounting land use change could improve the reliability of this methodology. Indeed, together with climatic factors, land use practices represent the main factors influencing FFR, as the expansion of agricultural fields and urban areas can significantly raise the fire ignition. Nevertheless, the methodology suffers for some drawbacks: (i) the approach is empirical and not driven by deterministic numerical models, thus the proposed weights cannot be straightforwardly applied in other regions if not calibrated with a local dataset and (ii) the calibration was performed for a single year, while multiple time series would provide more robust results. Ultimately, the methodology fulfills the obligation to produce clear and easily understandable maps, allowing local agencies to act properly and on time to prevent forest fires

6. Conclusions

Characterized by a Mediterranean climate and with at least 30% of the area covered by forests, the Campania region is highly susceptible to FFR. In the last 30 years over 3400 fires were registered, with 80% of cases occurred in the last three years. This moved the general audience attention to the issue of reducing FFR and protecting the ecosystem. In this study, a new FFR mapping methodology was proposed using a parametric linear combination of easily accessible data supported by an AHP application. The methodology divided the Campania region into five classes of risk, from very low to very high, and was validated using fire alerts from the MODIS database in three different areas. The FFR map that considers a decreasing risk approaching water courses and springs showed a good correlation with fire alert (0.79 R^2), confirming that the proposed methodology is reliable. Meanwhile, the standard literature assumption that considers an increasing risk approaching water courses and springs showed a poor correlation with fire alert (0.51 R^2). Besides, using the climate projections for 2050, a near-future FFR map was produced. The predicted FFR map highlighted a worsening situation with a spreading of higher risk classes over the territory due to mean temperature increases and a decrease in precipitation. Thus, the proposed methodology represents a reliable screening method for areas under FFR and can support authorities to direct preventive actions towards mitigating forest fires in cases where limited data are available.

Supplementary Materials:
Figure S1: Fire alerts and meteorological station location inside Campania Region, Figure S2: Temperature and precipitation risk classification: The 2018 classifications are shown in panel a for temperatures and panel c for precipitations, and those for the 2050 period are shown in panel b for temperatures and panel d for precipitations.

Author Contributions: Conceptualization, G.B. and E.G.; Methodology, G.B. and N.K.; Software, G.B. and N.K.; Validation, G.B. and N.C.; Formal analysis, E.G. and G.B.; Data curation, G.B. and E.G.; Writing—original draft preparation, G.B. and E.G.; Writing—review and editing, N.K. and N.C.

References

1. Pausas, J.G.; Keeley, J.E. Wildfires as an ecosystem service. *Front. Ecol. Environ.* **2019**, *17*, 289–295. [CrossRef]
2. Meddens, A.J.; Kolden, C.A.; Lutz, J.A. Detecting unburned areas within wildfire perimeters using Landsat and ancillary data across the northwestern United States. *Remote Sens. Environ.* **2016**, *186*, 275–285. [CrossRef]
3. Syphard, A.D.; Brennan, T.J.; Keeley, J.E. The role of defensible space for residential structure protection during wildfires. *Int. J. Wildland Fire* **2014**, *23*, 1165–1175. [CrossRef]
4. Syphard, A.D.; Keeley, J.E.; Pfaff, A.H.; Ferschweiler, K. Human presence diminishes the importance of climate in driving fire activity across the United States. *Proc. Natl. Acad. Sci. USA* **2017**, *114*, 13750–13755. [CrossRef]
5. Parisien, M.A.; Walker, G.R.; Little, J.M.; Simpson, B.N.; Wang, X.; Perrakis, D.D. Considerations for modeling burn probability across landscapes with steep environmental gradients: An example from the Columbia Mountains, Canada. *Nat. Hazards* **2013**, *66*, 439–462. [CrossRef]
6. Wuschke, K.; Clare, J.; Garis, L. Temporal and geographic clustering of residential structure fires: A theoretical platform for targeted fire prevention. *Fire Saf. J.* **2013**, *62*, 3–12. [CrossRef]
7. Fernandes, P.M.; Botelho, H.S. A review of prescribed burning effectiveness in fire hazard reduction. *Int. J. Wildland Fire* **2003**, *12*, 117–128. [CrossRef]

8. Battipaglia, G.; Strumia, S.; Esposito, A.; Giuditta, E.; Sirignano, C.; Altieri, S.; Rutigliano, F.A. The effects of prescribed burning on Pinus halepensis Mill. as revealed by dendrochronological and isotopic analyses. *For. Ecol. Manag.* **2014**, *334*, 201–208. [CrossRef]
9. Catalanotti, A.E.; Giuditta, E.; Marzaioli, R.; Ascoli, D.; Esposito, A.; Strumia, S.; Rutigliano, F.A. Effects of single and repeated prescribed burns on soil organic C and microbial activity in a Pinus halepensis plantation of Southern Italy. *Appl. Soil Ecol.* **2018**, *125*, 108–116. [CrossRef]
10. Erten, E.; Kurgun, V.; Musaoglu, N. Forest fire risk zone mapping from satellite imagery and GIS: A case study. In Proceedings of the XXth Congress of the International Society for Photogrammetry and Remote Sensing, Istanbul, Turkey, 12–23 July 2004; pp. 222–230.
11. Pourtaghi, Z.S.; Pourghasemi, H.R.; Aretano, R.; Semeraro, T. Investigation of general indicators influencing on forest fire and its susceptibility modeling using different data mining techniques. *Ecol. Indic.* **2016**, *64*, 72–84. [CrossRef]
12. Jaiswal, R.K.; Mukherjee, S.; Raju, K.D.; Saxena, R. Forest fire risk zone mapping from satellite imagery and GIS. *Int. J. Appl. Earth. Obs.* **2002**, *4*, 1–10. [CrossRef]
13. Jung, J.; Kim, C.; Jayakumar, S.; Kim, S.; Han, S.; Kim, D.H.; Heo, J. Forest fire risk mapping of Kolli Hills, India, considering subjectivity and inconsistency issues. *Nat. Hazard.* **2013**, *65*, 2129–2146. [CrossRef]
14. Amalina, P.; Prasetyo, L.B.; Rushayati, S.B. Forest Fire Vulnerability Mapping in Way Kambas National Park. *Procedia Environ. Sci.* **2016**, *33*, 239–252. [CrossRef]
15. Linn, R.; Reisner, J.; Colman, J.J.; Winterkamp, J. Studying wildfire behaviorusing FIRETEC. *Int. J. Wildland Fire* **2002**, *11*, 233–246. [CrossRef]
16. Sturtevant, B.R.; Scheller, R.M.; Miranda, B.R.; Shinneman, D.; Syphard, A. Simulating dynamic and mixed-severity fire regimes: A process-based fireextension for LANDIS-II. *Ecol. Model.* **2009**, *220*, 3380–3393. [CrossRef]
17. Duarte, L.; Teodoro, A.C.; Goncalves, J.A.; Soares, D.; Cunha, M. Assessing soil erosion risk using RUSLE through a GIS open source desktop and web application. *Environ. Monit. Assess.* **2016**, *188*, 351. [CrossRef]
18. Eugenio, F.C.; dos Santos, A.R.; Fiedler, N.C.; Ribeiro, G.A.; da Silva, A.G.; dos Santos, Á.B.; Schettino, V.R. Applying GIS to develop a model for forest fire risk: A case study in Espírito Santo, Brazil. *J. Environ. Manag.* **2016**, *173*, 65–71. [CrossRef]
19. Wang, L.; Zhao, Q.; Wen, Z.; Qu, J. RAFFIA: Short-term Forest Fire Danger Rating Prediction via Multiclass Logistic Regression. *Sustainability* **2018**, *10*, 4620. [CrossRef]
20. Oliveira, S.; Oehler, F.; San-Miguel-Ayanz, J.; Camia, A.; Pereira, J.M.C. Modeling spatial patterns of fire occurrence in Mediterranean Europe using multiple regression and random forest. *For. Ecol. Manag.* **2012**, *275*, 117–129. [CrossRef]
21. Hong, H.; Naghibi, S.; Dashtpagerdi, M.; Pourghasemi, H.; Chen, W. A comparative assessment between linear and quadratic discriminant analyses (LDA-QDA) with frequency ratio and weights-of-evidence models for forest fire susceptibility mapping in China. *Arab. J. Geosci.* **2017**, *10*, 167. [CrossRef]
22. Tien Bui, D.; Pradhan, B.; Lofman, O.; Revhaug, I.; Dick, O.B. Landslide susceptibility mapping at Hoa Binh province (Vietnam) using an adaptiveneuro-fuzzy inference system and GIS. *Comput. Geosci.* **2012**, *45*, 199–211. [CrossRef]
23. Short, K. A spatial database of wildfires in the United States, 1992–2011. *Earth Syst. Sci. Data* **2014**, *6*, 1–27. [CrossRef]
24. Zald, H.S.J.; Ohmann, J.L.; Roberts, H.M.; Gregory, M.J.; Henderson, E.B.; McGaughey, R.J.; Braaten, J. Influence of lidar, Landsat imagery, disturbance history, plotlocation accuracy, and plot size on accuracy of imputation maps of forest composition and structure. *Remote Sens. Environ.* **2014**, *143*, 26–38. [CrossRef]
25. You, W.; Lin, L.; Wu, L.; Ji, Z.; Yu, J.A.; Zhu, J.; He, D. Geographical information system-based forest fire risk assessment integrating national forest inventory data and analysis of its spatiotemporal variability. *Ecol. Indic.* **2017**, *77*, 176–184. [CrossRef]
26. González-Olabarria, J.; Rodríguez, F.; Fernández-Landa, A.; Mola-Yudego, B. Mapping fire risk in the model forest of Urbión (Spain) based on airborne LiDAR measurements. *For. Ecol. Manag.* **2012**, *282*, 149–156. [CrossRef]
27. Varela, V.; Vlachogiannis, D.; Sfetsos, A.; Karozis, S.; Politi, N.; Giroud, F. Projection of Forest Fire Danger due to Climate Change in the French Mediterranean Region. *Sustainability* **2019**, *11*, 4284. [CrossRef]

28. Oliveira, S.; Félix, F.; Nunes, A.; Lourenço, L.; Laneve, G.; Sebastián-López, A. Mapping wildfire vulnerability in Mediterranean Europe. Testing a stepwise approach for operational purposes. *J. Environ. Manag.* **2018**, *206*, 158–169. [CrossRef]
29. Petrakis, M.; Psiloglou, B.; Lianou, M.; Keramitsoglou, I.; Cartalis, C. Evaluation of forest fire risk and fire extinction difficulty at the mountainous park of Vikos-Aoos, Northern Greece: Use of remote sensing and GIS techniques. *IJRAM* **2005**, *5*, 50–65. [CrossRef]
30. Semerato, T.; Mastroleo, G.; Aretano, R.; Facchinetti, G.; Zurlini, G.; Petrosillo, I. GIS Fuzzy Expert System for the assessment of ecosystems vulnerability to fire in managing Mediterranean natural protected areas. *J. Environ. Manag.* **2016**, *168*, 94–103. [CrossRef]
31. Poldini, L.; Ganis, P.; Vidali, M.; Altobelli, A.; Bader, F.; Cantele, S. Inclusion of phytosociological data in an index of vegetation fire danger: Application and mapping on the Karst area around Trieste (Italy). *Plant Biosyst.* **2018**, *152*, 810–817. [CrossRef]
32. Wu, Z.; He, H.S.; Yang, J.; Liang, Y. Defining fire environment zones in the boreal forests of northeastern China. *Sci. Total Environ.* **2014**, *518*, 106–116. [CrossRef] [PubMed]
33. Pradhan, B.; Dini Hairi Bin Suliman, M.; Arshad Bin Awang, M. Forest fire susceptibility and risk mapping using remote sensing and geographical information systems (GIS). *Disaster Prev. Manag.* **2007**, *16*, 344–352. [CrossRef]
34. Xystrakis, F.; Koutsias, N. Differences of fire activity and their underlying factors among vegetation formations in Greece. *IForest* **2013**, *6*, 132. [CrossRef]
35. Bui, D.T.; Bui, Q.T.; Nguyen, Q.P.; Pradhan, B.; Nampak, H.; Trinh, P.T. A hybrid artificial intelligence approach using GIS-based neural-fuzzy inference system and particle swarm optimization for forest fire susceptibility modeling at a tropical area. *Agric. For. Meteorol.* **2017**, *233*, 32–44. [CrossRef]
36. Tian, X.; Zhao, F.; Shu, L.; Wang, M. Distribution characteristics and the influence factors of forest fires in China. *For. Ecol. Manag.* **2013**, *310*, 460–467. [CrossRef]
37. Carmo, M.; Moreira, F.; Casimiro, P.; Vaz, P. Land use and topography influences on wildfire occurrence in northern Portugal. *Landsc. Urban Plan.* **2011**, *100*, 169–176. [CrossRef]
38. Saura-Mas, S.; Paula, S.; Pausas, J.G.; Lloret, F. Fuel loading and flammability in the Mediterranean Basin woody species with different post-fire regenerative strategies. *Int. J. Wildland Fire* **2010**, *19*, 783–794. [CrossRef]
39. Price, O.; Bradstock, R. Countervailing effects of urbanization and vegetation extent on fire frequency on the wildland urban interface: Disentangling fuel and ignition effects. *Landscape Urban Plan.* **2014**, *130*, 81–88. [CrossRef]
40. Dillon, G.K.; Holden, Z.A.; Morgan, P.; Crimmins, M.A.; Heyerdahl, E.K.; Luce, C.H. Both topography and climate affected forest and woodland burn severity in two regions of the western US, 1984 to 2006. *Ecosphere* **2011**, *2*, 1–33. [CrossRef]
41. Lentile, L.B.; Smith, F.W.; Shepperd, W.D. Influence of topography and forest structure on patterns of mixed severity fire in ponderosa pine forests of the South Dakota Black Hills, USA. *Int. J. Wildland Fire* **2006**, *15*, 557–566. [CrossRef]
42. Verde, J.; Zêzere, J. Assessment and validation of wildfire susceptibility and hazard in Portugal. *Nat. Hazards Earth Syst. Sci.* **2010**, *10*, 485–497. [CrossRef]
43. Schmidt, D.A.; Taylor, A.H.; Skinner, C.N. The influence of fuels treatment and landscape arrangement on simulated fire behavior, Southern Cascade range, California. *For. Ecol. Manag.* **2008**, *255*, 3170–3184. [CrossRef]
44. Liu, Z.; Yang, J.; Chang, Y.; Weisberg, P.J.; He, H.S. Spatial patterns and drivers of fire occurrence and its future trend under climate change in a boreal forest of Northeast China. *Glob. Chang. Biol.* **2012**, *18*, 2041–2056. [CrossRef]
45. Zumbrunnen, T.; Menéndez, P.; Bugmann, H.; Conedera, M.; Gimmi, U.; Bürgi, M. Human impacts on fire occurrence: A case study of hundred years of forest fires in a dry alpine valley in Switzerland. *Reg. Environ. Chang.* **2012**, *12*, 935–949. [CrossRef]
46. Wu, Z.; He, H.S.; Yang, J.; Liu, Z.; Liang, Y. Relative effects of climatic and local factors on fire occurrence in boreal forest landscapes of northeastern China. *Sci. Total Environ.* **2014**, *493*, 472–480. [CrossRef] [PubMed]
47. Sirca, C.; Casula, F.; Bouillon, C.; García, B.F.; Ramiro, M.M.F.; Molina, B.V.; Spano, D. A wildfire risk-oriented GIS tool for mapping Rural-Urban Interfaces. *Environ. Model. Softw.* **2017**, *94*, 36–47. [CrossRef]

48. Ducci, D.; Tranfaglia, G. The Effect of Climate Change on the Hydrogeological Resources in Campania Region (Italy). In *Groundwater and Climatic Changes*; Dragoni, W., Ed.; Geological Society of London: London, UK, 2008; Volume 288, pp. 25–38.
49. Mastrocicco, M.; Busico, G.; Colombani, N. Deciphering Interannual Temperature Variations in Springs of the Campania Region (Italy). *Water* **2019**, *11*, 288. [CrossRef]
50. Busico, G.; Kazakis, N.; Colombani, N.; Mastrocicco, M.; Voudouris, K.; Tedesco, D. A modified SINTACS method for groundwater vulnerability and pollution risk assessment in highly anthropized regions based on NO3 and SO42 concentrations. *Sci. Total Environ.* **2017**, *609*, 1512–1523. [CrossRef]
51. JRC Forest Fires in Europe, Middle East and North Africa. 2017. Available online: https://effis.jrc.ec.europa.eu/applications/fire-news/ (accessed on 10 June 2019).
52. Ministero della difesa (Carabinieri.it). Available online: http://datiallefiamme.it (accessed on 24 May 2019).
53. Fire Global Forest Watch. Available online: https://fires.globalforestwatch.org/home/ (accessed on 2 May 2019).
54. MIPAAF. Agrometeorological Online Database. 2018. Available online: http://www.agricoltura.regione.campania.it/meteo/agrometeo.htm (accessed on 12 April 2019).
55. Arpac Database. 2019. Available online: http://www.sinanet.isprambiente.it/it (accessed on 11 March 2019).
56. Di Gennaro, A. *I sistemi di terra della Campania*; SELCA: Firenze, Italy, 2002.
57. Advance Spaceborn Thermal Emission and Reflection Radiometer. Available online: https://asterweb.jpl.nasa.gov/gdem.asp (accessed on 11 April 2019).
58. Copernicus, Land Monitoring Service. Available online: https://land.copernicus.eu/pan-european/corine-land-cover (accessed on 1 April 2019).
59. EURO-CORDEX. 2019. Available online: https://www.euro-cordex.net/index.php.en (accessed on 10 April 2019).
60. Saaty, T.H. A scaling method forms priorities in hierarchical structures. *J. Math. Psychol.* **1977**, *15*, 234–281. [CrossRef]
61. Napolitano, P.; Fabbri, A.G. Single Parameter Sensitivity Analysis for Aquifer Vulnerability Assessment Using DRASTIC and SINTACS. In *Hydrology GIS Application of Geographic Information Systems in Hydrology and Water Resources Management*; Kovar, K., Nachtnebel, H.P., Eds.; IAHS Publication: Wallingford, UK, 1996; pp. 559–566.
62. Kazakis, N.; Busico, G.; Colombani, N.; Mastrocicco, M.; Pavlou, A.; Voudouris, K. GALDIT-SUSI a modified method to account for surface water bodies in the assessment of aquifer vulnerability to seawater intrusion. *J. Environ. Manag.* **2019**, *235*, 257–265. [CrossRef]
63. Robinne, F.; Bladon, K.D.; Miller, C.; Parisien, M.; Mathieu, J.; Flannigan, M.D. A spatial evaluation of global wildfire-water risks to human and natural systems. *Sci. Total Environ.* **2018**, *610*, 1193–1206. [CrossRef] [PubMed]
64. Gontara, M.; Allouche, N.; Jmal, I.; Bouri, S. Sensitivity analysis for the GALDIT method based on the assessment of vulnerability to pollution in the northern Sfax coastal aquifer, Tunisia. *Environ. Earth Sci.* **2016**, *75*, 669. [CrossRef]
65. Bui, D.T.; Le, K.T.; Nguyen, V.; Le, H.; Revhaug, I. Tropical forest fire susceptibility mapping at the cat Ba national park area, Hai Phong city, Vietnam, using GIS-Based kernel logistic regression. *Remote Sens.* **2016**, *8*, 347. [CrossRef]
66. Chuvieco, E.; Aguado, I.; Yebra, M.; Nieto, H.; Salas, J.; Martín, M.P.; Vilar, L.; Martinez, J.; Martin, S.; Ibarra, P.; et al. Development of a framework for fire risk assessment using remote sensing and geographic information system technologies. *Ecol. Model.* **2010**, *221*, 46–58. [CrossRef]
67. Renard, Q.; Pelissier, R.; Ramesh, B.R.; Kodandapani, N. Environmental suscep-tibility model for predicting forest fire occurrence in the Western Ghats of India. *Int. J. Wildland Fire* **2012**, *21*, 368–379. [CrossRef]
68. Adab, H.; Devi Kanniah, K.; Solaimani, K. Modeling forest fire risk in the northeast of Iran using remote sensing and GIS techniques. *Nat. Hazards* **2013**, *65*, 1723–1743. [CrossRef]
69. Scholze, M.; Knorr, W.; Arnell, N.W. Prentice IC. A climate-change risk analysis for world ecosystems. *Proc. Natl. Acad. Sci. USA* **2006**, *103*, 13116–13120. [CrossRef]
70. Nunes, M.C.S.; Vasconcelos, M.J.; Pereira, J.M.C.; Dasgupta, N.; Alldredge, R.J.; Rego, F.C. Land-cover type and fire in Portugal: Do fires burn land cover selectively? *Landsc. Ecol.* **2005**, *20*, 661–673. [CrossRef]
71. Rajabi, M.; Alesheikh, A.; Chehreghan, A.; Gamzeh, H. An innovative methodfor forest fire risk zoning map using fuzzy inference system and GIS. *Int. J. Sci. Technol. Res.* **2013**, *2*, 57–64.

72. Calviño-Cancela, M.; Chas-Amil, M.L.; García-Martínez, E.D.; Touza, J. Interacting effects of topography, vegetation, human activities and wildland-urban interfaces on wildfire ignition risk. *For. Ecol. Manag.* **2017**, *397*, 10–17. [CrossRef]
73. Hong, H.; Tsangaratos, P.; Ilia, I.; Liu, J.; Zhu, A.X.; Xu, C. Applying genetic algorithms to set the optimal combination of forest fire related variables and model forest fire susceptibility based on data mining models. The case of Dayu County, China. *Sci Total Environ.* **2018**, *630*, 1044–1056. [CrossRef] [PubMed]
74. Michetti, M.; Pinar, M. Forest Fires Across Italian Regions and Implications for Climate Change: A Panel Data Analysis. *Environ. Resour. Econ.* **2019**, *72*, 207. [CrossRef]
75. Nadezhdina, N.; Ferreira, M.I.; Conceição, N.; Pacheco, C.A.; Häusler, M.; David, T.S. Water uptake and hydraulic redistribution under a seasonal climate: Long-term study in a rainfed olive orchard. *Ecohydrology* **2015**, *8*, 387–397. [CrossRef]
76. Giorgi, F.; Lionello, P. Climate change projections for the Mediterranean region. *Glob. Planet. Chang.* **2008**, *63*, 90–104. [CrossRef]
77. Bucchignani, E.; Montesarchio, M.; Zollo, A.L.; Mercogliano, P. High-resolution climate simulations with COSMO-CLM over Italy: Performance evaluation and climate projections for the 21st century. *Int. J. Climatol.* **2016**, *36*, 735–756. [CrossRef]
78. Turco, M.; von Hardenberg, J.; AghaKouchak, A.; Llasat, M.C.; Provenzale, A.; Trigo, R.M. On the key role of droughts in the dynamics of summer fires in Mediterranean Europe. *Sci. Rep.* **2017**, *7*, 81. [CrossRef]
79. Eskandari, S. A new approach for forest fire risk modeling using fuzzy AHP and GIS in hyrcanian forests of Iran. *Arab. J. Geosci.* **2017**, *10*, 190. [CrossRef]
80. Giuditta, E.; Coenders-Gerrits, A.M.J.; Bogaard, T.A.; Wenninger, J.; Greco, R.; Rutigliano, F.A. Measuring changes in forest floor evaporation after prescribed burning in southern Italy pine plantations. *Agric. For. Meteorol.* **2018**, *256*, 516–525. [CrossRef]
81. Lukenbach, M.C.; Hokanson, K.J.; Moore, P.A.; Devito, K.J.; Kettridge, N.; Thompson, D.K.; Wotton, B.M.; Petrone, R.M.; Waddington, J.M. Hydrological controls on deep burning in a northern forested peatland. *Hydrol. Process.* **2015**, *29*, 4114–4124. [CrossRef]
82. Hokanson, K.J.; Lukenbach, M.C.; Devito, K.J.; Kettridge, N.; Petrone, R.M.; Waddington, J.M. Groundwater connectivity controls peat burn severity in the boreal plains. *Ecohydrology* **2016**, *9*, 574–584. [CrossRef]
83. Battipaglia, G.; Tognetti, R.; Valese, E.; Ascoli, D.; De Luca, P.F.; Basile, S.; Ottaviano, M.; Marchetti, M.; Esposito, A. Incendi 2017: Un'importante lezione. *J. Silvic. For. Ecol.* **2017**, *14*, 231. [CrossRef]
84. Faggian, P. Estimating fire danger over Italy in the next decades. *J. Environ. Integr.* **2018**, *3*, 15. [CrossRef]

2

GIS-Based Multicriteria Evaluation of Land Suitability for Grasslands Conservation in Chihuahua, Mexico

Griselda Vázquez-Quintero [1], Jesús A. Prieto-Amparán [2], Alfredo Pinedo-Alvarez [2], María C. Valles-Aragón [1], Carlos R. Morales-Nieto [2] and Federico Villarreal-Guerrero [2,*]

1 Facultad de Ciencias Agrotecnológicas, Universidad Autónoma de Chihuahua, Ave. Pascual Orozco s/n, Campus 1, Chihuahua, Chihuahua 31350, Mexico; gquintero@uach.mx (G.V.-Q.); valles.cecilia@gmail.com (M.C.V.-A.)

2 Facultad de Zootecnia y Ecología, Universidad Autónoma de Chihuahua, Periférico Francisco R. Almada Km 1, Chihuahua, Chihuahua 31453, Mexico; jesus_prieto06@hotmail.com (J.A.P.-A.); apinedo@uach.mx (A.P.-A.); cnieto@uach.mx (C.R.M.-N.)

* Correspondence: fvillarreal@uach.mx

Abstract: This study developed a GIS-based framework for the zoning of land suitability for grassland conservation (LSGC) in the Central Valleys of Chihuahua, México. For that, a Geographic Information Systems (GIS)-based multicriteria evaluation techniques with weighted overlay (MCE-WO), and a fragmentation analysis were performed. The framework for LSGC consisted in the development of four scenarios: Nonintensive Agriculture, Intensive Agriculture, Urban, and Rural. The LSGC classes defined with the MCE-WO technique were: Very high, High, Moderate, Low, and Very low land suitability. Results showed that the zone with a high suitability covered the largest area in the four scenarios with a surface of 44,264 km^2. The zones with low and very low suitability were concentrated mainly in the central region of the study area. At the landscape level, fragmentation of LSGC showed the Nonintensive Agriculture and the Rural scenarios with the highest Number of patches (54,640 and 46,210, respectively). The fragmentation of LSGC, under the scenarios evaluated, was mainly due to land opening for agriculture and to the influence of rural communities. The integration of GIS with MCE-WO is useful and effective for the evaluation of LSGC. This tool can provide a solid source of information for decision-makers regarding planning of land use to mitigate grasslands degradation.

Keywords: conservation zones; MCE; weighted overlay; fragmentation

1. Introduction

Grasslands cover about one-third of the world's land area, distributed mainly in semiarid and arid areas [1]. These ecosystems maintain various ecological functions such as water conservation [2], carbon storage [3], and soil stability [4]. Grasslands provide habitat for wildlife, serve as carbon reservoirs, and help mitigate the impacts of global climate change [5]. They also provide environmental services of recreation and appreciation with great views of the landscape [6].

Deforestation, overgrazing, agriculture, displacement of native species, and urbanization accompanied by population growth have induced global desertification and degradation of grasslands, putting at risk the services and functions of these ecosystems [7,8]. Grassland degradation could have a significant impact on the carbon cycle, the regional economy and climate [9,10]. In addition, changes in the precipitation regimes due to global environmental change could produce heavy rains with soil saturation in grasslands, which has been proven to have a negative effect on the ecosystem, with a decrease in the reproductive and germination rates in grasslands [11,12].

Grasslands in Mexico are distributed over an area of ≈100,000 km^2 [13]. Specifically, the grasslands of northern Mexico are extensively used for grazing and have also experienced an intensive change in soil due to agricultural activities, resulting in changes in the composition of species in these ecosystems [14,15]. Agricultural and livestock activities contribute to more than 75% of the recent changes in land use, followed by deforestation and urbanization [16].

The establishment of conservation zones for grasslands is of great importance to avoid new disturbances as well as the restoration of degraded grasslands [10]. Many species of both flora and fauna require large areas to maintain their viable populations. The conservation of healthy and functional ecosystems in which biodiversity is maintained in human presence requires the integration of reserve design rules and ecosystem management approaches at the species, ecosystem, and landscape levels [17–19].

Land decision-makers face the problem of selecting the appropriate framework for the establishment of land suitability for grasslands conservation. These zoning decisions are strongly influenced by the type and amount of information available, the land surface area, legislation, and the possibilities of new data collection. The elaboration and establishment of the zoning scheme is possibly the most relevant process in the planning of ecosystem protection zones. Through zoning, specific uses are assigned to the units of the territory to be conserved. The provision of spatial methods and computational technologies to assist land decision-makers in making decisions has had a major influence on land zoning [20–22].

Multicriteria Evaluation (MCE) is a method commonly implemented by decision support systems to compare alternative courses of action based on multiple factors and to identify the solution with the best performance [23,24]. MCE has received attention in the context of decision-making based on geographic information systems (GIS) [25–27], which are very useful for solving issues requiring a large set of variables covering extensive territories, which are sometimes inaccessible. For this reason, GIS-based MCE techniques have been used in different studies mainly for decision-making on land use [28,29]. Some authors have used MCE techniques to find potential areas for agriculture [22], silvicultural management [30], integrated watershed management [31], settlements development [32], among others. Weighted Overlay (WO) analysis is a component of spatial modeling that uses MCE, placing importance on the criteria in a differentiated manner [33,34]. In other words, WO includes different weights in the multicriteria spatial analysis and it is considered different from conventional MCE techniques [35], evaluating the influence of each criterion and assigning values within a defined range of classes [36]. In addition, simulation of scenarios serves to project what will probably happen and may also be used to develop support decision-making processes based on hypothesis of interest [37].

Different criteria at the landscape level, such as roads, agriculture, and urban development, have been shown to have significant negative effects on the spatial configuration of grasslands [38]. The monitoring and quantification of the spatial configuration of grassland ecosystems has been carried out using fragmentation indices, which are of great importance for understanding ecosystem degradation caused by natural and anthropogenic factors [15].

The Central Valleys of the state of Chihuahua is a transition zone between the mountainous area and the Chihuahuan Desert in northern Mexico. Their topography goes from moderate slopes to large plains covered mostly by grasslands, covering 24% of the State. These grasslands are one of the most suitable ecosystems to sustain domestic herbivorous animals, which, in turn, serve for human consumption. The diversity found in these ecosystems is seriously threatened by changes in land use/land cover, demographic pressure, and the accelerated growth of the surface area occupied by agriculture [39]. Thus, the implementation of a zoning scheme for grassland conservation is of great importance for this region.

The objective was to develop a GIS-based framework for grassland conservation zoning in the Central Valleys of Chihuahua, México. For that, four scenarios of land suitability for grasslands conservation (LSGC) were generated. Scenarios included intensive agriculture, nonintensive agriculture, urban, and rural. Scenario generation was carried out using multicriteria evaluation with weighted overlay (MCE-WO) analysis, while the evaluation was performed through a fragmentation analysis. This study was based on GIS and multicriteria evaluation methods and is intended as a tool for local and regional authorities, as well as for land decision-makers. The proposed zoning for LSGC through MCE-WO allows for locating spatial opportunities (i.e., priority areas) for Nature-Based Solutions (NBS) [40–43]. Conservation actions, incorporating nature into decision-making [41], could then be performed in these areas. Focusing on land-use management, through zoning, enables one to optimize NBS through an efficient use of natural resources, and helps in facing future challenges such as the ecosystems' degradation and the global environmental change [44].

2. Materials and Methods

2.1. Study Area

The Central Valleys have an area of ≈59,952.8 km^2, distributed in elevations from 1176 m to 2978 m. The region was delimited based on the map of the terrestrial ecoregion of Mexico [45] and was adjusted to the limits of the state of Chihuahua (Figure 1). The main urban settlements of the area are Chihuahua, Cuauhtemoc, Parral, Casas Grandes, and Nuevo Casas Grandes. The climate of the region is semiarid-temperate, with an average annual temperature range from 12 °C to 18 °C, and an annual precipitation ranging on average from 300 mm to 400 mm [46]. In the grasslands of this region, we can find shrub species and dune vegetation [40]. In some of these grassland ecosystems, mammals such as *Bison bison* and *Cynomys ludovicanus* may be found. These species have multiple effects on the ecosystem and both of them are considered key for maintaining the grassland's habitats [47]. Mammal species that can also be found include *Puma concolor*, which is under pressure, and its populations are in a decreasing status, according to the Red List of threatened species of the International Union for the Conservation of Nature [48].

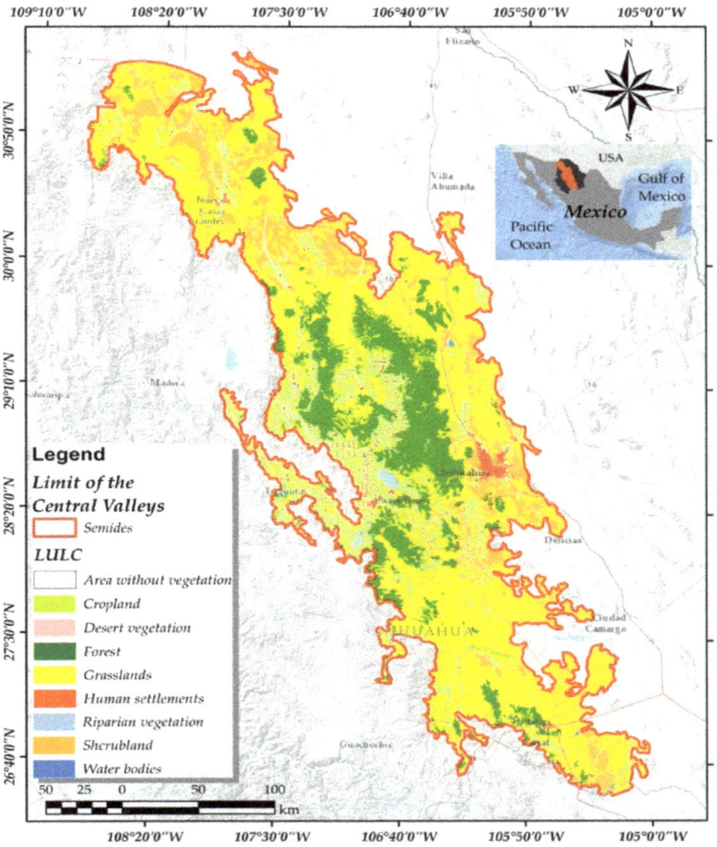

Figure 1. Location of the study area. LULC = Land use/land cover.

2.2. Summary of the Framework

Figure 2 summarizes the methodology used for zoning based on LSGC and for the analysis of the land-suitability configuration in the Central Valleys of Chihuahua, Mexico.

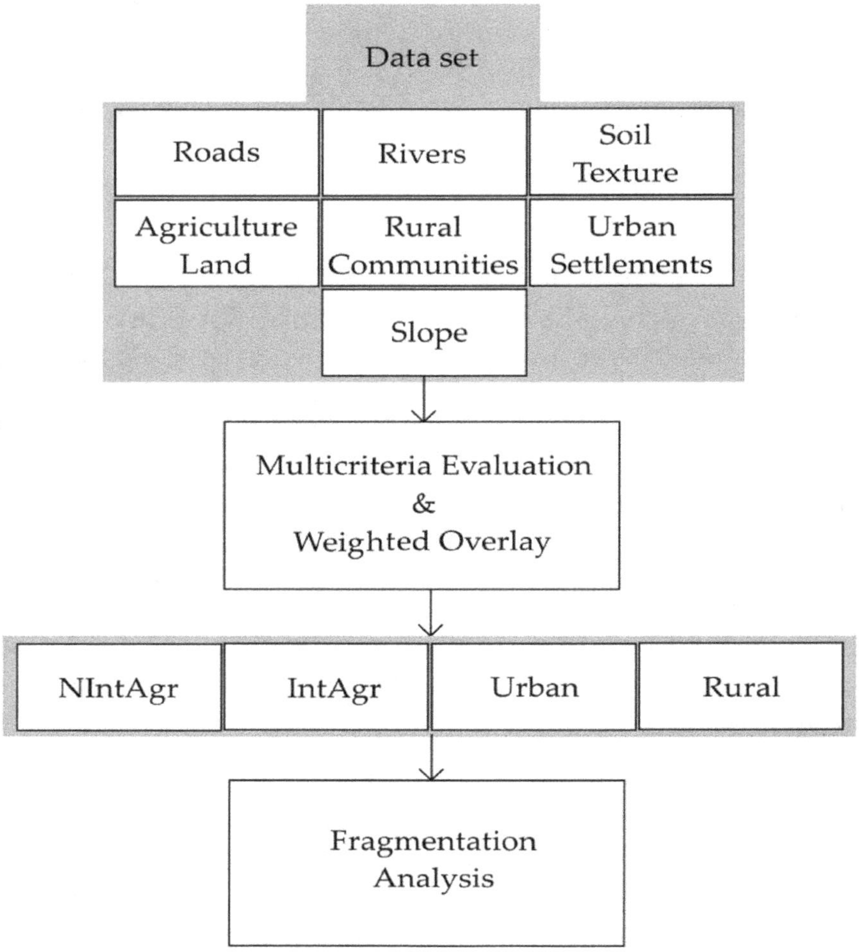

Figure 2. Flowchart summarizing the methodology for the analysis of land suitability for grasslands conservation. NIntAgr = Nonintensive Agriculture, IntAgr = Intensive Agriculture.

2.3. Data Set and Zoning

Areas of LSGC were defined by using a GIS-based multicriteria evaluation with weighted overlay (MCE-WO) technique. This quantitative method allowed for including qualitative considerations to establish relationships among a set of relevant criteria influencing grassland conservation. Based on expert knowledge [49] and data availability, seven criteria related to LSGC were selected. The criteria, represented in maps and used for the analysis, were: distance to roads, distance to rivers, distance to agricultural areas, distance to urban areas, distance to rural localities (Euclidean distance), slope, and soil texture (Table 1). All these criteria are related, to a greater or lesser extent, to the alteration of areas suitable for grassland conservation.

In addition, areas corresponding to urban settlements and water bodies were excluded for the analysis, since these areas are 100% not able to be converted into grassland conservation areas. Regarding agriculture lands, they were kept for the analysis since agriculture in the study area usually takes place on the best lands for grasslands conservation and they could be reconverted if policy modifications happened to occur in the future. In the case of the other land use/land cover types, e.g., shrublands and forests, we considered that it was appropriate to include them in the analysis because, in the study area, these land use/land covers are associated with grassland vegetation (transition areas), which contribute to maintaining the diversity of the flora and fauna of these ecosystems.

Table 1. Criteria used in the analysis of land suitability for grasslands conservation.

Criteria	Unit	Source
Distance to agricultural land	km	INEGI [50]
Distance to roads	km	INEGI [51]
Distance to rural communities	km	INEGI [51]
Distance to urban settlements	km	INEGI [51]
Distance to rivers	km	INEGI [52]
Slope	Degree	INEGI [53]
Texture	Unitless	INEGI [54]

The maps were obtained from various sources and presented different degrees of generalization and scales in vector format. They were then converted to a raster format and homogenized to a cell size of 100 m by using the ArcMap 10.3© GIS software (ESRI, Redlands, CA, USA; https://www.esri.com/en-us/home).

2.4. Weighted Overlay Analysis

The MCE-WO was used to reclassify the criteria from Table 1 to the same scale according to their influence on LSGC. Values from 1 to 5 (criteria standardization) were applied to each criterion, as it has been performed on a previous study [55]. According to the degree of suitability for grassland conservation, a value of 1 was assigned to the areas with the lowest suitability, while a value of 5 was assigned to the most suitable areas. The classes manually assigned in this study were: 1 (very low), 2 (low), 3 (moderate), 4 (high), and 5 (very high), as shown in Table 2.

Table 2. Reclassification of the criteria influencing land suitability for grasslands conservation.

Distance to										Slope		Texture	
Agric. Land		Roads		Rural Communities		Urban Settlem.		Rivers					
km	C	km	C	km	C	km	C	km	C	°	C	Type	C
0–0.1	1	0–0.5	1	0–0.5	1	0–1	1	0–0.5	1	0–2	1	Fine	1
0.1–0.5	2	0.5–5	2	0.5–1	2	1–20	2	0.5–1	2	2–5	2	Med.	5
0.5–1	3	5–10	3	1–5	3	20–30	3	1–2	3	5–10	3	Coarse	3
1–10	4	10–20	4	5–10	4	30–40	4	2–5	4	10–20	4		
10->	5	20->	5	10->	5	40->	5	5->	5	20->	5		

Agric. = Agriculture; Settlem = settlements; C = Class; Med. = Medium.

Maps of distances to agricultural lands, roads, rural communities, urban settlements, and rivers were classified into the five aforementioned classes. These are relevant because the distance to anthropogenic activities, or the distance to incentives for anthropogenic activities (i.e., rivers), affects the grassland's suitability for conservation. As distances are small, susceptibility decreases.

There is no consensus in the literature about the proximities or the criteria used in this study. Hence, we proposed the distance ranges stated in Table 2. The range used for each class was defined by expert knowledge and by literature review, as it has been performed previously [56]. In addition, the first class of each criterion was defined with the shortest range because we assumed that the nearest distance has the strongest influence.

The distances to agricultural lands were classified into 5 classes, the lowest value being assigned to the lowest distances (0–0.1 km), because we assumed that the shortest distances have the strongest influences on changing the configuration of the landscape. The distances to roads and rural communities were also classified into 5 classes by assigning the lowest value to the shortest distances (0–0.5 km).

Likewise, urban areas were classified into 5 classes, assigning the lowest value to the nearest distance (0–1.0 km). The proximity to rivers can attract anthropogenic activities and, consequently, increase soil erosion. The map of distance to rivers was classified into 5 classes, assigning the lowest suitability to the shortest distances. Regarding slope, low values (0–2°) make the land more suitable for the establishment of anthropogenic activities, such as cropping or grazing. These criteria specify the horizontal distance at which the anthropogenic activity could cause an alteration.

The soil textures were classified into 3 classes. Fine textures are associated with small pore sizes, which do not favor infiltration, and the soil tends to become saline. Conversely, when soil textures are coarse, pores are large and the soil gets dry quickly, preventing appropriate root development of grasses. Therefore, the highest suitability value was assigned to the medium texture, while the coarse and fine textures got suitability values of 3 and 5, respectively.

Based on expert knowledge and on the literature, the technique would result in a spatially explicit index to evaluate the suitability of areas for grassland conservation.

2.5. Scenarios

After reclassifying the criteria, four scenarios were generated. A scenario was produced by ranking, with the highest weight being a criterion of interest to reflect its importance for LSGC. The result was then represented in a map. Four weighting scenarios were considered for this study (Table 3). The first scenario corresponded to Nonintensive agriculture, where it is assumed that agriculture land, rural communities, and urban settlements (criteria) produce the same pressure on grassland conservation. It was represented by assigning a weight of 20 to each one of these criteria. The second scenario considered intensive agriculture as the criterion emitting the highest pressure. Under this case, agriculture is the main driver for the modification of the landscape and, consequently, the alteration of grasslands. That was represented by giving a weight of 40 to this criterion. The Urban criterion, with a weight of 40, was considered as the main alteration driver for the third scenario. Similarly for the fourth scenario, rural communities, with a weight of 40, were considered as the main landscape modifier. Under this scenario, dispersed localities of the rural type have the greatest pressure on the territory.

Table 3. Proposed scenarios and weighting (%) of the criteria based on their influence on the land suitability for grasslands conservation.

Criteria \ Scenario	Nonintensive Agriculture	Intensive Agriculture	Urban	Rural
Agriculture land	20	40	15	20
Roads	10	10	10	10
Rural communities	20	20	10	40
Urban settlements	20	5	40	5
Slope	10	10	10	10
Rivers	10	5	5	10
Texture	10	10	10	5
Total	100	100	100	100

The weights assigned to the criteria in each scenario represent the strength or influence of each criterion within the scenario. The values of each pixel from each criterion are multiplied by its corresponding weight (Figure 3). This is repeated for each criterion, the resulting values are added and then rounded to the nearest integer before they are assigned to each pixel to obtain a final map for one scenario. Within each scenario, the lower the cumulative value for each pixel, the lower the LSGC. Conversely, the higher the cumulative value for each pixel, the higher the suitability.

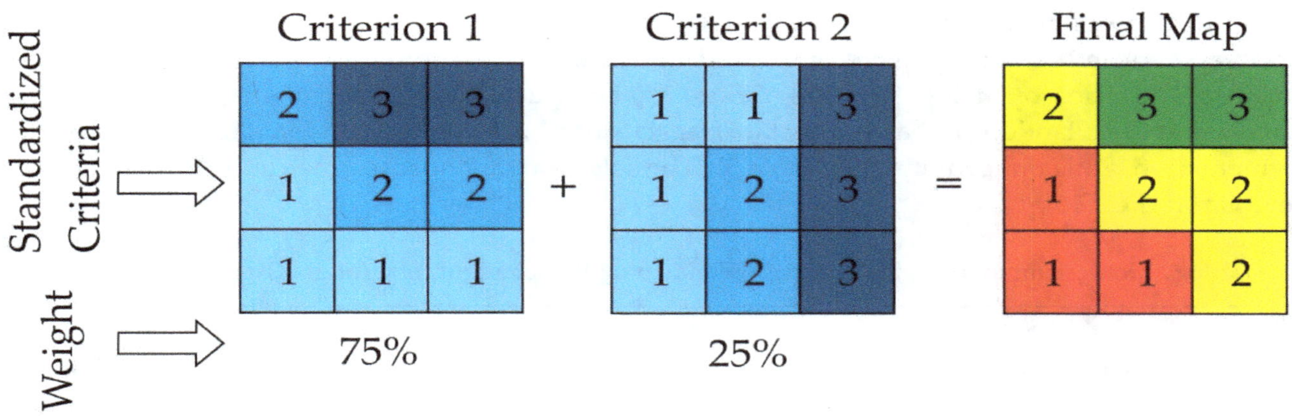

Figure 3. A schematic example of the final map generation of a scenario with the MCE-WO technique. MCE-WO: multicriteria evaluation techniques with weighted overlay.

Finally, the categories assigned to the cumulative values were: Very high, High, Moderate, Low and Very low, as can be seen in Table 4. The LSGC considered as Very high includes the greatest distance to roads, agricultural areas, rural communities, urban settlements, rivers, moderate slopes, and medium soil textures. Meanwhile, areas with Very low LSGC are those showing proximity to agriculture lands, roads, rural communities, urban settlements, rivers, low slopes, and fine soil textures.

Table 4. Definition of land-suitability classes for grasslands conservation.

Land-Suitability Class	Value	Description
Very high	5	Optimal lands for the development of areas for grasslands conservation activities together with other land cover types. This allows for maintaining diversity.
High	4	Lands with good characteristics for grasslands conservation together with other land cover types, which allows for maintaining diversity. However, some patches corresponding to human activities could be found close by.
Moderate	3	Lands, which are moderately appropriate for the implementation of grasslands conservation activities because of the close presence of anthropogenic activities.
Low	2	Lands that are not good for grasslands conservation activities, either because anthropogenic activities take place nearby or because natural conditions do not favor conservation of grasslands.
Very low	1	Lands with very bad conditions for the development of grasslands conservation activities. There exists a direct influence from anthropogenic activities.

The reclassification of the input maps and the MCE-WO was carried out using the Reclassify and Weighted Overlay tool, respectively. The software used for that purpose was ArcMap 10.3© software (ESRI, Redlands, CA, USA; https://www.esri.com/en-us/home).

2.6. *Scenarios Comparison through Landscape Indexes*

Fragmentation analysis was used to understand the land-suitability configuration for grasslands conservation within each scenario [57]. The metrics of the fragmentation analysis show numerical information about the land-suitability composition, configuration, and dimension. They also allow comparisons of different times and even help recreate future scenarios [58,59]. Consequently, landscape metrics have been widely used to study large natural areas, the evolution of grasslands, urban expansion, among others [15].

Landscape metrics can be applied at three different scales: landscape, class, and patch [57]. Landscape and class metrics were only used in this study. Nine metrics were calculated for the proposed scenarios by using FRAGSTATS v4.2© software (UMASS, Amherst, MA, USA; https://www.umass.edu/landeco/research/fragstats/fragstats.html), calculating land-suitability heterogeneity, fragmentation, and connectivity [60] (Table 5). The selection of landscape metrics was made based on previous studies [15,61].

Table 5. Metrics employed in the analysis of land-suitability configuration for grasslands conservation.

Index	Acronym	Level of Analysis	Type
Number of Patches	NumP	L/C	Fragmentation
Patch Density	PD	L/C	Fragmentation
Edge Density	ED	L/C	Fragmentation
Mean Patch Size	MPS	L/C	Fragmentation
Perimeter-Area Fractal Dimension	PAFRAC	L	Fragmentation
Shannon's Diversity Index	SHDI	L	Heterogeneity
Percentage of Landscape	PLAND	C	Fragmentation
Euclidean Nearest-Neighbor Distance	ENN_MN	C	Connectivity
Patch Cohesion Index	COHESION	C	Connectivity

L = Landscape; C = Class.

3. Results

3.1. Spatial Distribution of LSGC

The application of the MCE-WO technique made it possible to identify areas of LSGC based on seven criteria selected and applied to the Central Valleys of the state of Chihuahua. That allowed for zoning and spatially representing the land suitability based on five classes within four scenarios. The surface areas of the land-suitability categories are shown in Table 6. The very high suitability zone resulted with a surface that went from 1050 km^2 (rural) up to 5763 km^2. The high-suitability zone was the one covering most of the surface for the four scenarios with areas ranging from 34,619 km^2 (intensive agriculture) to 44,264 km^2 (urban). The moderate suitability zone has an area ranging from 6653 km^2 (urban) to 19,559 km^2 (nonintensive agriculture). The low and very low zones showed surfaces ranging from 60 km^2 (nonintensive agriculture) to 5565 km^2.

Table 6. Surface area occupied by five land-suitability classes for grasslands conservation in the Central Valleys of Chihuahua in four different scenarios.

LSC	Nonintensive Agriculture		Intensive Agriculture		Urban		Rural	
	Surface (km^2)	%	Surface (km^2)	%	Surface (km^2)	%	Surface (km^2)	%
Very high	1178	1.96	1141	1.90	5763	9.61	1050	1.75
High	36,072	60.17	34,619	57.74	44,264	73.83	38,752	64.64
Moderate	19,559	32.62	18,426	30.73	6653	11.10	17,456	29.12
Low	3084	5.14	5565	9.28	3085	5.15	2533	4.23
Very low	60	0.10	202	0.34	188	0.31	161	0.27
Total	59,953	100	59,953	100	59,953	100	59,953	100

LSC = Land-suitability class.

In the nonintensive agriculture scenario (Figure 4), the high land-suitability area predominated (60%), followed by the moderate land-suitability zone (32.62%). In the intensive agriculture scenario, high and moderate land-suitability areas also dominated with 57.74% and 30.73%, respectively. The same behavior was registered for the urban and rural scenarios.

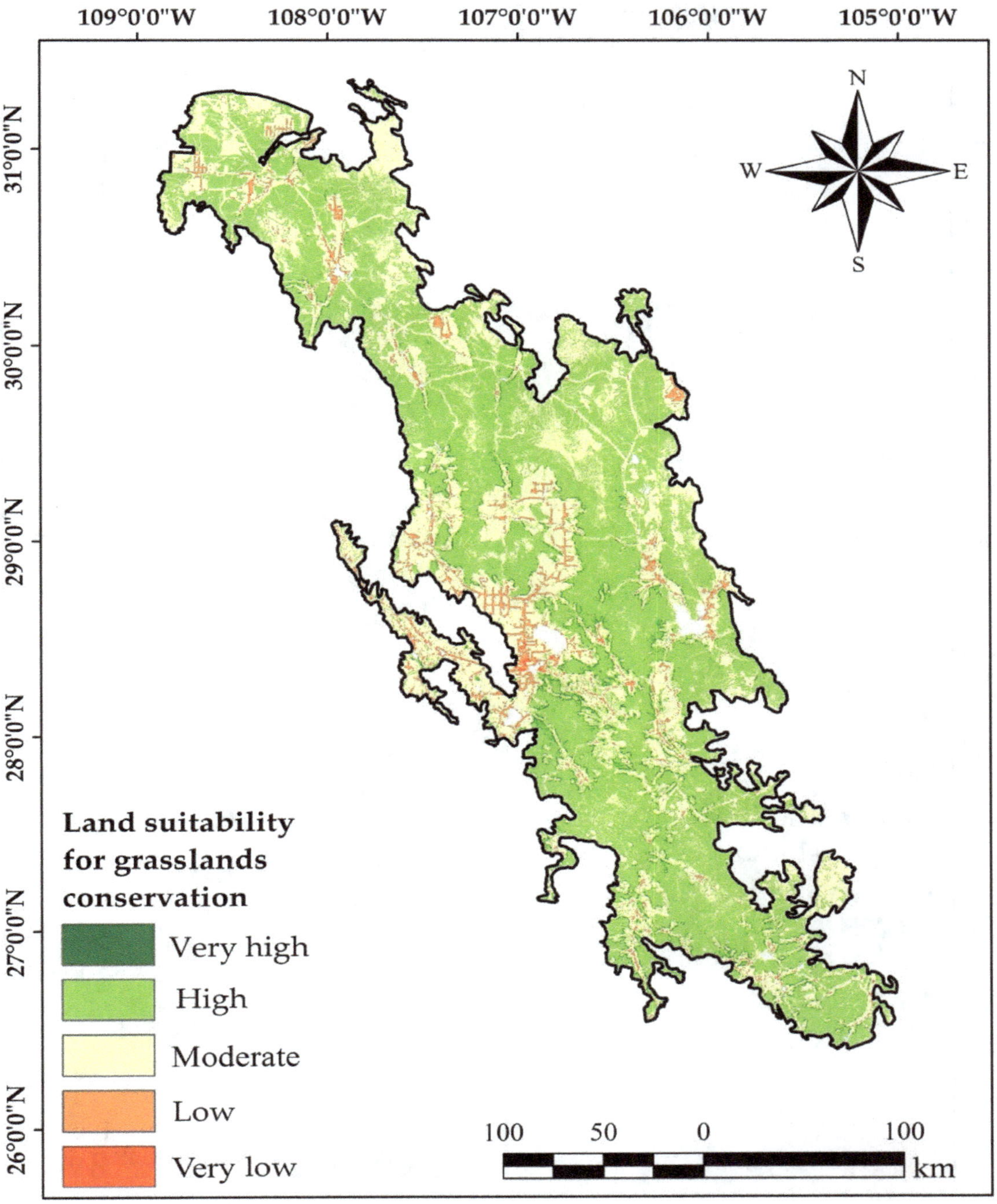

Figure 4. Spatial distribution of land suitability for grasslands conservation in the Central Valleys of the state of Chihuahua. Nonintensive Agriculture scenario.

Regarding land-suitability distribution, in the nonintensive agriculture scenario (Figure 4), very high suitability areas for grasslands conservation are small and distributed throughout the study area. Instead, the high suitability areas are large and distributed throughout the central valleys. The low and very low zones are distributed mainly in the central part of the study area, with some isolated surfaces along the central valleys.

For the intensive agriculture scenario (Figure 5), very high suitability areas resulted with a reduced distribution, mainly located at the edge of the high and moderate suitability areas. The high suitability zones had a distribution similar to the previous scenario. The low and very low suitability zones showed a reduction in their surface areas, which became concentrated mainly to the central and northern central valleys.

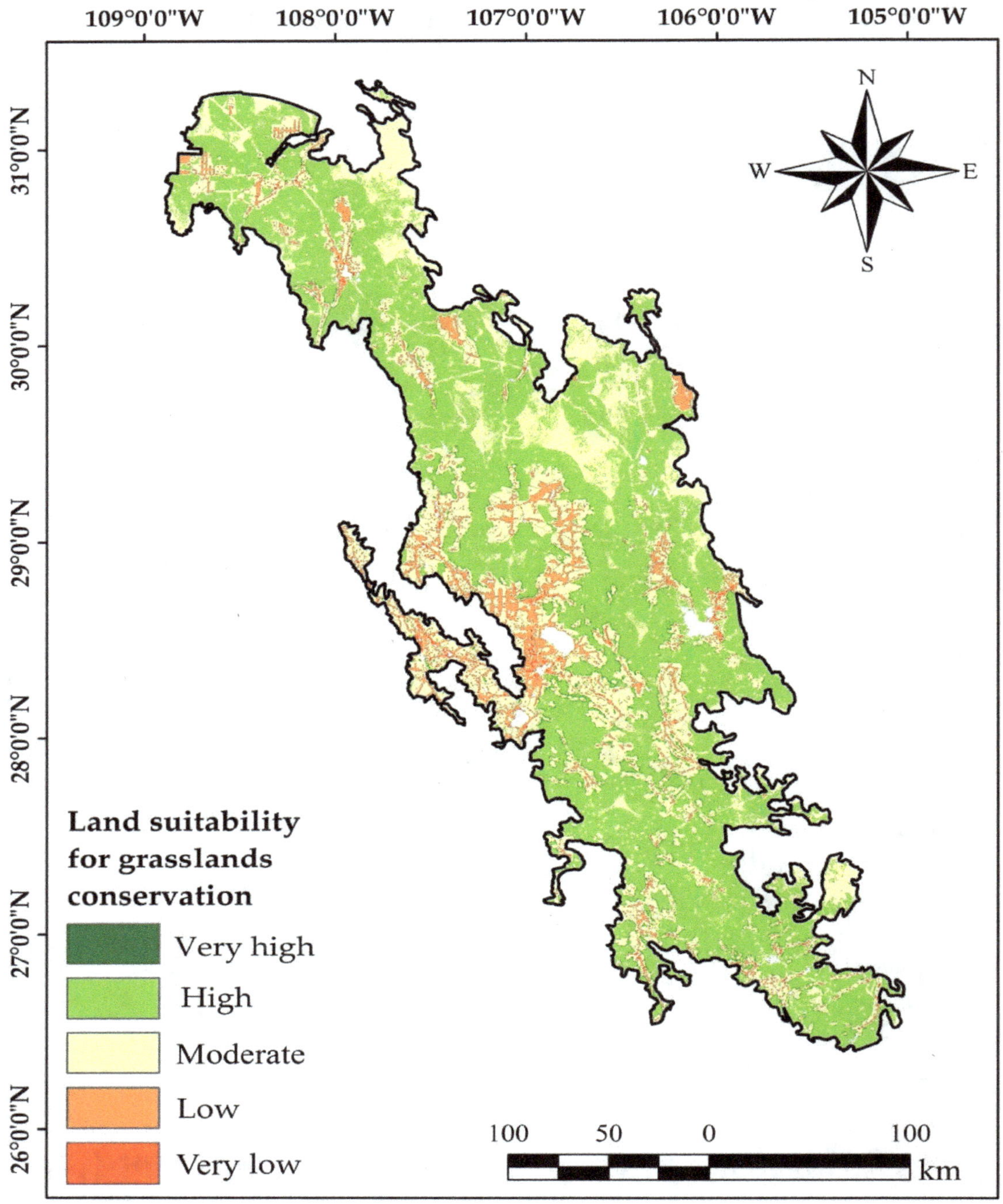

Figure 5. Spatial distribution of land suitability for grasslands conservation in the Central Valleys of the state of Chihuahua. Intensive Agriculture scenario.

For the urban scenario (Figure 6), the areas with very high LSGC have a larger surface area compared to the previous scenarios. The high suitability areas are continuously distributed throughout the central valleys, without presenting isolations. The low and very low areas show a reduction in their surface areas, concentrated in urban areas, as the main driver of change for this scenario.

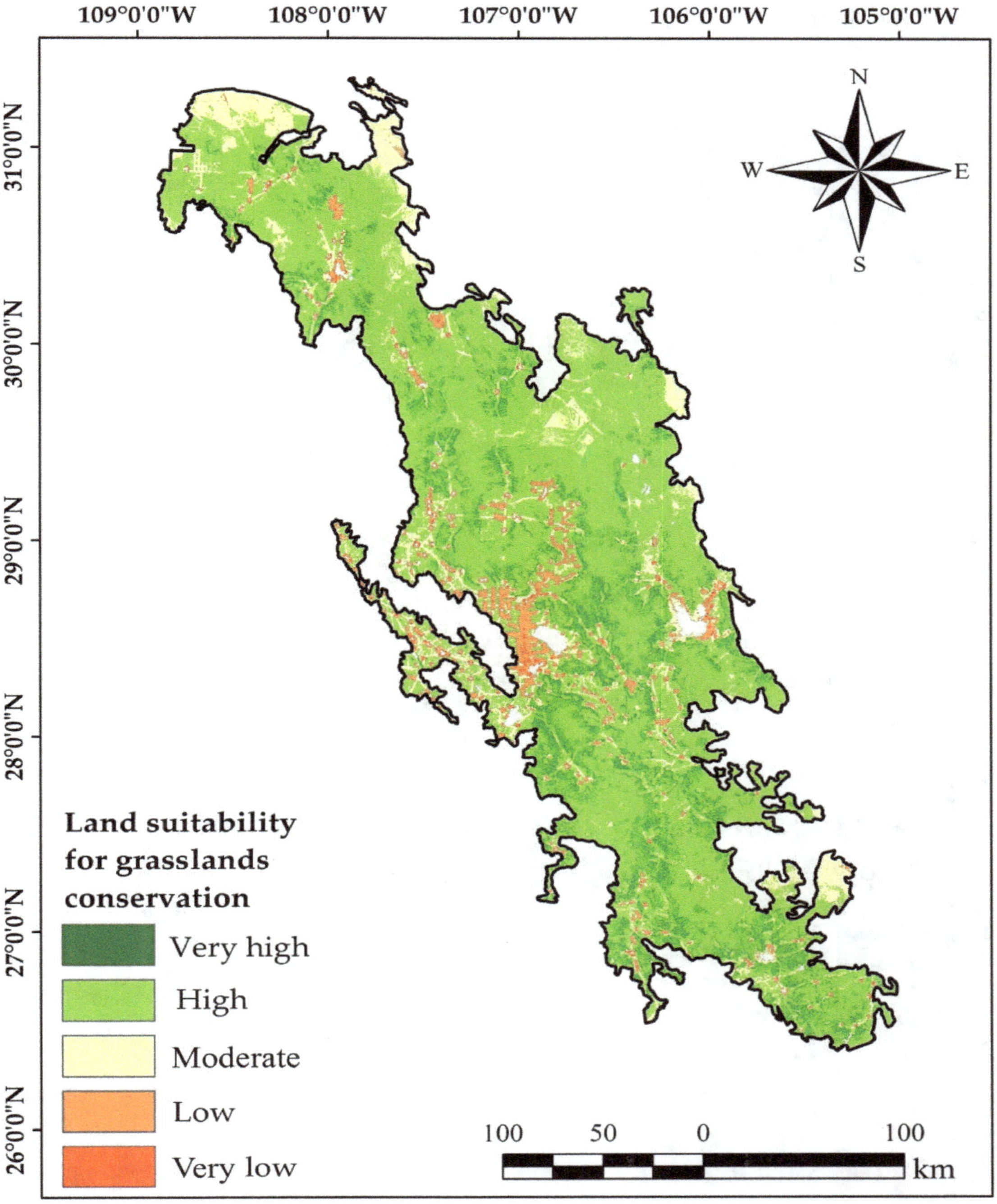

Figure 6. Spatial distribution of land suitability for grasslands conservation in the Central Valleys of the state of Chihuahua. Urban scenario.

For the rural scenario (Figure 7), areas with very high LSGC were small with a dispersed distribution. The high-suitability areas are widely distributed through the central valleys. The low and very low suitability areas show a reduction in surface area compared to the previous scenario, concentrated in the central part of the central valleys. For this scenario, the moderate suitability zones increased their area, but with a dispersed distribution of the rural communities.

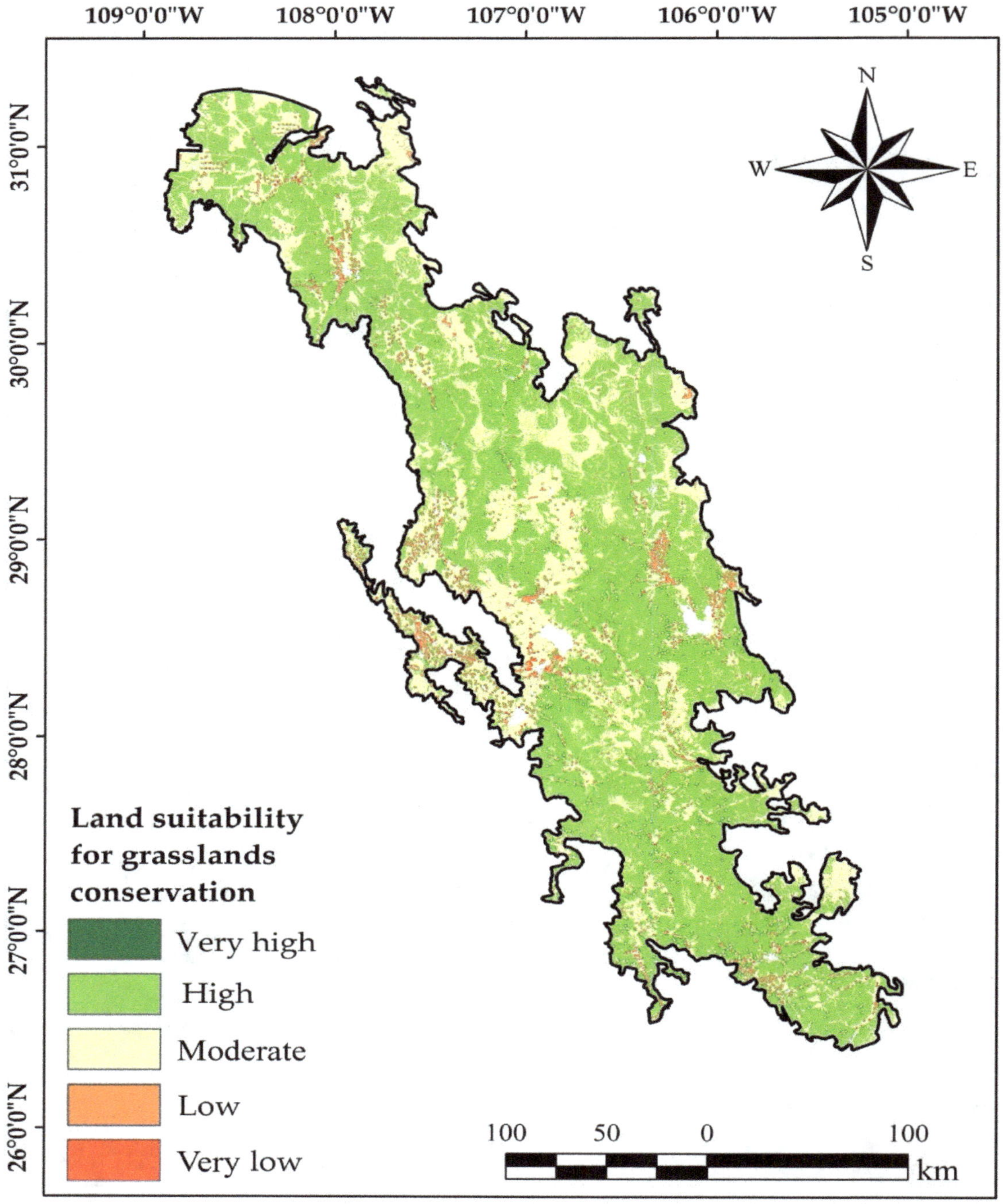

Figure 7. Spatial distribution of land suitability for grasslands conservation in the Central Valleys of the state of Chihuahua. Rural scenario.

3.2. Land Suitability Configuration for Grasslands Conservation

Landscape metrics provided information on the spatial configuration of the land-suitability classes for grasslands conservation. The metrics provide insights on fragmentation, connectivity, and heterogeneity measured at the landscape or class levels. At the landscape level, the number of patches and the density of patches were the highest for nonintensive agriculture (54,640 and 0.92), followed by the rural scenario (46,210 and 0.77). The nonintensive agriculture and urban scenarios showed the highest edge density with values of 20.71 and 18.67, respectively. The mean patch size showed the highest value (155.86) for the intensive agriculture scenario. The perimeter-area fractal dimension did not have large fluctuations among the scenarios, with the highest values corresponded to the urban and nonintensive agriculture scenarios. The intensive agriculture scenario presented the highest value of the Shannon's diversity index (1.0), followed by the nonintensive agriculture scenario with a value of 0.92 (Table 7).

Table 7. Landscape-level metrics used in the land-suitability configuration for grasslands conservation analysis for the Central Valleys of the state of Chihuahua, Mexico.

Scenario	NumP	PD	ED	MPS	PAFRAC	SHDI
Nonintensive Agriculture	54,640	0.92	20.71	108.59	1.43	0.92
Intensive Agriculture	38,070	0.64	15.36	155.86	1.38	1.00
Urban	45,069	0.75	18.67	131.66	1.44	0.88
Rural	46,210	0.77	17.55	128.41	1.36	0.87

NumP = Number of patches; PD = Patch density; ED = Edge density; MPS = Mean patch size; PAFRAC = Perimeter-area fractal dimension; SHDI = Shannon's diversity index.

The class level metrics for each scenario are summarized in Table 8. Class level fragmentation indices show differences within classes or land-suitability classes. An increase in the number of patches was registered for the moderate suitability class in the nonintensive agriculture and rural scenarios. Meanwhile, a smaller number of patches was shown for the Intensive agriculture and urban scenarios. The Euclidean nearest-neighbor distance reflected a decrease in class connectivity for all the scenarios, except for the nonintensive agriculture scenario.

Table 8. Class level metrics used in the land-suitability configuration for grasslands conservation analysis for the Central Valleys of the state of Chihuahua, Mexico.

LSC	PLAND				NumP				ED			
	IntAgr	NIntAgr	Rural	Urban	IntAgr	NIntAgr	Rural	Urban	IntAgr	NIntAgr	Rural	Urban
Very high	2.0	2.1	1.8	10.0	5178	7783	6984	22,936	2.5	3.0	2.6	12.5
High	57.2	59.5	64.1	73.3	11,058	11,584	9820	10,083	10.6	17.7	15.5	17.4
Moderate	31.1	33.1	29.6	11.3	16,247	30,577	23,886	9853	12.9	17.6	13.9	5.5
Low	9.4	5.2	4.3	5.2	4755	4346	4634	1580	4.4	3.0	2.8	1.6
Very low	0.3	0.1	0.3	0.3	832	350	886	617	0.3	0.1	0.3	0.3

LSC	MPS				ENN_MN				COHESION			
	IntAgr	NIntAgr	Rural	Urban	IntAgr	NIntAgr	Rural	Urban	IntAgr	NIntAgr	Rural	Urban
Very high	22.5	15.8	15.7	25.7	359.3	331.5	378.5	263.9	88.4	86.4	84.4	95.4
High	307.1	304.6	387.1	431.1	269.2	275.3	270.5	237.3	99.9	100.0	100.0	100.0
Moderate	113.5	64.3	73.4	67.9	319.0	288.4	307.9	358.7	99.1	99.2	98.8	97.8
Low	116.9	71.3	54.5	195.0	417.0	516.8	725.6	556.1	98.8	97.3	92.1	97.9
Very low	24.7	17.7	18.6	30.9	1057.0	2074.8	1161.3	1231.8	86.5	84.5	83.7	87.6

PLAND = Percentage of landscape; NumP = Number of patches; ED = Edge density; LSC = Land-suitability class; MPS = Mean patch size; ENN_MN = Euclidean nearest-neighbor distance; COHESION = Patch cohesion index; IntAgr = Intensive agriculture; NIntAgr = Nonintensive agriculture.

4. Discussions

In several regions of the world, habitat fragmentation, caused by the increase in grazing areas, as well as by the increase in the agricultural surface area, are considered the main factors causing the loss of the grasslands structure and surface [62,63]. The demand for producing sufficient food from crops for public consumption is leading to the encroachment of agriculture over productive grasslands, where these converted grasslands are often premium-quality lands used by ranchers to graze their livestock during dry years to produce meat [64,65]. This represents a land use/land cover conflict, where, in many cases, grasslands get relegated to areas with low and very low suitability for the implementation of conservation activities. The Central Valleys of the state of Chihuahua do not escape from these pressures [14,15]. Therefore, zoning based on multiple criteria and subsequent conservation schemes are necessary [66]. This study contributes with knowledge about the zoning of grassland areas based on multiple criteria. Moreover, it provides insight on the spatial configuration of the LSGC in the Central Valleys of the state of Chihuahua, Mexico.

The advantages of the MCE and WO techniques with a spatial approach have been reported previously [24]. These techniques were useful to identify the suitability, at different levels, of areas for conservation under four pressure scenarios. The manuscript also integrated a fragmentation analysis at the landscape and class levels, which provided a practical evaluation framework for prioritizing conservation decisions. The scenarios were constructed with seven criteria, which incorporated both natural attributes and proximity to anthropogenic disturbances. These criteria offered several advantages, such as the simplicity of obtaining or generating them at the chosen spatial scale, and the ease of modifying or extracting the desired data.

The results of this work, as well as previous studies [24,49], were useful for identifying different classes of land suitability. In the four scenarios generated, the areas with very low suitability for grasslands conservation are practically null. Likewise, the very high suitability class is also represented by small areas, except for the urban scenario. Regarding the total areas represented by each of the five land-suitability classes for grasslands conservation, the four scenarios showed similar values. However, the spatial distribution of such classes showed notorious differences.

The fragmentation analysis provided an understanding of the configuration of LSGC for different scenarios. Although the nonintensive agriculture scenario is dominated by the areas of high and moderate land-suitability classes, it shows the greatest number of patches, the greatest patch density, edge density, Shannon's diversity index and the smallest medium patch size (Table 5), which may be due to the influence of the small patches represented by plots of nonintensive agriculture, indicating that this scenario is fragmented. This was corroborated by the results from the metrics applied to the scenarios at five land-suitability classes (Table 6), where the number of patches, the percentage of the landscape, the edge density, and the cohesion were also higher for the moderate land suitability in the nonintensive agriculture scenario. For this scenario, the average patch size was the second smallest in the moderate class, followed by the rural scenario. In areas located nearby the study area, fragmentation has been assessed with similar metrics in previous studies, and the results show the presence of fragmentation [67,68].

The connectivity indices indicate, in general, a low landscape connectivity for all the scenarios, especially for the low and very low suitability classes. In particular, for the nonintensive agriculture scenario, the patches of high suitability for grasslands conservation showed the greatest distance within them. This is due to the similar weights given to agriculture and rural communities (Table 2), which caused these two criteria to be drivers of alterations in this scenario. The opposite resulted for the urban scenario, where the area of high suitability proved to be the largest with the shortest distance between patches.

The heterogeneity for the four scenarios showed to be similar. The Shannon's diversity index for the rural scenario showed a value of 0.87, being the least heterogeneous, while the scenario of intensive agriculture showed a value of 1.0, being the most heterogeneous. This is consistent with the findings reported by Sutti et al. [69] in a study on conservation areas for grassland birds.

This study integrated MCE and WO analysis with landscape fragmentation over four scenarios, which served to identify different classes of land suitability and their spatial configuration. This was possible by evaluating the fragmentation and its spatial representation in the four scenarios, providing information about the configuration of the landscape under different pressures. Previous studies have shown that drivers such as agriculture have altered the configuration of grassland ecosystems [63,70,71]. Such alteration has contributed to the decline of grassland bird species, which are distributed in the Central Valleys of Chihuahua [72]. Thus, our study highlights the importance of including agricultural as a driver of disturbance.

The results indicate the difference between the pressures given to the grassland ecosystems in each scenario, maintaining or decreasing the structure of the zones of LSGC and the patterns of the landscape. This is the first study of this kind, which integrates MCE-WO methodologies and fragmentation analysis, providing a starting point for grassland conservation planning.

For the sake of practicality, repeatability, and simplicity of the model, the criteria were not exhaustive and only some criteria known for the definition of LSGC zones were included. However, additional criteria, such as information on actual management regimes, established conservation zones in the study area, socioeconomic criteria that may influence decisions on grassland habitat selection, could be incorporated. In this study, the proposed scenarios were generated by ranking, with the highest weight, the criterion, or criteria assumed to have the strongest influence on each scenario. It is worth mentioning that there exist additional methodologies for the assignment of weights [73]; however, we adhered to this particular one given its simplicity.

The choice of criteria with a knowledge-based approach, such as that of this study, is commonly applied in land-use planning studies [74,75]. Although this approach may become somewhat subjective compared to automated or data-driven approaches, it allows for more realistic results because it is based on local knowledge and from a review of a study previously developed in the study area [15].

Scenarios of LSGC, based on multicriteria evaluation, have a minimum of entry requirements and they can be easily modified [76]. Based on the availability of data from our interest, the modeling of land suitability for conservation can be applied for other ecosystems [77,78]. Several weights can be tested to explore different scenarios with the aim of developing policies that address specific problems [36].

If new rangeland zoning policies are put in place, it is easy to redesign the zoning exercise within a new context. To do this, various land demand scenarios can be constructed, modifying the area of the five conservation classes and repeating the whole analysis. This will make it possible to visualize and compare the effects of the different allocation policies under the different scenarios.

5. Conclusions

The integration of MCE-WO and fragmentation analysis is a valuable approach to estimate land suitability for grasslands conservation. The maps generated should be considered as a guiding tool for detecting lands with a high suitability for grasslands conservation. Managers and stakeholders can then make efforts in promoting conservation, as well as management practices on these lands, where they may have a greater chance of success.

In the Central Valleys of the state of Chihuahua, the areas with a high land suitability for grasslands conservation were distributed mainly in the central part of the study area. The seven criteria chosen served to assess the variability of the land suitability when a higher relative importance is assigned to different pressures, as it was represented by the several scenarios evaluated. Land suitability had a similar behavior for the intensive agriculture, urban, and rural scenarios.

Future research could include the central valleys, not just from Chihuahua. Grasslands do not respond based on geopolitical boundaries but on disturbances. Additional criteria, such as socioeconomic activities, actual zoning plans, and other landscape modifiers than the ones employed in this study, could be considered. This could lead to more comprehensive studies on the best locations to implement conservation management plans and activities.

Although the technique has been applied in this study for grasslands, it has the potential to be applicable for a multitude of other ecosystems or regions.

Author Contributions: Conceptualization, G.V.-Q. and F.V.-G.; Data curation, J.A.P.-A.; Formal analysis, A.P.-A.; Investigation, M.C.V.-A.; Software, G.V.-Q.; Supervision, A.P.-A.; Writing—original draft, G.V.-Q. and J.A.P.-A.; Writing—review & editing, C.R.M.-N. and F.V.-G. All authors have read and agreed to the published version of the manuscript.

References

1. Xu, D.; Guo, X. Some insights on grassland health assessment based on remote sensing. *Sensors* **2015**, *15*, 3070–3089. [CrossRef] [PubMed]
2. Blanco, H.; Lal, R. Soil and water conservation. In *Principles of Soil Conservation and Management*, 1st ed.; Blanco, H., Lal, R., Eds.; Springer Science & Business Media: New York, NY, USA, 2010; Volume 1, pp. 1–16.
3. Fan, J.; Zhong, H.; Harris, W.; Yu, G.; Wang, S.; Hu, Z.; Yue, Y. Carbon storage in the grasslands of China based on field measurements of above-and below-ground biomass. *Clim. Chang.* **2008**, *86*, 375–396. [CrossRef]
4. Zuazo, V.H.D.; Pleguezuelo, C.R.R. Soil-erosion and runoff prevention by plant covers: A review. In *Sustainable Agriculture*, 1st ed.; Lichtfouse, E., Navarrete, M., Debaeke, P., Véronique, S., Alberola, C., Eds.; Springer: Dordrecht, The Netherlands; Heidelberg, Germany; London, UK; New York, NY, USA, 2009; pp. 785–811.
5. Jurado, G.P.; Saucedo, T.R.A.; Morales, N.C.R.; Martínez, S.M. *Almacén y Captura de Carbono en Pastizales y Matorrales de Chihuahua*; INIFAP: Chihuahua, Mexico, 2013.
6. Lamarque, P.; Tappeiner, U.; Turner, C.; Steinbacher, M.; Bardgett, R.D.; Szukics, U.; Scherme, M.; Lavorel, S. Stakeholder perceptions of grassland ecosystem services in relation to knowledge on soil fertility and biodiversity. *Reg. Environ. Chang.* **2011**, *11*, 791–804. [CrossRef]
7. Cheng, X.; An, S.; Chen, J.; Li, B.; Liu, Y.; Liu, S. Spatial relationships among species, above-ground biomass, N, and P in degraded grasslands in Ordos Plateau, northwestern China. *J. Arid Environ.* **2007**, *68*, 652–667. [CrossRef]
8. Gang, C.; Zhou, W.; Chen, Y.; Wang, Z.; Sun, Z.; Li, J.; Qi, J.; Odeh, I. Quantitative assessment of the contributions of climate change and human activities on global grassland degradation. *Environ. Earth Sci.* **2014**, *72*, 4273–4282. [CrossRef]
9. Facelli, J.M.; Temby, A.M. Multiple effects of shrubs on annual plant communities in arid lands of South Austrlia. *Austral Ecol.* **2002**, *27*, 422–432. [CrossRef]
10. He, C.; Zhang, Q.; Li, Y.; Li, X.; Shi, P. Zoning grassland protection area using remote sensing and cellular automata modeling—A case study in Xilingol steppe grassland in northern China. *J. Arid Environ.* **2005**, *63*, 814–826. [CrossRef]
11. Gellesch, E.; Khan, M.A.A.; Jentsch, A.; Beierkuhnlein, C. Grassland experiments under climatic extremes: Reproductive fitness versus biomass. *Environ. Exp. Bot.* **2017**, *144*, 68–75. [CrossRef]
12. Concilio, A.L.; Prevéy, J.S.; Omasta, P.; O'Connor, J.; Nippert, J.B.; Seastedt, T.R. Response of a mixed grass prairie to an extreme precipitation event. *Ecosphere* **2015**, *6*, 1–12. [CrossRef]
13. Mas, J.F.; Velázquez, A.; Díaz-Gallegos, J.R.; Mayorga-Saucedo, R.; Alcántara, C.; Bocco, G.; Castro, R.; Fernández, T.; Pérez-Vega, A. Assessing land use/cover changes: A nationwide multidate spatial database for Mexico. *Int. J. Appl. Earth Obs. Geoinf.* **2004**, *5*, 249–261. [CrossRef]
14. Estrada-Castillón, E.; Scott-Morales, L.; Villarreal-Quintanilla, J.A.; Jurado-Ybarra, E.; Cotera-Correa, M.; Cantú-Ayala, C.; García-Pérez, J. Clasificación de los pastizales halófilos del noreste de México asociados con perrito de las praderas (*Cynomys mexicanus*): Diversidad y endemismo de especies. *Rev. Mex. Biodivers.* **2010**, *81*, 401–416. [CrossRef]
15. Manjarrez-Dominguez, C.; Pinedo-Alvarez, A.; Pinedo-Alvarez, C.; Villarreal-Guerrero, F.; Cortes-Palacios, L. Vegetation landscape analysis due to land use changes on arid lands. *Pol. J. Ecol.* **2015**, *63*, 167–175. [CrossRef]
16. Instituto Nacional de Estadistica, Geografía en Informatica (INEGI). *Síntesis de información geográfica del Estado de Chihuahua*, 1st ed.; Instituto Nacional de Estadística, Geografía e Informática: Aguascalientes, Mexico, 2003; p. 113.
17. Margules, C.R.; Pressey, R.L. Systematic conservation planning. *Nature* **2000**, *405*, 243–253. [CrossRef]
18. Meffe, G.K.; Nielsen, L.A.; Knight, R.L.; Schenborn, D.A. *Ecosystem Management: Adaptive, Community-Based Conservation*; Island Press: Washington, DC, USA, 2002; pp. 169–184.
19. Knight, A.T.; Cowling, R.M. Embracing opportunism in the selection of priority conservation areas. *Conserv. Biol.* **2007**, *21*, 1124–1126. [CrossRef]
20. Walke, N.; Reddy, G.O.; Maji, A.K.; Thayalan, S. GIS-based multicriteria overlay analysis in soil-suitability evaluation for cotton (*Gossypium* spp.): A case study in the black soil region of central India. *Comput. Geosci.* **2012**, *41*, 108–118. [CrossRef]
21. Zolekar, R.B.; Bhagat, V.S. Multi-criteria land suitability analysis for agriculture in hilly zone: Remote sensing and GIS approach. *Comput. Electron. Agric.* **2015**, *118*, 300–321. [CrossRef]

22. Aldababseh, A.; Temimi, M.; Maghelal, P.; Branch, O.; Wulfmeyer, V. Multi-criteria evaluation of irrigated agriculture suitability to achieve food security in an arid environment. *Sustainability* **2018**, *10*, 803. [CrossRef]
23. Massam, B.H. Multi-criteria decision making techniques in planning. *Prog. Plan.* **1988**, *30*, 1–84. [CrossRef]
24. Geneletti, D.; van Duren, I. Protected area zoning for conservation and use: A combination of spatial multicriteria and multiobjective evaluation. *Landsc. Urban Plan.* **2008**, *85*, 97–110. [CrossRef]
25. Pereira, J.M.C.; Duckstein, L. A multiple criteria decision-making approach to GIS based land suitability evaluation. *Int. J. Geogr. Inf. Sci.* **1993**, *7*, 407–424. [CrossRef]
26. Malczewski, J. GIS-based land-use suitability analysis: A critical overview. *Prog. Plan.* **2004**, *62*, 3–65. [CrossRef]
27. Eastman, R. Multi-criteria evaluation and GIS. Chapter 35. In *Geographical Information Systems*; Longley, P.A., Goodchild, M.F., Maguire, D.J., Rhind, D.W., Eds.; Wiley: New York, NY, USA, 1999; pp. 493–502.
28. Cengiz, T.; Akbulak, C. Application of analytical hierarchy process and geographic information systems in land-use suitability evaluation: A case study of Dümrek village (Çanakkale, Turkey). *Int. J. Sustain. Dev. World Ecol.* **2009**, *16*, 286–294. [CrossRef]
29. Mendas, A.; Delali, A. Integration of MultiCriteria Decision Analysis in GIS to develop land suitability for agriculture: Application to durum wheat cultivation in the region of Mleta in Algeria. *Comput. Electron. Agric.* **2012**, *83*, 117–126. [CrossRef]
30. Huth, A.; Drechsler, M.; Köhler, P. Using multicriteria decision analysis and a forest growth model to assess impacts of tree harvesting in Dipterocarp lowland rain forests. *For. Ecol. Manag.* **2005**, *207*, 215–232. [CrossRef]
31. Pourghasemi, H.R.; Pradhan, B.; Gokceoglu, C. Application of fuzzy logic and analytical hierarchy process (AHP) to landslide susceptibility mapping at Haraz watershed, Iran. *Nat. Hazards* **2012**, *63*, 965–996. [CrossRef]
32. Torrieri, F.; Batà, A. Spatial multi-criteria decision support system and strategic environmental assessment: A case study. *Buildings* **2017**, *7*, 96. [CrossRef]
33. Hailegebriel, S. Irrigation Potential Evaluation and Crop Suitability Analysis Using GIS and Remote Sensing Technique in Beles Sub Basin, Beneshangul Gumez Region. Master's Thesis, Addis Ababa Universty, Addis Ababa, Ethiopia, 2007.
34. Zelalem, A. Land Use/Land Cover Dynamics and Vegetation Vulnerability Analysis: A Case Study of Arsi Negele Wereda. Master's Thesis, Addis Ababa Universty, Addis Ababa, Ethiopia, 2007.
35. Girard, L.F.; Cerreta, M.; De Toro, P. Integrated spatial assessment: A multidimensional approach for sustainable planning. In Proceedings of the MTISD 2008—Methods, Models and Information Technologies for Decision Support System, Lecce, Italy, 18–20 September 2008; pp. 277–280.
36. Mayfield, C.J. Automating the Classification of Thematic Rasters for Weighted Overlay Analysis in GeoPlanner for ArcGIS. Master's Thesis, University of Redlands, Redlands, CA, USA, 2016.
37. Camacho Olmedo, M.T.C.; Paegelow, M.; Mas, J.F.; Escobar, F. *Geomatic Approaches for Modeling Land Change Scenarios*; Springer: Cham, Switzerland, 2018; pp. 27–30.
38. Rodewald, A.D. The importance of land uses within the landscape matrix. *Wildl. Soc. Bull.* **2003**, *31*, 586–592.
39. Comisión Nacional para el Conocimiento y Uso de la Biodiversidad (CONABIO). *La Biodiversidad en Chihuahua: Estudio de Estado*; Comisión Nacional para el Conocimiento y Uso de la Biodiversidad: Distrito Federal, Mexico, 2014.
40. International Union for Conservation of Nature (IUCN). Nature-Based Solutions. 2016. Available online: http://www.iucn.org/regions/europe/our-work/nature-based-solutions (accessed on 16 December 2019).
41. Guerrero, P.; Haase, D.; Albert, C. Locating spatial opportunities for nature-based solutions: A river landscape application. *Water* **2018**, *10*, 1869. [CrossRef]
42. Stürck, J.; Schulp, C.J.; Verburg, P.H. Spatio-temporal dynamics of regulating ecosystem services in Europe—The role of past and future land use change. *Appl. Geogr.* **2015**, *63*, 121–135. [CrossRef]
43. Cohen-Shacham, E.; Walters, G.; Janzen, C.; Maginnis, S. *Nature-Based Solutions to Address Global Societal Challenges*; IUCN: Gland, Switzerland, 2016; pp. 2–10.
44. Chen, F.; Li, L.; Niu, J.; Lin, A.; Chen, S.; Hao, L. Evaluating Ecosystem Services Supply and Demand Dynamics and Ecological Zoning Management in Wuhan, China. *Int. J. Environ. Res. Public Heath* **2019**, *16*, 2332. [CrossRef]

45. Comisión Nacional para el Conocimiento y Uso de la Biodiversidad (CONABIO). Sierra del Nido-Pastizal de Flores Magón. Available online: http://www.conabio.gob.mx/conocimiento/regionalizacion/doctos/rtp_047.pdf (accessed on 20 January 2019).
46. García, E. *Climas (Clasificación de Köeppen, Modificado por García) Escala 1:1000,000*; CONABIO: Distrito Federal, Mexico, 1998.
47. Samson, F.B.; Knopf, F.L.; Ostlie, W.R. Great Plains ecosystems: Past, present, and future. *Wildl. Soc. Bull.* **2004**, *32*, 6–15. [CrossRef]
48. International Union for Conservation of Nature (IUCN). Available online: http://www.iucnredlist.org/details/18868/0 (accessed on 20 January 2019).
49. Davies, A.L.; Bryce, R.; Redpath, S.M. Use of multicriteria decision analysis to address conservation conflicts. *Conserv. Biol.* **2013**, *27*, 936–944. [CrossRef] [PubMed]
50. Instituto Nacional de Geografía e Informática. Uso del Suelo y Vegetación. Available online: https://www.inegi.org.mx/temas/usosuelo/default.html#Descargas (accessed on 20 February 2018).
51. Instituto Nacional de Geografía e Informática. Marco Geoestadistico. Available online: https://www.inegi.org.mx/temas/mg/default.html#Descargas (accessed on 20 February 2018).
52. Instituto Nacional de Geografía e Informática. Simulador de Flujos de Agua de Cuencas Hidrográficas. Available online: http://antares.inegi.org.mx/analisis/red_hidro/siatl/# (accessed on 20 February 2018).
53. Instituto Nacional de Geografía e Informática. Continuo de elevaciones mexicano (CEM). Available online: https://www.inegi.org.mx/app/geo2/elevacionesmex/ (accessed on 20 February 2018).
54. Instituto Nacional de Geografía e Informática. Edafología. Available online: https://www.inegi.org.mx/temas/edafologia/default.html#Descargas (accessed on 20 February 2018).
55. Food and Agriculture Organization of the United Nations. *A Framework for Land Evaluation*; FAO: Rome, Italy, 1977.
56. Gorsevski, P.V.; Donevska, K.R.; Mitrovski, C.D.; Frizado, J.P. Integrating multi-criteria evaluation techniques with geographic information systems for landfill site selection: A case study using ordered weighted average. *Waste Manag.* **2012**, *32*, 287–296. [CrossRef] [PubMed]
57. McGarigal, K.; Marks, B. FRAGSTATS: Spatial Pattern Analysis Program for Quantifying Landscape Structure. USDA Forest Service General Technical Report PNW-GTR-351. 1995. Available online: https://www.fs.fed.us/pnw/pubs/pnw_gtr351.pdf (accessed on 13 February 2019).
58. Mas, J.F.; Pérez-Vega, A.; Clarke, K.C. Assessing simulated land use/cover maps using similarity and fragmentation indices. *Ecol. Complex* **2012**, *11*, 38–45. [CrossRef]
59. Sohares-Filho, B.; Silvestrini, R.; Nepstad, D.; Brando, P.; Rodrigues, H.; Alencar, A.; Coe, M.; Locks, C.; Lima, L.; Hissa, L.; et al. Forest fragmentation, climate change and understory fire regimes on the Amazonian landscapes of the Xingu headwaters. *Lands. Ecol.* **2012**, *27*, 585–598. [CrossRef]
60. McGarigal, K.; Cushman, S.A.; Neel, M.C.; Ene, E. FRAGSTATS: Spatial Pattern Analysis Program for Categorical Maps. 2002. Available online: www.umass.edu/landeco/research/fragstats/fragstats.html (accessed on 20 March 2019).
61. De León Mata, G.D.; Pinedo Álvarez, A.; Martínez Guerrero, J.H. Aplicación de sensores remotos en el análisis de la fragmentación del paisaje en cuchillas de la zarca, México. *Investigaciones Geográficas* **2014**, *84*, 42–53. [CrossRef]
62. Otuoma, J.; Kinyamario, J.; Ekaya, W.; Kshatriya, M.; Nyabenge, M. Effects of human-livestock-wildlife interactions on habitat in an eastern Kenya rangeland. *Afr. J. Ecol.* **2009**, *47*, 567–573. [CrossRef]
63. Reino, L.; Beja, P.; Araujo, M.B.; Dray, S.; Segurado, P. Does local habitat fragmentation affect large-scale distributions? The case of a specialist grassland bird. *Divers. Distrib.* **2013**, *19*, 423–432. [CrossRef]
64. Dodd, J. Desertification and degradation of Africa's rangelands. *Rangelands* **1994**, *16*, 180–183.
65. Rashford, B.S.; Walker, J.A.; Bastian, C.T. Economics of grassland conversion to cropland in the Prairie Pothole Region. *Conserv. Biol.* **2011**, *25*, 276–284. [CrossRef]
66. Janišová, M.; Bartha, S.; Kiehl, K.; Dengler, J. Advances in the conservation of dry grasslands: Introduction to contributions from the seventh European Dry Grassland Meeting. *Plant Biosyst. Int. J. Deal. All Asp. Plant Biol.* **2011**, *145*, 507–513. [CrossRef]
67. Martínez-Estévez, L.; Balvanera, P.; Pacheco, J.; Ceballos, G. Prairie dog decline reduces the supply of ecosystem services and leads to desertification of semiarid grasslands. *PLoS ONE* **2013**, *8*, e75229. [CrossRef] [PubMed]

68. Pool, D.B.; Panjabi, A.O.; Macias-Duarte, A.; Solhjem, D.M. Rapid expansion of croplands in Chihuahua, Mexico threatens declining North American grassland bird species. *Biol. Conserv.* **2014**, *170*, 274–281. [CrossRef]
69. Sutti, F.; Strong, A.; Perlut, N. A multicriteria decision analysis for identifying priority conservation areas 472 for grassland birds. *Northeast Nat.* **2017**, *24*, 99–119. [CrossRef]
70. Hoekstra, J.M.; Boucher, T.M.; Ricketts, T.H.; Roberts, C. Confronting a biome crisis: Global disparities of habitat loss and protection. *Ecol. Lett.* **2005**, *8*, 23–29. [CrossRef]
71. Zhou, X.; Helmers, M.J.; Asbjornsen, H.; Kolka, R.M.; Tomer, M.D. Perennial filter strips reduce nitrate levels in soil and shallow groundwater after grassland-to-cropland conversion. *J. Environ. Qual.* **2010**, *39*, 2006–2015. [CrossRef]
72. Macias-Duarte, A.; Panjabi, A.O.; Pool, D.; Youngberg, E.; Levandoski, G. Wintering Grassland Bird Density in Chihuahuan Desert Grassland Priority Conservation Areas, 2007–2011. Rocky Mountain Bird Observatory, Brighton, CO, RMBO Technical, Report I-NEOTROP-MXPLAT-10-2. Available online: https://birdconservancy.org/wp-content/uploads/2014/06/2011_Chihuahuan_Desert_Grassland_Bird_report_with_appendixAB.pdf (accessed on 18 February 2019).
73. Malczewski, J. *GIS and Multicriteria Decision Analysis*; John Wiley & Sons: New York, NY, USA, 1999; pp. 177–192.
74. Krueger, T.; Page, T.; Hubacek, K.; Smith, L.; Hiscock, K. The role of expert opinion in environmental modelling. *Environ. Model. Softw.* **2012**, *36*, 4–18. [CrossRef]
75. Prieto-Amparán, J.A.; Villarreal-Guerrero, F.; Martínez-Salvador, M.; Manjarrez-Domínguez, C.; Vázquez-Quintero, G.; Pinedo-Alvarez, A. Spatial near future modeling of land use and land cover changes in the temperate forests of Mexico. *PeerJ* **2019**, *7*, e6617. [CrossRef]
76. Bottero, M.; Comino, E.; Duriavig, M.; Ferretti, V.; Pomarico, S. The application of a Multicriteria Spatial Decision Support System (MCSDSS) for the assessment of biodiversity conservation in the province of Varese (Italy). *Land Use Policy* **2013**, *30*, 730–738. [CrossRef]
77. Garmendia, E.; Gamboa, G. Weighting social preferences in participatory multi-criteria evaluations: A case study on sustainable natural resource management. *Ecol. Econ.* **2012**, *84*, 110–120. [CrossRef]
78. Nguyen, T.T.; Verdoodt, A.; Van Y, T.; Delbecque, N.; Tran, T.C.; Van Ranst, E. Design of a GIS and multi-criteria based land evaluation procedure for sustainable land-use planning at the regional level. *Agric. Ecosyst. Environ.* **2015**, *200*, 1–11. [CrossRef]

3

Evaluation of Groundwater Potential by GIS-Based Multicriteria Decision Making as a Spatial Prediction Tool

Recep Çelik

Eng. Faculty Civil Eng. Dpt, Dicle University, 21280 Diyarbakır, Turkey; recep.celik@dicle.edu.tr

Abstract: The Tigris River Batman-Hasankeyf region sub-basin drainage area is in the Upper Tigris basin and lies between the area where the Batman stream joins the river and the Yanarsu stream flows into the river. Intensive agricultural activities are carried out in this region, and irrigation is generally obtained from groundwater just as it moves away from the riverfront. The study area is a valuable basin for both Turkey and the Middle East. In this study, the effectiveness of the Geographic Information System (GIS)-based multicriteria decision-making (MCDM) analytic hierarchy process (AHP) as a spatial prediction tool was utilized in exploring the groundwater potential of the drainage area. In the analysis, eight hydrological and hydrogeological criteria were considered as influencing factors, namely, geomorphology, geology, rainfall, drainage density, slope, lineament density, land use, and soil properties. The weights of these criteria were determined through the AHP method; the Arc GIS 10.2.2 program and its submodules were used. The major findings of the study were that groundwater-potential index values of the basin were derived. Groundwater-potential-zone evaluation of the basin was obtained as follows: very poor (19%), poor (17%), moderate (34%), good (17%), very good (13%); and groundwater potential zone (GWPZ) maps of the sub-basin were created.

Keywords: groundwater-potential zone; multicriteria decision making; GIS; analytic hierarchy process (AHP); Upper Tigris sub-basin

1. Introduction

Groundwater is an important alternative water resource for the development of urban and rural areas with increasing needs. Demand for fresh water is increasing due to factors such as population growth, irrigated agricultural activities, and climate change [1]. Climate change and the threat of droughts, surface-water potential, and the risk of pollution of surface waters caused by human activities such as agriculture and industrialization [2,3] necessitate the identification of alternative water resources. Being less polluted than surface water makes groundwater a valuable essential resource. Groundwater is also an essential water source during alternative-surface-water shortages, especially in regions with a semiarid climate such as the Tigris Basin. However, unplanned groundwater use disrupts the natural feeding balance of aquifers. Excessive use of aquifers could have adverse consequences on the environment, ecology, and economy of the region [4,5]. For this reason, groundwater-discharge zones need to be determined with a focus on appropriate water management. While it is always possible to identify groundwater potential with observation-purpose water wells, it is uneconomical in extensive areas.

Several techniques have been adopted by various researchers that do not require fieldwork, e.g., the decision-tree model [6], principal component analysis (PCA) [7], and the logistic-regression model [8]. Most of these methods are based on multivariate statistical techniques [9]. In contrast, the analytical hierarchy process (AHP) is considered to be a simple, effective, transparent, and reliable technique [10]. The AHP method is easy to identify in a useful way by integrating geographic information system (GIS) and remote sensing (RS) data. Groundwater parameters such as precipitation, aquifers, land use, and soil type can easily be defined as spatial data in a GIS environment. Thus, the method is very useful for identifying potential groundwater areas and is well-suited for the useful identification of groundwater-potential regions [11–13]. AHP is an approach that was developed by Thomas L. Saaty in the early 1980s [14]. It was developed with planned criteria according to a certain stage, evaluating the weights of these criteria, and comparing the criteria with other similar studies [15,16]. It is possible and preferable to determine hydrological and hydrogeological parameters that influence groundwater formation and their effects on groundwater formation and generate potential groundwater maps with GIS-based Multicriteria Decision-Making (MCDM) techniques. GIS-based multicriteria decision analysis (MCDA) has been extensively used in the field of hydrological and water resources [16,17]. It has gained popularity in the last decade as it provides GIS-based mapping [18].

Most current studies [19–22] widely applied RS- and GIS-based AHP–MCDM techniques for the assessment of Groundwater Potential Zones (GWPZs) with effective results. When these studies were examined, it was found that land use (LU), geomorphology (GM), geology (G), recharge rate (RR), drainage density (DD), DEM, rainfall (RF), slope (SL), surface water body (SW), lineament density (LD), lithology (LI), and water table depth (WTD) parameters were used in almost all studies. groundwater potential index (GWPI) values of the basins were conditioned by overlaying all those criteria weights with AHP methods [23–26]. These GWPI values were then classified, and GWPZ maps of different global regions were created [27–30]. However, the thematic layers used to identify GWPZs differ according to studies and regions, and the selection of qualitative layers is arbitrary. In almost all studies, the LU, Soil, GM, DD, SL parameters were mainly used. Murthy (2000) added RF in GWPZ studies of the semiarid Andhra Pradesh region [29]. İbrahim Bathis and Ahmed (2016) [30] also added the RF and LD parameters. Sikdar et al. (2004) [31], Prasad et al. (2007) [32], Madrucci et al. (2008) [33], Chowdhury et al. (2009) [34], Senanayake et al. (2016) [13], Zaidi et al. (2015) [35] also took into account the G parameter. Jha et al. (2010) [36] considered SW even though it is a substrate of Land Use. Machiwal et al. (2011) [10] utilized SW, RR, and WTD. Senanayake et al. (2016), and Agarwal and Garg (2016) [37] added the DEM criterion. In pre-2012 studies, LI was also taken into account (Jaiswal et al., 2003 [38] by Srivastava and Bhattacharya in 2006 [39] and Magesh et al. in 2012 [40]).

In this study, the GIS-based MCDM AHP method was used to produce groundwater-potential maps of the Tigris River Batman-Hasankeyf drainage area using meteorological, hydrological, and hydrogeological characteristics without drilling water wells (test drilling) in the area. In this study, the ArcInfo 10.2.2 GIS program and extensions such as spatial analyst and arc hydro were used to define groundwater potential. Eight criteria that affect groundwater discharge, namely, land use, slope, geology, geomorphology, soil map, drainage density, fault density, and precipitation were taken into account. The relative weights of each criterion were determined by the AHP–MCDM method in a GIS environment. GWPZ maps of the basin were obtained as a result of this study. The acquired potential-groundwater-discharge maps properly represent a significant source that helps water-resource managers to make accurate management plans and much better use of groundwater resources.

2. Materials and Methods

2.1. Study Area

The Tigris river is an important source of water that originates within the borders of Turkey and discharges into the Persian Gulf. The total length of the Tigris River is 1750 km, about 500 km of which is within the borders of Turkey. The primary source of the river is in the southeast of Turkey, from Hazar lake and Hazarbaba mountain (2290 m). Within the borders of Turkey, after crossing the

Diyarbakir plain and south of the Batman city center, the river passes through Gercüş and historical Hasankeyf districts. This area is also upstream of the Ilısu dam, which is under the Southeastern Anatolia Project (GAP). The total discharge area of the sub-basin is 1140 km^2 (Figure 1). The Tigris Batman Discharge area is geographically at between 41°01′–41°38′ East longitude and 37°51′–37°32′ North latitude.

Tigris River is an important resource that is used for potable and utility water, as well as agricultural irrigation, with dams constructed in the regions of the river basin. The river has significant agricultural potential in the drainage area, which is why humans have been living in these basins for 10,000 years [41,42]. The study area comprises the ancient Hasankeyf settlement (which is located southeast of Turkey) and Tigris valleys. Although some regions along the river basin have rich surface-water resources, it is solely operated for groundwater use in agricultural irrigation due to incomplete irrigation projects [43]. Although the study area is a valuable basin for Turkey and the Middle East, the assessment of water resources has transcended locality and has reached significant value on a global scale, just like oil and other natural resources.

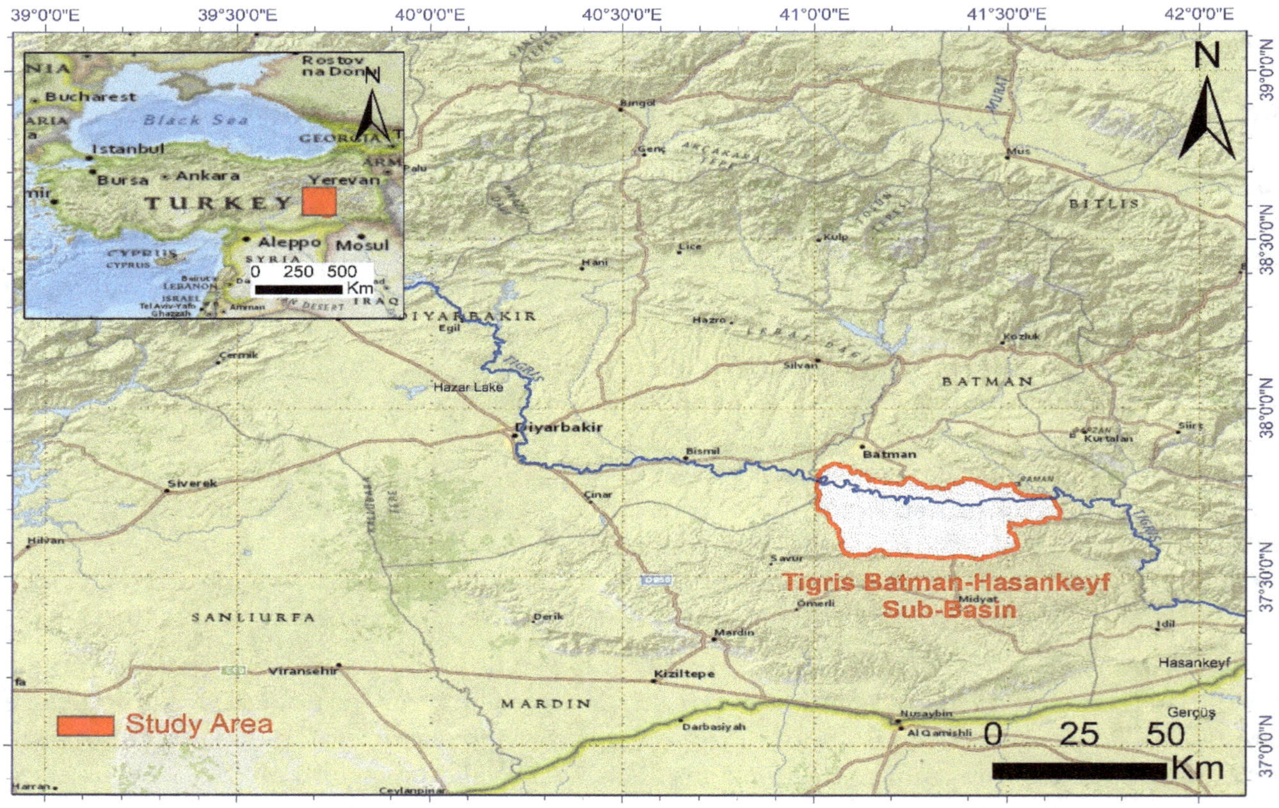

Figure 1. Tigris River Batman-Hasankeyf sub-drainage basin (study area) and its location.

2.2. Method

The GIS–MCDM combination is an excellent evaluation tool that analyzes multiple-criteria methodologies under a GIS environment [44,45]. It simplifies solving problems and improves the use of multiple criteria [46]; the Spatial Analyst extension of the Arc GIS program is perfectly suitable to implement this model. The analysis process was performed with the Spatial Analysis and Overlay tools taking into account the relative values, and grading and interpolating the final thematic maps. To determine the groundwater potential of the region, 8parameters were taken into account: GM, LU, G, LD, DD, S, R, and ST. The relative weight of each parameter was determined by the AHP method. The binary comparison matrices of these parameters are indicated in Table 1, and the weights of each parameter obtained as a result of the normalized values of these parameters are indicated in Table 2. Finally, thematic maps of the groundwater potential of the Upper Tigris Basin Batman subdrainage area were acquired with the Arc GIS spatial analyst overlay method. The methodology flowchart of the entire study is shown in Figure 2.

To obtain groundwater-potential maps of the Batman subdrainage area, 1/25,000 M45 and M46 topographic maps with 5 X 5 m resolution were first digitized; then, DEM maps of the region were produced (Figure 3a). Hydrological land characteristics were extracted by using drainage-density maps. Drainage density, as shown in Equation (1), was acquired by dividing the length of the stream by the drainage area. DD was prepared with DEM data by using the Arc Hydro program. A flow direction (Fdr) map was obtained (Figure 3b); then, the drainage-density map was extracted by dividing thematic river maps that were derived from a flow accumulation (Fac) and stream definition (Figure 3c) operation in the basin area.

$$DD = L/A \tag{1}$$

where DD is Drainage Density, L is drainage length, and A is per-unit area [47].

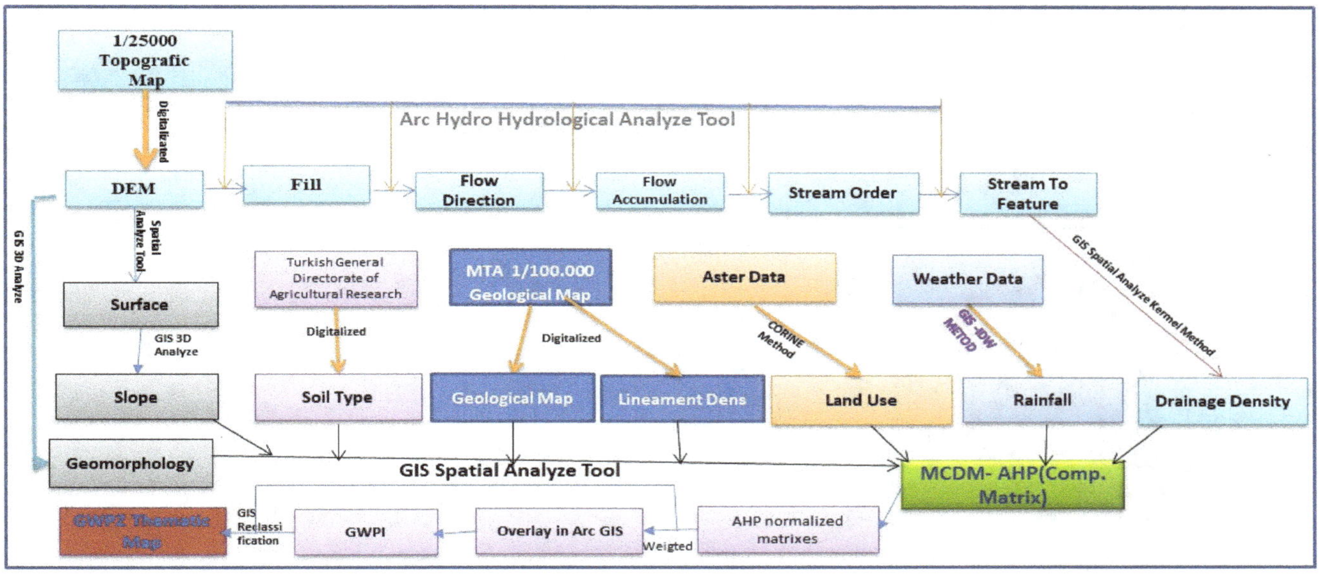

Figure 2. Study methodology flowchart.

From DEM maps, geomorphological maps were produced with the Arc GIS program Spatial Analyst, and S maps were created with the 3D Analysis program. First, geological maps and lineament (fault) lines were converted into kml (Google Earth) format, then into shp format by using the Data Interoperability command via the Turkish General Directorate of Mineral Research and Exploration website [48]. Geological formations were converted into raster format with the related representation, and the geological status of the region was analyzed. Lineament (fault) density (LL) was calculated as per Equation (2) [49]:

$$LL = \sum_{i=1}^{n} Li/A \tag{2}$$

where LL is lineament density, Li is lineament length, i is lineament number, and A is area.

Precipitation (rainfall) data were acquired from the Diyarbakır regional office of the Turkish State Meteorological Service. The average annual-precipitation figures of the last 30 years, observed by meteorological stations at the Diyarbakır–Batman regions, were analyzed by the Spatial Analyst Inverse Distance Weighting IDW interpolation method; then, precipitation maps were formed.

The soil-type study was derived from the Turkish General Directorate of Agricultural Research and Policies. LU data were extracted from the Global Land Cover Facility site as Erdas images. These data were classified with the Coordination of Information on the Environment (CORINE) method. CORINE are land-cover/-use data generated by the computer-assisted visual-interpretation method over satellite imagery and RS data according to land-cover/-use classifications set by the European Environment

Agency [50]. CORINE is employed to utilize the terrain-classification degree of RS data in a GIS environment [51]. In satellite imagery, the coordinates of each pixel were defined according to the coordinates of the ground control points (rectification).

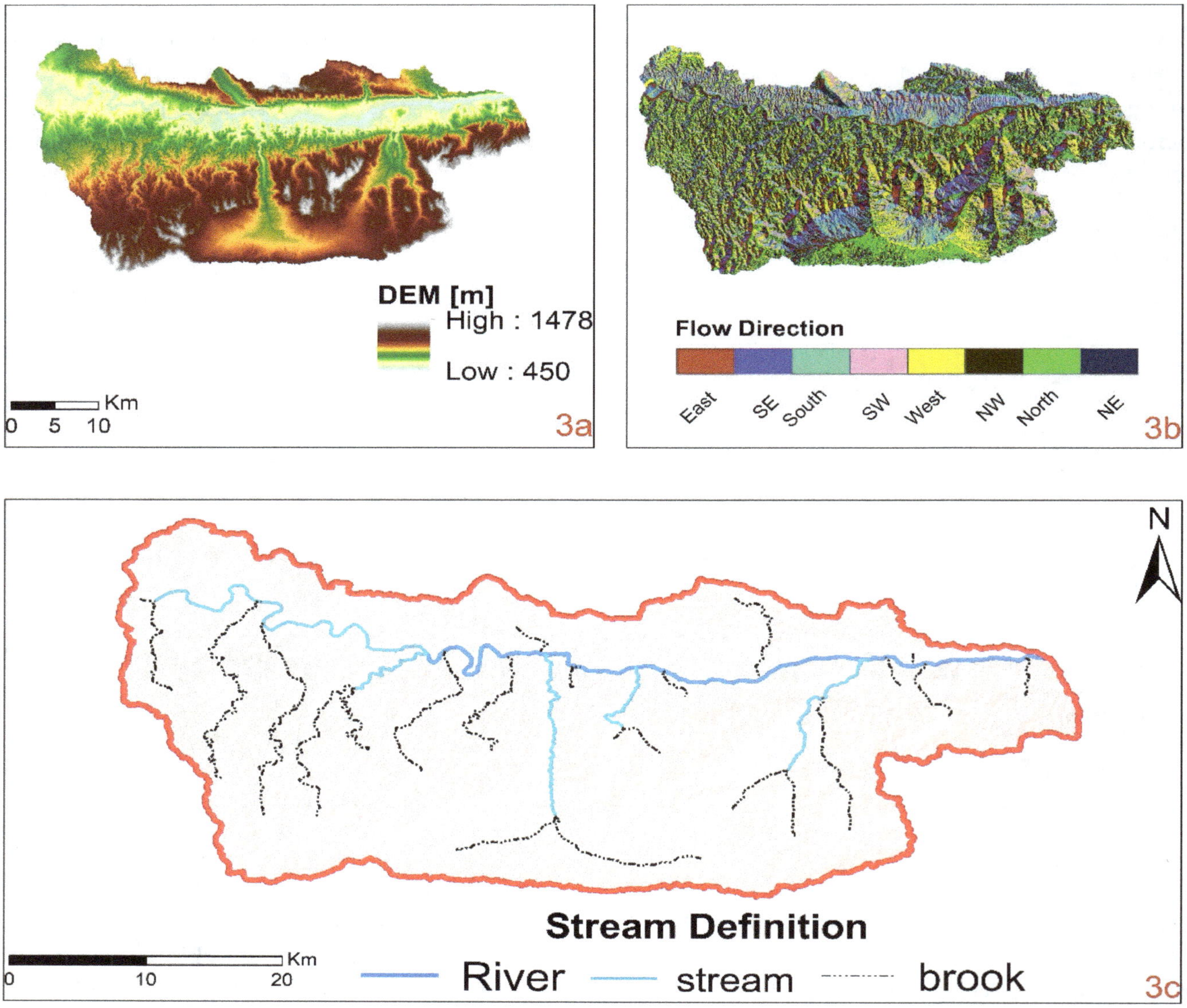

Figure 3. Drainage-density creation process; 3 (**a**): DEM map, 3 (**b**): Flow Direction (Fdr) map; 3 (**c**): Stream Definition.

GWPI is a dimensionless magnitude formed by the sum of the relative weights of the parameters affecting groundwater potential. This coefficient was obtained by calculating the weights of the criteria that made up the groundwater potential by the AHP and GIS-based MCDM method as shown in Equation (3) [52]. This indicator gives information about groundwater potential in different regions [53].

It is calculated according to the AHP method as follows:

$$GWPI = Sr.Sw + LDr.LDw + Gr.Gw + GMr.GMw + LUr.LUw + STr.STw + Rr.Rw + DDr.DDw \quad (3)$$

where GWPI is groundwater potential index, S is Slope, LD is lineament density, G is geology, GM is geomorphology, LU is land use, ST is soil type, R is rainfall, and DD is drainage density. In addition, subscripts r and w refer to the rating and weight of the parameter, respectively.

2.3. GIS-Based AHP Method

The AHP is one of the most widely used MCDM methods. It was developed by Thomas L. Saaty in 1980, and it is an approach created by planned criteria (identification of selection criteria) at a certain stage that involves evaluating criteria weights (relative importance), comparing the alternatives for each criterion, and determining an overall ranking of the alternatives. The AHP is widely used in areas such as reducing confusion/complexity in problems, facilitating decisions, finding priorities by applying expert provisions with comparisons [54], distinguishing resources and solving complexities/confusion, uncovering best options, and planning [55]. Here, the AHP method was applied in rating the selected factors. In order to determine the relative importance of selection factors and weight formation, the AHP methodology was applied, and objective weight assessment was carried out. There had to be at least 2 criteria and 2 alternatives to the technique [56]. Relative importance was given to these criteria, so information and experience are particularly important for accurate assessment and analysis. Criteria could be qualitative and quantitative [57,58]. The GIS-based MCDM as an expert-knowledge-based approach that is very useful for solving complex problems.

The application of the GIS technique and multicriteria decision analysis provides more flexible solutions for the prediction of groundwater-potential zones.

Accordingly, a decision matrix was formed on the basis of 1–9 scaling for parameters affecting a decision. The criteria in the scoring scale were as follows: (1) extremely unimportant, (2) very unimportant, (3) unimportant, (4) moderately unimportant, (5) equally important, (6) moderately important, (7) more important, (8) very important, (9) extremely important [59].

In this study, the weighting of various criteria was carried out through field experiences and literature review. The basic steps for determining the system's normalized weight and consistency ratio (CR) were as follows:

Step 1. Establishment of judgment matrices (p) by pairwise comparison:

$$\mathbf{p} = \begin{vmatrix} p_{11} & p_{12} & \cdots & p_{1n} \\ p_{21} & p_{22} & \cdots & p_{2n} \\ \cdots & \cdots & \cdots & \cdots \\ p_{n1} & p_{n2} & \cdots & p_{nn} \end{vmatrix} \tag{4}$$

where p_n displays the n th indicator unit and p_{nn} is the judgment matrix element.

Step 2. Calculation of normalized weight:

$$Wn = GMn / \sum_{n=1}^{N} GMn \tag{5}$$

where W is weight vector (column) and GMn is the geometric mean of the i th row of the judgment.

Step 3. CR calculation to verify the coherence of the judgements:

$$CR = CI/RCI \tag{6}$$

Consistency index (CI) is denoted as follows:

$$CI = \frac{\lambda\text{max} - N}{N - 1} \tag{7}$$

where λ is the eigenvalue of judgment matrix and it is calculated as follows:

$$\lambda = \frac{\sum_{i=1}^{n} (P_i W)n}{N \cdot W} \tag{8}$$

Random consistency index (RCI) could be obtained from standard tables [60]. The *CR* value had to be about 0.10 or less to be accepted.

3. Results and Discussion

In this study, slope, lineament-density, geology, geomorphology, land-use, soil, rainfall, and drainage-density criteria were taken into consideration. The pairwise comparison matrix is depicted in Table 1. The relative normalized weight (W) of the criteria was then calculated and is shown in Table 2. RCI was calculated as 0.011. Hence, the weight criteria within the criteria were quite consistent. The distribution in the basin and relative weights of these thematic maps are shown in Table 3. In order to determine the groundwater-potential zones in the Tigris River Batman drainage area, the derived parameters of the thematic map are shown in Figure 4. These maps are GM, G, LD, S, R, ST, and LU. As a result, the GWPZ map extracted by the weights of these parameters is shown in Figure 5.

Table 1. Analytic hierarchy process (AHP) pairwise comparison matrix.

Parameters	Slope	Lineament Density	Geology	Geomorphology	Land Use	Soil	Rainfall	Drainage Density
Slope	1.00	1.80	1.29	1.13	1.50	1.29	1.00	1.29
Lineament dens	0.56	1.00	0.71	0.63	0.83	0.71	0.56	0.71
Geology	0.78	1.40	1.00	0.88	1.17	1.00	0.78	1.00
Geomorphology	0.89	1.60	1.14	1.00	1.33	1.14	0.89	1.14
Land use	0.56	1.20	0.86	0.75	1.00	0.86	0.67	0.86
Soil	0.78	1.40	1.00	0.88	1.17	1.00	0.78	1.00
Rainfall	0.89	1.80	1.29	1.13	1.50	1.29	1.00	1.29
Drainage dens	0.49	1.40	1.00	0.88	1.17	1.00	0.78	1.00

Table 2. Normalized AHP matrix.

Parameters	Slope	Lineament Density	Geology	Geomorphology	Land Use	Soil	Rainfall	Drainage Density	Normalized Weight
Slope	0.17	0.16	0.16	0.16	0.16	0.16	0.16	0.16	0.16
Lineament dens	0.09	0.09	0.09	0.09	0.09	0.09	0.09	0.09	0.12
Geology	0.13	0.12	0.12	0.12	0.12	0.12	0.12	0.12	0.11
Geomorphology	0.15	0.14	0.14	0.14	0.14	0.14	0.14	0.14	0.13
Land use	0.09	0.10	0.10	0.10	0.10	0.10	0.10	0.10	0.12
Soil	0.13	0.12	0.12	0.12	0.12	0.12	0.12	0.12	0.11
Rainfall	0.15	0.16	0.16	0.16	0.16	0.16	0.16	0.16	0.14
Drainage Dens	0.08	0.12	0.12	0.12	0.12	0.12	0.12	0.12	0.13

$\lambda_{max} = 8.12$, CI = 0.02, RCI = 1.41, CR = 0.011 < 0.1: acceptable.

3.1. Geomorphology

Geomorphology maps (Figure 4A) are topographic maps showing mountainous, lowland, hilly, and alluvial regions of the region. Alluvial regions have the highest groundwater potential, while mountainous regions have the lowest. The geomorphological distribution of the study area is as follows: mountainous areas (23%), hilly areas (21%), alluvial areas (1%), plains (43%), and plateaus (12%).

3.2. Geology

The geological parameter is one of the most important parameters in terms of groundwater potential. Geological characteristics are important in terms of reflecting aquifer status, which shows groundwater storage. The geological thematic map of the basin is shown in Figure 4B. Limestone, and sandstone and conglomerate ratios were the highest in the basin with 43% and 17%, respectively. Water-retention capacity is higher in these formations. Evaporite rocks and moraine formations are at 22% in the region. The percentage of other formations is shown in Table 3.

Table 3. Summarized AHP assessment of parameter sub-properties.

Weight	Criteria	Sub-Feature	Area Cover (%)	Sub-Rank
7	Geomorphology	Alluvion	1	9
		Plain	43	7
		Plateau	12	5
		Hill	21	3
		Mountain	23	1
7	Geology	Shale	10.9	1
		Sandstone–conglomerate	17.1	7
		Evaporites/moraine	21.7	3
		Clayey limestone/basalt/quartz	0.6	5
		Pebble–sandstone–conglomerate	5.0	7
		Limestone	43.7	8
		Alluvion	1.0	9
5	Lineament density (%)	0–0.01	93.7	1
		0.01–1.5	1.4	3
		1.5–3.5	2.6	5
		3.5–6.50	2.3	7
9	Slope (%)	0–2	9	9
		2–4	6	7
		4–8	12	5
		8–15	25	3
		>15	48	1
7	Rainfall (mm/year)	448–478	43.57	5
		479–508	32.16	6
		509–548	17.08	7
		549–592	7.18	8
6	Soil	Alluvial soil	0.10	9
		Brown forestry soil	56.90	6
		Brown Soil	41.30	4
		Reddish-brown soil	1.70	3
5	Drainage density (%)	0–0.01	55.5	1
		0.1–2	4.1	7
		2–4.06	40.5	9
		Bare rock	7.23	1
		Discrete rural building	0.13	3
		Non-irrigated agricultural field/vineyards	28.92	4
		Agricultural areas with natural vegetation/non-irrigated orchard	12.32	5
		Sparse plant areas	30.23	6
		Grassland	19.79	7
		Irrigated area	0.17	8
		Water bodies	1.21	9

3.3. Lineament Density

Lineament density is an important geological feature that shows rainfall penetration into the ground [61]. Feeding is more common in underground fractured areas. Although fractured areas create discontinuity between regions, they enable rainfall to be fed to underground aquifers in less time. In areas where lineament density is high, groundwater feeding is higher; in areas where density is low, groundwater feeding is less.

Results of the lineament-density map of the basin (Figure 4C) are summarized in Table 3. Accordingly, there is no fault density in 94% of the basin. The area covered by a density of <1.5 part in the basin is 1%, with a density 1.5–3.5 is 3%, and with a density of 3.5–6.5 is 2%.

3.4. Slope

Precipitation infiltration (leakage) is an effective hydrological parameter in obtaining groundwater potential [62]. In low-slope locations, infiltration into the ground is more common as surface flow will be less. As slope ratio increases, surface flow increases, and infiltration into the ground is less. As seen in Figure 4D, the slope is greater than 15% in 48% of the basin. The slope at 25% of the study area is between 8%–15%. Areas having the ideal slope rate (less than 2%) for groundwater formation are 9%. A further 2%–4% of the slope rate areas are 6% in the basin.

3.5. Rainfall

Rainfall (precipitation) is one of the most important groundwater parameters. Groundwater is formed by rainfall infiltrating the ground. Accordingly, 43.57% of the basin has a rainfall rate of 448–478 mm/year and 32.16% has 479–508 mm/year. Only 7% of the basin has a rainfall rate of 549–592 mm/year (Figure 4E).

3.6. Soil

Soil is an important parameter that influences rainfall infiltration into the ground. The rate and amount of infiltration vary depending on grain structure. In coarse-grained porous structures, groundwater infiltration is higher. In fine-grained soils such as clay, infiltration is slower and less. Porosity and permeability both play an important role in groundwater movement and recovery. In the soil map of the study area (Figure 4F), we can see that, in the basin, alluvial soil: 0.10%, brown forestry soil: 56.90%, brown soil: 41.30%, and reddish-brown soil: 1.70% (Table 3).

3.7. Drainage Density

Drainage density is one of the parameters affecting the groundwater-potential index. In river flows, drainage density is low in areas with low infiltration [63]. Where drainage density is small, infiltration occurs at a higher speed because the flow is slow. Hence, the performance coefficient of ratios with low drainage density was taken as a higher value. Drainage density is generally low in the basin (Figure 4G).

3.8. Land Use

Land use is directly related to rainfall leaking into the ground. It is therefore an important criterion for groundwater formation. Due to urbanization in construction sites, leakage is rare and surface flow rate is therefore higher. Wetlands and water bodies are considered to have the highest evaluation score. Similarly, in rivers, alluvial deposits have a high leakage rate. In addition, irrigated agricultural areas' water infiltration into the ground is also high. In a similar way, there is a high rate of leakage in plant zones, partly in land form, because surface flow is prevented by plantations. Land-use rate and utilization classification are shown in Table 4. As depicted in the table, water mass and irrigated areas in the Batman discharge area are very small. By majority, there are non-irrigated agricultural areas (28%), pastures (19%), and areas with different vegetation.

Table 4. Tigris Batman drainage field GWPZ evaluation.

Evaluation	km^2	%
Very poor	262.45	19
Poor	241.71	17
Moderate	484.22	34
Good	234.53	17
Very good	177.09	13

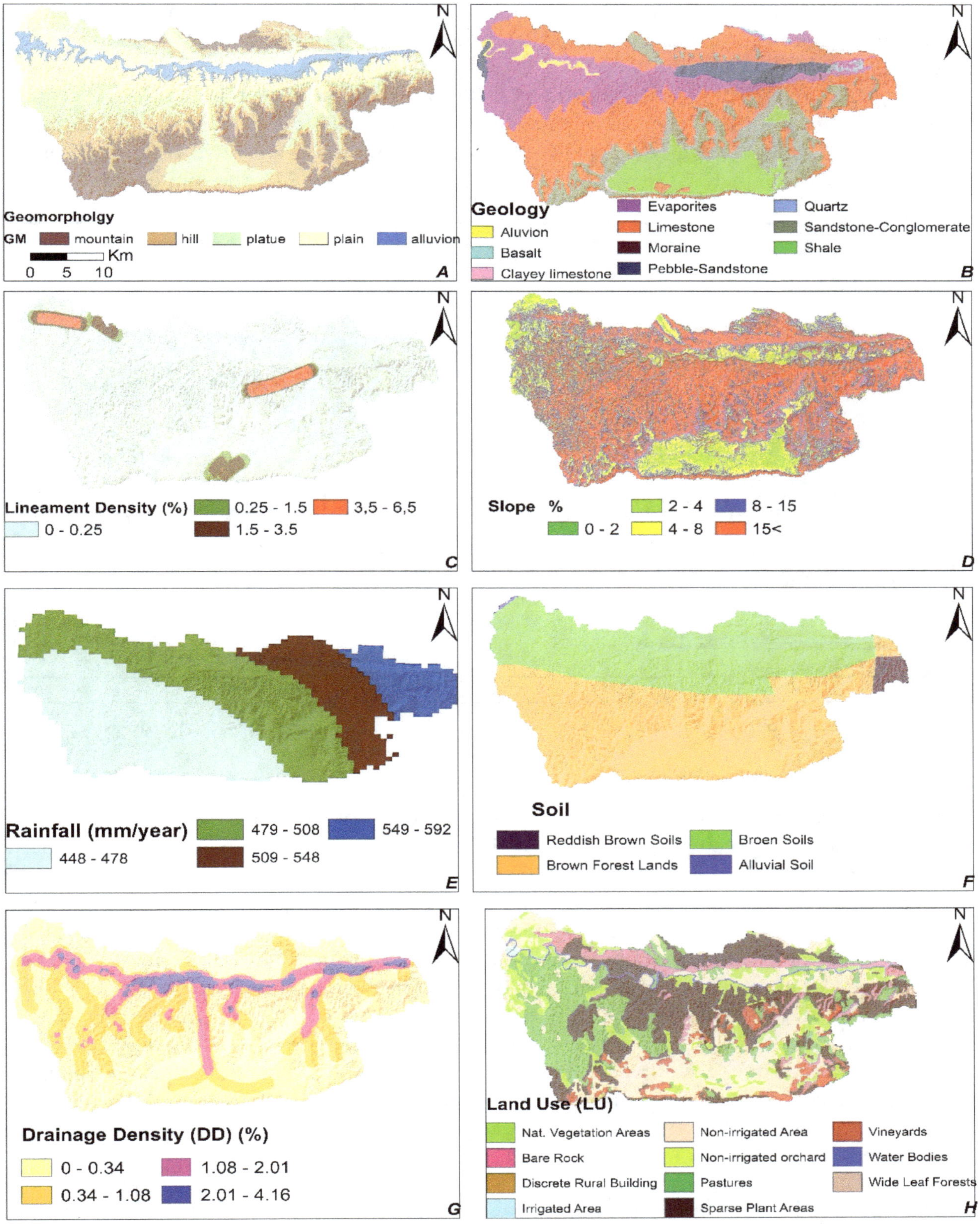

Figure 4. Main criteria affecting reclassified Groundwater Potential Index (GWPI) raster maps (**A**): Geomorphology, (**B**): Geology, (**C**): Lineament Density, (**D**): Slope, (**E**): Rainfall, (**F**): Soil, (**G**): Drainage Density, (**H**): Land Use map.

3.9. GWPZ Delination

GWPI is a coefficient showing the groundwater potential of the region. This coefficient is defined on the probability of obtaining groundwater resources for that area. The GWPI of an area varies according to the variability of properties that make up groundwater potential and, predominantly, the coefficient of significance. This final map shows the low–medium–high rate GW discharge regions for potential.

The GWPZ map was derived by periodic GWPI classification, shown in Figure 5. Accordingly, the GWPZ evaluation of the basin is summarized in Table 4 as follows: very poor (262.45 km^2; 19%), poor (241.71 km^2; 17%), moderate (484.22 km^2; 34%), good (234.53 km^2; 17%), and very good (177.09 km^2; 13%). As depicted, the maximum weight of the basin is 34% moderate GWPZ.

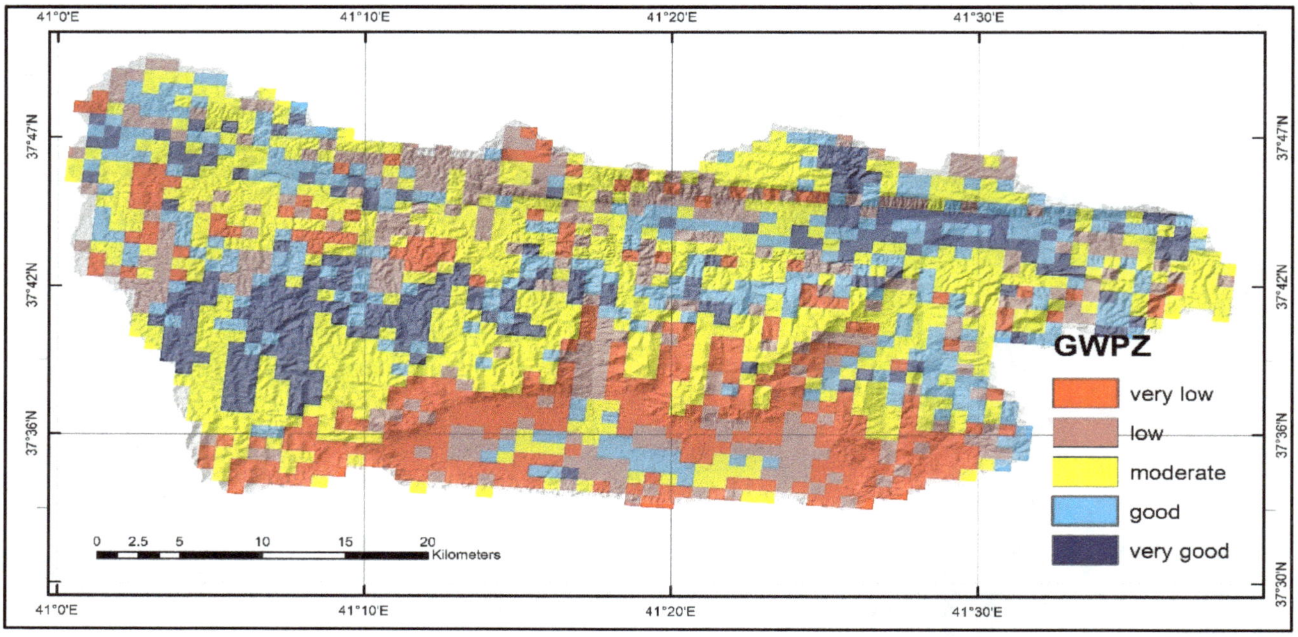

Figure 5. Tigris Batman-Hasankeyf drainage field Groundwater Potential Zone (GWPZ) map.

3.10. Discussion

GIS-based multicriteria decision analysis is a flexible tool that supports decision makers in identifying groundwater potential. Therefore, a major challenge in the proposed GIS-based multicriteria decision analysis is the selection of criteria for groundwater-potential-zone mapping. Criteria selection requires good knowledge of the site and watershed data, the accurate weighting of criteria by hydrogeologist experts, and accurate cooperation between selected factors. One of the most important stages of the decision-making process is the determination of the criteria weight, in other words, their importance. Weights given to the criteria directly affect the outcome of the decision analysis. Analysis accuracy depends on the determination of the weights with sufficient accuracy. There are several methods for determining criteria weights, and since these methods differ from each other in terms of accuracy, ease of use, clarity, and theoretical structure, which method to use in determining weights varies according to the priorities of the decision maker. Although ranking and scoring methods are easier to use, binary comparison and preference priority analysis methods give more accurate results.

The GIS-based MCDM method is a good alternative in this research area. MCDM techniques are used very effectively to solve problems in various and different areas [64–66]. Despite the many methods used in MCDM applications, the most common is AHP. In the selection of the appropriate areas, considering the relative impact of the various criteria in relation to each other [67] gives useful results in groundwater-discharge assessment [68–71]. The model used in this study allowed us to identify areas with high, medium, and low potential for groundwater charging zones. The presence of the thematic data obtained with this approach requires that accuracy is checked on the field. This is not always an economical and fast method, but it is possible to compare the water budgets of the basins with the field test. However, results achieved by this approach play an important role in this type of groundwater-management study.

In this study, it was concluded that places with the highest groundwater potential are in river crossings and the alluvial zone near the Tigris river. Findings showed that this method is suitable for the Batman-Hasankeyf sub-basin. Another feature of this region is that the slope is very low. Similar studies using the same method in different regions had varying results. In an Indian study, the Indian Hooghly district was classified into three groundwater-potential zones, 'good', 'moderate', and 'poor', covering 71%, 23%, and 6%, respectively. 'Good' GWPZ indicated that the alluvial plain has excellent groundwater potential [22]. Madrucci et al. studied groundwater-favorability mapping on a fractured terrain in the eastern portion of the state of São Paulo, Brazil [33].

They determined a "moderate favorability" GW potential level in almost half of the basin. Results obtained from the study are acceptable when compared with well logs on the field. GWPZs were calculated by a similar method in the Yamuna basin, another basin of India, and it was found that about half of the basin had good–very good value, and 24% had weak potential [25]. In a study on the Manuba aquifers in NE Tunisia with the GIS-based MCDM method [66], it was found that only 24% of the basin had a low GW potential. GWPZ values of a region vary according to hydrological and hydrogeological conditions. However, it is crucial to evaluate the GW potential of a basin by correctly using these variables.

4. Conclusions

In this study, groundwater potential was explored in the Tigris river Batman region sub-basin drainage area by integrating spatial-analyzer tools with MCDM. The MCDM–AHP technique was successfully applied to the normalized weight and rank of each parameter and its sub-properties affecting the groundwater potential of the basin. In total, eight layers were used in system integration. As a result of the overlay of these maps, GWPI values of the basin were extracted. These GWPI values were reclassified into five classes: very poor, poor, moderate, good, and very good. By utilizing this method, local GWPZ maps of the basin were derived.

The final maps showed that 34% of the basin has moderate groundwater potential. However, 36% of the basin demonstrated poor potential, while 30% of the basin has good potential. This GWPZ map should be taken into account in watershed planning by managers. On the basis of the findings, controlled groundwater use should be made in the poor-potential areas of the basin. In these regions, groundwater should only be used as potable water. The use of groundwater can be realized in agricultural areas at moderate-potential zones, but large-scale agricultural irrigation should be prevented. In areas with good potential, groundwater can be used in agricultural irrigation. However, water-well drillings must be properly controlled and opened at wide intervals on the field.

It is more accurate to learn the groundwater potential of a narrow area by drilling several observation wells. However, in large-scale basins, this is both expensive and time-consuming. It is purely logical and valuable to preferably discover the groundwater potential of large-scale basins in the macroplan with the MCDM method. Hydrological and hydrogeological data are now easier and relatively cheaper to obtain with RS and satellite methods. This method could be applied not only to the upstream Tigris basin but to anywhere else in the world as well. The most important topic to consider is that specific hydrological and meteorological criteria that affect groundwater and their weight might change depending on each region's condition.

References

1. Okello, C.; Tomasello, B.; Greggio, N.; Wambiji, N.; Antonellini, M. Impact of population growth and climate change on the freshwater resources of Lamu Island, Kenya. *Water* **2015**, *7*, 1264–1290. [CrossRef]
2. Diamantino, C.; Henriques, M.J.; Oliveira, M.M.; Ferreira, J.P.L. Methodologies for pollution risk assessment of water resources systems. *IAHS Publ.* **2007**, *310*, 298.
3. Alessa, L.; Kliskey, A.; Lammers, R.; Arp, C.; White, D.; Hinzman, L.; Busey, R. The arctic water resource vulnerability index: An integrated assessment tool for community resilience and vulnerability with respect to freshwater. *Environ. Manag.* **2008**, *42*, 523. [CrossRef] [PubMed]
4. Danielopol, D.L.; Griebler, C.; Gunatilaka, A.; Notenboom, J. Present state and future prospects for groundwater ecosystems. *Environ. Conserv.* **2003**, *30*, 104–130. [CrossRef]
5. De Stefano, L.; Lopez-Gunn, E. Unauthorized groundwater use: Institutional, social and ethical considerations. *Water Policy* **2012**, *14*, 147–160. [CrossRef]
6. Lee, S.; Lee, C.W. Application of decision-tree model to groundwater productivity-potential mapping. *Sustainability* **2015**, *7*, 13416–13432. [CrossRef]
7. Helena, B.; Pardo, R.; Vega, M.; Barrado, E.; Fernandez, J.M.; Fernandez, L. Temporal evolution of groundwater composition in an alluvial aquifer (Pisuerga River, Spain) by principal component analysis. *Water Res.* **2000**, *34*, 807–816. [CrossRef]
8. Pourtaghi, Z.S.; Pourghasemi, H.R. GIS-based groundwater spring potential assessment and mapping in the Birjand Township, southern Khorasan Province, Iran. *Hydrogeol. J.* **2014**, *22*, 643–662. [CrossRef]
9. Thapa, R.; Gupta, S.; Guin, S.; Kaur, H. Assessment of groundwater potential zones using multi-influencing factor (MIF) and GIS: A case study from Birbhum district, West Bengal. *Appl. Water Sci.* **2017**, *7*, 4117–4131. [CrossRef]
10. Machiwal, D.; Jha, M.K.; Mal, B.C. Assessment of groundwater potential in a semi-arid region of India using remote sensing, GIS and MCDM techniques. *Water Resour. Manag.* **2011**, *25*, 1359–1386. [CrossRef]
11. Yin, H.; Shi, Y.; Niu, H.; Xie, D.; Wei, J.; Lefticariu, L.; Xu, S. A GIS-based model of potential groundwater yield zonation for a sandstone aquifer in the Juye Coalfield, Shangdong, China. *J. Hydrol.* **2018**, *557*, 434–447. [CrossRef]
12. Elbeih, S.F. An overview of integrated remote sensing and GIS for groundwater mapping in Egypt. *Ain Shams Eng. J.* **2015**, *6*, 1–15. [CrossRef]
13. Senanayake, I.P.; Dissanayake, D.M.D.O.K.; Mayadunna, B.B.; Weerasekera, W.L. An approach to delineate groundwater recharge potential sites in Ambalantota, Sri Lanka using GIS techniques. *Geosci. Front.* **2016**, *7*, 115–124. [CrossRef]
14. Saaty, T.L. *The Analytic Hierarchy Process*; McGrawHill: New York, NY, USA, 1980.
15. Saaty, T.L. *Fundamentals of Decision Making and Priority Theory*; RWS Publications: Pittsburgh, PA, USA, 2000.
16. Guo, J.-Y.; Zhang, Z.-B.; Sun, Q.-Y. Applications of AHP method in safety science. *J. Saf. Sci. Technol.* **2008**, *2*, 69–73.
17. Wang, Y.; Li, Z.; Tang, Z.; Zeng, G. A GIS-based spatial multi-criteria approach for flood risk assessment in the Dongting Lake Region, Hunan, Central China. *Water Resour. Manag.* **2011**, *25*, 3465–3484. [CrossRef]
18. Tang, Z.; Zhang, H.; Yi, S.; Xiao, Y. Assessment of flood susceptible areas using spatially explicit, probabilistic multi-criteria decision analysis. *J. Hydrol.* **2018**, *558*, 144–158. [CrossRef]
19. Chen, H.; Wood, M.D.; Linstead, C.; Maltby, E. Uncertainty analysis in a GIS-based multi-criteria analysis tool for river catchment management. *Environ. Model. Softw.* **2011**, *26*, 395–405. [CrossRef]
20. Nair, N.C.; Srinivas, Y.; Magesh, N.S.; Kaliraj, S. Assessment of groundwater potential zones in Chittar basin, Southern India using GIS based AHP technique. *Remote Sens. Appl. Soc. Environ.* **2019**, *15*, 100248. [CrossRef]
21. Abrams, W.; Ghoneim, E.; Shew, R.; LaMaskin, T.; Al-Bloushi, K.; Hussein, S.; AbuBakr, M.; Al-Mulla, E.; Al-Awar, M.; El-Baz, F. Delineation of groundwater potential (GWP) in the northern United Arab Emirates and Oman using geospatial technologies in conjunction with Simple Additive Weight (SAW), Analytical Hierarchy Process (AHP), and Probabilistic Frequency Ratio (PFR) techniques. *J. Arid Environ.* **2018**, *157*, 77–96. [CrossRef]
22. Patra, S.; Mishra, P.; Mahapatra, S.C. Delineation of groundwater potential zone for sustainable development: A case study from Ganga Alluvial Plain covering Hooghly district of India using remote sensing, geographic information system and analytic hierarchy process. *J. Clean. Prod.* **2018**, *172*, 2485–2502. [CrossRef]

23. Gdoura, K.; Anane, M.; Jellali, S. Geospatial and AHP-multicriteria analyses to locate and rank suitable sites for groundwater recharge with reclaimed water. *Resour. Conserv. Recycl.* **2015**, *104*, 19–30. [CrossRef]
24. Saidi, S.; Hosni, S.; Mannai, H.; Jelassi, F.; Bouri, S.; Anselme, B. GIS-based multi-criteria analysis and vulnerability method for the potential groundwater recharge delineation, case study of Manouba phreatic aquifer, NE Tunisia. *Environ. Earth Sci.* **2017**, *76*, 511. [CrossRef]
25. Kaur, L.; Rishi, M.S.; Singh, G.; Thakur, S.N. Groundwater potential assessment of an alluvial aquifer in Yamuna sub-basin (Panipat region) using remote sensing and GIS techniques in conjunction with analytical hierarchy process (AHP) and catastrophe theory (CT). *Ecol. Indic.* **2020**, *110*, 105850. [CrossRef]
26. Gupta, M.; Srivastava, P.K. Integrating GIS and remote sensing for identification of groundwater potential zones in the hilly terrain of Pavagarh, Gujarat, India. *Water Int.* **2010**, *35*, 233–245. [CrossRef]
27. Rahmati, O.; Samani, A.N.; Mahdavi, M.; Pourghasemi, H.R.; Zeinivand, H. Groundwater potential mapping at Kurdistan region of Iran using analytic hierarchy process and GIS. *Arab. J. Geosci.* **2015**, *8*, 7059–7071. [CrossRef]
28. Pinto, D.; Shrestha, S.; Babel, M.S.; Ninsawat, S. Delineation of groundwater potential zones in the Comoro watershed, Timor Leste using GIS, remote sensing and analytic hierarchy process (AHP) technique. *Appl. Water Sci.* **2017**, *7*, 503–519. [CrossRef]
29. Murthy, K.S.R. Ground water potential in a semi-arid region of Andhra Pradesh—A geographical information system approach. *Int. J. Remote Sens.* **2000**, *21*, 1867–1884. [CrossRef]
30. Ibrahim-Bathis, K.; Ahmed, S.A. Geospatial technology for delineating groundwater potential zones in Doddahalla watershed of Chitradurga district, India. *Egypt. J. Remote Sens. Space Sci.* **2016**, *19*, 223–234. [CrossRef]
31. Sikdar, P.K.; Chakraborty, S.; Adhya, E.; Paul, P.K. Land use/Land cover changes and groundwater potential zoning in and around Raniganj coal mining area, Bardhaman District, West Bengal—A GIS and Remote Sensing Approach. *J. Spat. Hydrol.* **2004**, *4*, 1–24.
32. Prasad, R.K.; Mondal, N.C.; Banerjee, P.; Nandakumar, M.V.; Singh, V.S. Deciphering potential groundwater zone in hard rock through the application of GIS. *Environ. Geol.* **2007**, *55*, 467–475. [CrossRef]
33. Madrucci, V.; Taioli, F.; De Araújo, C.C. Groundwater favorability map using GIS multicriteria data analysis on crystalline terrain, Sao Paulo State, Brazil. *J. Hydrol.* **2008**, *357*, 153–173. [CrossRef]
34. Chowdhury, A.; Jha, M.K.; Chowdary, V.M.; Mal, B.C. Integrated remote sensing and GIS-based approach for assessing groundwater potential in West Medinipur district, West Bengal, India. *Int. J. Remote Sens.* **2009**, *30*, 231–250. [CrossRef]
35. Zaidi, F.K.; Nazzal, Y.; Ahmed, I.; Naeem, M.; Jafri, M.K. Identification of potential artificial groundwater recharge zones in Northwestern Saudi Arabia using GIS and Boolean logic. *J. Afr. Earth Sci.* **2015**, *111*, 156–169. [CrossRef]
36. Jha, M.K.; Chowdary, V.M.; Chowdhury, A. Groundwater assessment in Salboni Block, West Bengal (India) using remote sensing, geographical information system and multi-criteria decision analysis techniques. *Hydrogeol. J.* **2010**, *18*, 1713–1728. [CrossRef]
37. Agarwal, R.; Garg, P.K. Remote sensing and GIS based groundwater potential and recharge zones mapping using multi-criteria decision making technique. *Water Resour. Manag.* **2016**, *30*, 243–260. [CrossRef]
38. Jaiswal, R.K.; Mukherjee, S.; Krishnamurthy, J.; Saxena, R. Role of remote sensing and GIS techniques or generation of groundwater prospect zones towards rural developmentean approach. *Int. J. Remote Sens.* **2003**, *24*, 993–1008. [CrossRef]
39. Srivastava, P.K.; Bhattacharya, A.K. Groundwater assessment through an integrated approach using remote sensing, GIS and resistivity techniques: A case study from a hard rock terrain. *Int. J. Remote Sens.* **2006**, *27*, 4599–4620. [CrossRef]
40. Magesh, N.S.; Chandrasekar, N.; Soundranayagam, J.P. Delineation of groundwater potential zones in Theni district, Tamil Nadu, using remote sensing, GIS and MIF techniques. *Geosci. Front.* **2012**, *3*, 189–196. [CrossRef]
41. Ahunbay, Z.; Balkiz, Ö. Outstanding universal value of Hasankeyf and the Tigris valley. *Doğa Deneği.* **2009**, 22. Available online: https://goo.gl/qZ9tqX (accessed on 10 August 2019).
42. Miyake, Y.; Maeda, O.; Tanno, K.; Hongo, H.; Gündem, C.Y. New excavations at Hasankeyf Höyük: A 10th millennium cal. BC site on the upper Tigris, southeast Anatolia. *Neo Lithics* **2012**, *1*, 3–7.

43. Çelik, R. Temporal changes in the groundwater level in the Upper Tigris Basin, Turkey, determined by a GIS technique. *J. Afr. Earth Sci.* **2015**, *107*, 134–143. [CrossRef]
44. Rikalovic, A.; Cosic, I.; Lazarevic, D. GIS based multi-criteria analysis for industrial site selection. *Procedia Eng.* **2014**, *69*, 1054–1063. [CrossRef]
45. Mahmoud, S.H. Delineation of potential sites for groundwater recharge using a GIS-based decision support system. *Environ. Earth Sci.* **2014**, *72*, 3429–3442. [CrossRef]
46. Chenini, I.; Mammou, A.B.; El May, M. Groundwater recharge zone mapping using GIS-based multi-criteria analysis: A case study in Central Tunisia (Maknassy Basin). *Water Resour. Manag.* **2010**, *24*, 921–939. [CrossRef]
47. Tarboton, D.G.; Bras, R.L.; Rodriguez-Iturbe, I. A physical basis for drainage density. *Geomorphology* **1992**, *5*, 59–76. [CrossRef]
48. General Directorate of Mineral Research on Exploration. Available online: http://yerbilimleri.mta.gov.tr/anasayfa.aspx (accessed on 8 August 2019).
49. Mandal, U.; Sahoo, S.; Munusamy, S.B.; Dhar, A.; Panda, S.N.; Kar, A.; Mishra, P.K. Delineation of groundwater potential zones of coastal groundwater basin using multi-criteria decision making technique. *Water Resour. Manag.* **2016**, *30*, 4293–4310. [CrossRef]
50. Gallego, F.J.; Batista, F.; Rocha, C.; Mubareka, S. Disaggregating population density of the European Union with CORINE land cover. *Int. J. Geogr. Inf. Sci.* **2011**, *25*, 2051–2069. [CrossRef]
51. Ateşoğlu, A. Havza çalışmalarında kullanılan CORINE 2006 arazi sınıflandırma verilerinin doğruluğunun araştırılması. *İstanb. Üniv. Orman Fakültesi Derg.* **2016**, *66*, 173–183. [CrossRef]
52. Rao, B.V.; Briz-Kishore, B.H. A methodology for locating potential aquifers in a typical semi-arid region in India using resistivity and hydrogeologic parameters. *Geoexploration* **1991**, *27*, 55–64.
53. Shekhar, S.; Pandey, A.C. Delineation of groundwater potential zone in hard rock terrain of India using remote sensing, geographical information system (GIS) and analytic hierarchy process (AHP) techniques. *Geocarto Int.* **2015**, *30*, 402–421. [CrossRef]
54. Punniyamoorty, M.; Ponnusamy, M.; Lakshmi, G. A combined application of structural equation modeling (SEM) and analytic hierarchy process (AHP) in supplier selection. *Benchmarking Int. J.* **2012**, *19*, 70–92. [CrossRef]
55. Kannan, V. Benchmarking the service quality of ocean container carriers using AHP. *Benchmarking Int. J.* **2010**, *17*, 637–656. [CrossRef]
56. Greene, R.; Devillers, R.; Luther, J.E.; Eddy, B.G. GIS-based multiple-criteria decision analysis. *Geogr. Compass* **2011**, *5*, 412–432. [CrossRef]
57. Triantaphyllou, E.; Sanchez, A. A sensitivity analysis approach for some deterministic multi-criteria decision making methods. *Decis. Sci.* **1997**, *28*, 151–194. [CrossRef]
58. Ho, W.; Xu, X.; Dey, P.K. Multi-criteria decision making approaches for supplier valuation and selection: A literature review. *Eur. J. Oper. Res.* **2010**, *202*, 16–24. [CrossRef]
59. Saaty, T.L. Decision-making with the AHP: Why is the principal eigenvector necessary. *Eur. J. Oper. Res.* **2003**, *145*, 85–91. [CrossRef]
60. Alonso, J.A.; Lamata, M.T. Consistency in the analytic hierarchy process: A new approach. *Int. J. Uncertain. Fuzziness Knowl. Based Syst.* **2006**, *14*, 445–459. [CrossRef]
61. Magowe, M.; Carr, J.R. Relationship between lineaments and ground water occurrence in western Botswana. *Groundwater* **1999**, *37*, 282–286. [CrossRef]
62. Batelaan, O.; De Smedt, F. Wet Spass: A flexible, GIS based, distributed recharge methodology for regional groundwater modelling. *IAHS Publ.* **2001**, *269*, 11–18.
63. Tucker, G.E.; Bras, R.L. Hillslope processes, drainage density, and landscape morphology. *Water Resour. Res.* **1998**, *34*, 2751–2764. [CrossRef]
64. Mendoza, G.A.; Martins, H. Multi-criteria decision analysis in natural resource management: A critical review of methods and new modelling paradigms. *For. Ecol. Manag.* **2006**, *230*, 1–22. [CrossRef]
65. Makropoulos, C.K.; Butler, D. Spatial ordered weighted averaging: Incorporating spatially variable attitude towards risk in spatial multi-criteria decision-making. *Environ. Model. Softw.* **2016**, *21*, 69–84. [CrossRef]
66. Karnatak, H.C.; Saran, S.; Bhatia, K.; Roy, P.S. Multicriteria spatial decision analysis in web GIS environment. *Geoinformatica* **2017**, *11*, 407–429. [CrossRef]

67. Adiat, K.A.N.; Nawawi, M.N.M.; Abdullah, K. Assessing the accuracy of GIS-based elementary multicriteria decision analysis as a spatial prediction tool—A case of predicting potential zones of sustainable groundwater resources. *J. Hydrol.* **2012**, *440*, 75–89. [CrossRef]
68. Azimi, S.; Moghaddam, M.A.; Monfared, S.A.H. Spatial assessment of the potential of groundwater quality using fuzzy AHP in GIS. *Arab. J. Geosci.* **2018**, *11*, 142. [CrossRef]
69. Singh, P.K.; Dahiphale, P.; Yadav, K.K.; Singh, M. Delineation of groundwater potential zones in Jaisamand Basin of Udaipur District. In *Groundwater*; Springer: Singapore, 2018; pp. 3–20.
70. Gnanachandrasamy, G.; Zhou, Y.; Bagyaraj, M.; Venkatramanan, S.; Ramkumar, T.; Wang, S. Remote sensing and GIS based groundwater potential zone mapping in Ariyalur District, Tamil Nadu. *J. Geol. Soc. India* **2018**, *92*, 484–490. [CrossRef]
71. Souissi, D.; Msaddek, M.H.; Zouhri, L.; Chenini, I.; El May, M.; Dlala, M. Mapping groundwater recharge potential zones in arid region using GIS and Landsat approaches, southeast Tunisia. *Hydrol. Sci. J.* **2018**, *63*, 251–268. [CrossRef]

4

Evaluation of Geological and Ecological Bearing Capacity and Spatial Pattern along Du-Wen Road based on the Analytic Hierarchy Process (AHP) and the Technique for Order of Preference by Similarity to an Ideal Solution (TOPSIS) Method

Zhoufeng Wang [1,*], Xiangqi He [1], Chen Zhang [1], Jianwei Xu [1] and Yujun Wang [2]

1 School of Geoscience and Technology, Southwest Petroleum University, Chengdu 610500, China; 201922000062@stu.swpu.edu.cn (X.H.); 201721000107@stu.swpu.edu.cn (C.Z.); 201822000076@stu.swpu.edu.cn (J.X.)

2 Hefei Surveying and mapping Institute, Hefei 230031, China; wangyujun@hfchy.com

* Correspondence: wangzf@swpu.edu.cn.

Abstract: As China is a mountainous country, a large quantity of the population has to live in mountainous areas due to limited living space. Most of them cluster along roads in areas with relatively poor traffic conditions. In view of the spatial-temporal change of complex geological and ecological environment along the roads in the mountains, this paper takes the Dujiangyan- Wenchuan (Du-Wen) Road as the research object, and puts forward a method to evaluate the bearing capacity of regional geological and ecological environment based on the evaluation of quality and spatial coupling of bearing capacity. For the needs of the current research, a total number of 20 indicators from three aspects of geological, ecological, and social attributes were selected to carry out the assessment. Based on the GIS platform and evaluation index system, the weight of each evaluation index factor is determined by Analytic Hierarchy Process (AHP). The comprehensive quality of bearing capacity is calculated by the Technique for Order of Preference by Similarity to an Ideal Solution (TOPSIS) algorithm through weighted superposition, which comes to result in the evaluation of geological, ecological, and social environment. Afterward, the bearing capacity of the study area is classified, combining the results of hot spot analysis. The study shows that the spatial distribution of geological, ecological, and socio-economic bearing capacity in this area is highly aggregated, with 31.12% of the area to be classified as suitable construction area, 31.98% as backup reserve area, and 36.79% as unsuitable construction area. The studied triangle area, composed of Yingxiu Town, Xuankou Town, and Dujiangyan City, presents a large area of a high-valued aggregation area, with comprehensive high quality bearing capacity and spatial aggregation, which is better for planning and construction.

Keywords: bearing capacity; analytic hierarchy process; geographical survey of national conditions; hotspot analysis; topsis algorithm

---

1. Introduction

China is a mountainous country, with mountainous areas accounting for 69% of the country's total land area; and whose flat areas, suitable for human habitation and production activities, such as plains and basins, are extremely limited. At the same time, China has a large population and a growing shortage of living space. For example, 56% of the population has to live in mountainous areas, where transportation is inconvenient and production activities are not suitable. Sichuan Province is located in the southwest of China, with a large area of mountains. Except for the Sichuan Basin, which is dominated by low and medium mountains and hills, the surrounding areas of the basin are dominated by medium and high mountains, with steep terrain and widely geological disasters. In particular, the Wenchuan earthquake, in 2008, made the geological environment very fragile. According to the data from Sichuan Provincial Department of Transportation in 2016, the total mileage of national roads and provincial roads in Sichuan Province has reached 22,526 kilometers, of which, 11,793 kilometers are located at the edge of the basin or the surrounding plateau area, accounting for 52.4% of the total mileage. A violent debris flow hazard occurred in Wenchuan, Sichuan Province in August 20, 2019; the flood disaster caused by the rainstorm killed 16 people, and 22 people lost contact. The direct economic loss was 3.6 billion yuan.

With the sustained and rapid development of China's economy, the interference of mankind's social activities on the natural ecosystem is also intensifying, leading to the continuous deterioration of the natural environment. Solving the contradiction among resources, population, and environment, achieving harmony between man and nature, and achieving coordinated development, become the inherent requirements to ecological civilization construction. The adaptation of ecological civilization construction and economic development is the prerequisite for sustainable development. Ma and Ma [1] have proposed that environmental carrying capacity refers to the ability of the environmental system to undergo no structural changes during a period of time and in a certain geographical area, while maintaining its normal functioning, the total amount of human beings and various economic activities that the environmental system can withstand.

In recent years, Chinese scholars have carried out various research on the comprehensive bearing capacity evaluation of different regions. Some studies are carried out on a national scale, such as Liu's [2] comprehensive evaluation of China's water resources carrying capacity. Some are provincial; Liu et al. [3] have studied dynamic evaluation of regional comprehensive carrying capacity in Henan Province. Others are about specific cities, such as Lu et al.'s [4] analysis of ecological environment resources carrying capacity of Guiyang City. Studies on single element bearing capacity mainly include the following aspects: Li [5] studied the land ecological security of Dujiangyan City, while Wang et al. [6] compared the ecological bearing capacity of the upper reaches of Shiyang River in time evolution. Liu et al. [7] studied the evaluation of geo-environmental carrying capacity of the Wulonggou mine, based on the spatial statistical analysis method. What was used mainly include: Yang et al. [8] used the Analytic Hierarchy Process (AHP) comprehensive evaluation method, Feng et al. [9] used the Technique for Order of Preference by Similarity to an Ideal Solution (TOPSIS) algorithm, Zheng et al. [10] used the ecological footprint method. The research direction of bearing capacity mainly focuses on the following three aspects: Gu [11] elaborated the development of bearing capacity concept in different periods, Xu et al. [12] evaluated the bearing capacity of traffic environment in Beijing, and Wang et al. [13] planed the development area based on the bearing capacity. Taking the planning area as an example, Ni [14] divided the quality of land in Qushui region of Tibet; Tong [15] divided the land use of Wuhan city; Wang et al. [16] divided the area along Du-wen Road into comprehensive areas, including the agricultural development area, ecological restoration area, construction suitable area, fault avoidance area, and conversion of farmland to forest area.

After entering the 21st century, based on the research on the basic concepts and theories of bearing capacity, domestic scholars have carried out rich researches on regional "ecological environment bearing capacity" and "geological environment bearing capacity" for the perspective of "nature-economy-society" coordinated development. Ma and Jiang [17,18], respectively, took the Beijing Tianjin Hebei and Yangtze River economic belt as research objects, analyzed the land, resources, and environmental conditions, and other major geological factors of this economic belt. Zhang et al. [19] carried out the risk-based assessment of geological environment bearing capacity, put forward the concept of a risk-based geological environment bearing capacity and ultimate bearing capacity, and distinguished the bearing capacity status into three grades: safe bearing, allowable overload, and unacceptable overload, which has important reference value for the delineation of geological disaster red line.

Meng et al. [20] conducted a background test evaluation on the bearing capacity of the geological environment in China.

In the existing research system of an earthquake mountainous area, the system model method is often used to evaluate the bearing capacity of regional resources and environment, and the comprehensive quality evaluation of bearing capacity is used to calculate regional population and carry out industrial planning. Based on the Vigor-Organization-Resilience model [21] coupled with ecosystem services, Zhu [22] quantitatively evaluates the ecosystem health in Wenchuan earthquake area while Tang [23] built the model of the bearing capacity of resources and environment in the mountainous earthquake area, according to the derivation from the stress mechanism. Fan [24] evaluated the bearing capacity of resources and environment in the reconstruction in the Lushan area after the earthquake by using the AHP comprehensive evaluation method; there is still lack of research on the spatial-temporal dynamic evolution of the bearing capacity of resources and the environment in the mountainous earthquake area. Mountainous area accounts for a large proportion of the land area in China where earthquakes, along with other kinds of secondary disasters, occurred frequently and often caused serious damage. Reconstruction and recovery in these areas are becoming more difficult; meanwhile, the basic conditions for economic and social development are more arduous. Given to the situation stated above, it is of great practical significance to carry out the research on the bearing capacity of the seismic mountainous area.

The above research enriches the theories and methods of bearing capacity, but the existing research mainly focuses on the quantitative evaluation of bearing capacity. Due to different natural, social, and economic conditions in each study area, as well as the diversity of the evaluation factor selection and the application of evaluation methods in bearing capacity research, a mature, complete, and recognized research system has not been formed in the process of quantitative measurement. There are few studies on the spatial distribution pattern in the evaluation of regional bearing capacity. It is difficult to find correlation among bearing capacity quality, scale, and spatial distribution. The comprehensive evaluation of bearing capacity needs to be considered from aspects of comprehensive development suitability and spatial stability. It is necessary to consider the suitability of development from the perspective of comprehensive quality, and to consider the spatial stability of sustainable development from the perspective of spatial layout.

The main objectives of current research were (1) Establishing three evaluation index systems for geological environment, ecological environment, and socioeconomics, which are suitable for the three major geomorphic areas of Sichuan Basin edge area, Southwest Sichuan mountain area, and Northwest Sichuan Plateau area. (2) Studying the integration between the national geographical monitoring data and ecological environment bearing capacity evaluation index system, based on the background database of the geo-ecological conditions. (3) Using AHP to reasonably determine the weight of each index, combining with the comprehensive evaluation method and establishing the mathematical model of geological ecological environment bearing capacity. (4) Based on this, the comprehensive quality score of bearing capacity is used as a spatial variable, and the hotspot analysis tools are used to analyze the spatial agglomeration characteristics of the comprehensive quality of bearing capacity in the study area. The research results realize the coupling of bearing capacity quality, scale, and spatial distribution pattern, provide scientific basis for the development and planning along Du-Wen Road.

2. Materials and Methods

2.1. Study Area

The study area is located in Wenchuan County, Sichuan Province, China. The geographical range is 103°14′~103°45′ east longitude, 30°54′~31°36′ north latitude, and the area is 925 km^2. There are the Maowen Fault and Yingxiu Fault along the Du-Wen Road in this study area, and Yingxiu Fault is the seismogenic fault of the "5. 12" earthquake (see Figure 1).

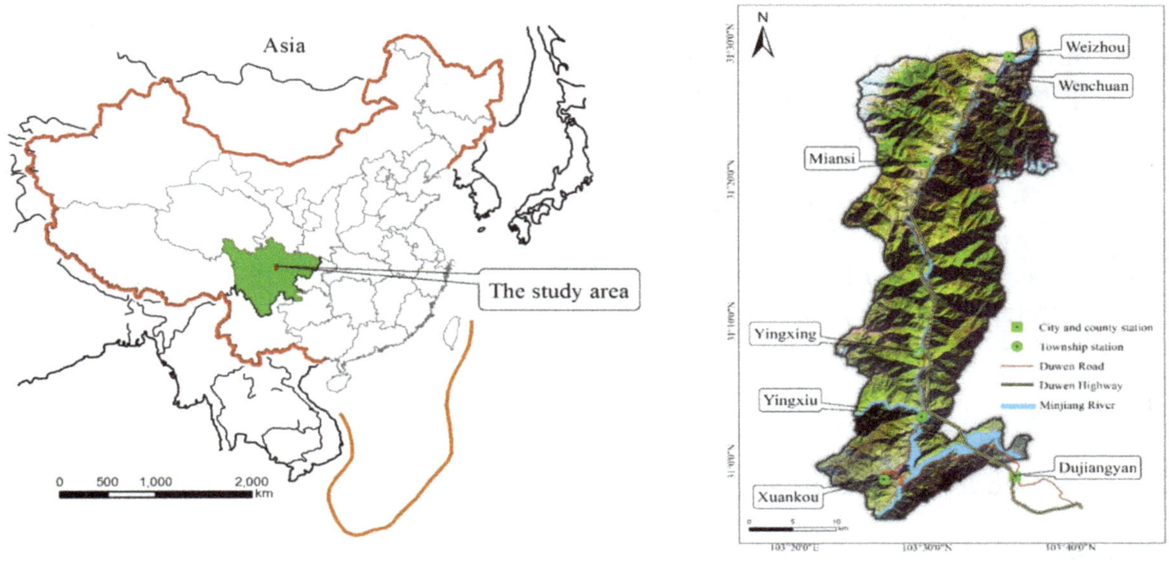

(a) (b)

Figure 1. (**a**) Location of the study area, Sichuan Province, China; (**b**) remote sensing image map of Dujiangyan-Wenchuan Road.

This area is located between the Longmen Mountains and the Qionglai Mountains in the highland plateau area of western Sichuan. It belongs to the transitional zone between the Chengdu Plain and the northwestern plateau of Sichuan. It has steep internal terrain, high mountains, and steep valleys. The average slope of mountains along the road is 30°~35°, with the Maowen Fault and Yingxiu Fault being the main geologic constructions. The Yingxiu Fault was the tremor of the "5 12" Wenchuan Earthquake. The study area is an important source of water conservation of the Yellow River and the Yangtze River. It belongs to the continental monsoon climate. The precipitation is unevenly distributed in the region, affected by the mountain blockage in the territory. It gradually decreases from northwest to southeast as the terrain decreases, with the most in August and the least in January. The natural rivers in the area belong to the main stream of the Lancang River and its tributaries. The larger tributaries include Yuzi River, Caopo River, Shouxi River, etc. There is also an artificial river, the Dujiangyan Irrigation River. As located in the area of the Longmen Mountain seismic belt, the geological structure is complex, with completed tectonic development and the stratigraphic excavation, diversified lithology. Meanwhile, the summer rainfall is abundant; it is easy to generate landslides and other geological disasters. The research on bearing capacity evaluation here will play a great practical and exemplary role.

The basic data used in this paper is derived from the first national survey of geographic conditions in Sichuan Province. With wide coverage, various types, and high quality, the data can nearly completely reflect the situation of the research area. It contents: 1) the basic status, such as type, quantity, location, scope and area, of natural geographical elements, including the topography, the distribution of disaster points, the coverage of vegetation, the waters and the land. 2) The basic data of human geography elements, including categories, locations, and scopes of transportation facilities, residential areas, public service facilities and energy supply, which are closely related to human activities. This national survey on geographic conditions has comprehensive geospatial information. By transforming the results, we can better serve the local economic construction and promote regional social development.

2.2. Data

The data involved in this paper mainly include the census data of the first national geographical conditions in Sichuan province (including types of land cover, roads, waters, structures, and geographical units). It also includes fault zone data and formation lithology data (from 1:200,000 digital geological map), seismic intensity data (from the Wenchuan 8.0 earthquake intensity distribution map of the China Seismological Bureau). There are data of geological hazard points, data of multi-year average precipitation, soil erosion data, Digital Elevation Model data (30 meters), raster data of land use type (30 meters), data of geological hazard points in 2018, and Point of Information data for 2018, from the Resources and Environment Data Center, Chinese Academy of Sciences. It also includes road data at all levels, water system data (reservoirs, rivers, ditches, etc.), data on settlements, place-name data of settlement place, administrative boundary data (dot, line, plane) from the National Geomatic Center of China; statistical yearbook of Wenchuan County People's Government, including the statistics of society, population, and GDP.

2.3. Methodology

2.3.1. Overall Methodological Framework

First, research on the bearing capacity constructs the evaluation research framework of the geological ecological environment bearing capacity. Then, the regional bearing capacity status evaluation index system is constructed, which includes three target levels of environment, geologic environment, ecological environment, and social economy, according to the first national geographical survey data of Sichuan Province. It is divided into several evaluation indicators. GIS and Remote Sensing technology are applied to preprocess the data of evaluation indicators, quantify each index factor, and divide the quantified results into five grades. AHP is then used to calculate and assign weights to each indicator. Finally, the TOPSIS method is used to calculate the comprehensive index of geological and ecological environmental bearing capacity of the study area through weighted superposition. The zoning rules, high-value clustering area, low-valued aggregation area, and no significant point area, are used to classify the bearing capacity of the study area through weighted superposition. Thus, the bearing capacity of the study area is comprehensively evaluated, the spatial agglomeration characteristics of the bearing capacity is analyzed, the temporal and spatial distribution of the bearing capacity of the study area is explored, and the priority area of the bearing capacity is delineated. The Overall Methodological Framework is shown as Figure 2.

2.3.2. Basic Theory of AHP

The Analytic Hierarchy Process (AHP) developed by Satty, provides a framework for dealing with decision-making problems and complex problems [25]. This paper chooses analytic hierarchy process (AHP) to determine the weight of the index. The basic idea of AHP to solve the problem is consistent with people's thinking of a multi-level, multi factor, and complex decision-making problem. Its most prominent feature is layer comparison and comprehensive optimization, which can provide a simple decision-making method for regional bearing capacity evaluation.

The basic principles of AHP can be summarized as defining and determining the problem; decomposing the problem in a hierarchy from top through the intermediate levels; constructing a set of pair-wise comparison matrices; testing the consistency index; synthesizing the hierarchy to find out the ranks of the alternatives [26]. The implementation steps are listed as follows.

(1) In the AHP, nine scales are usually used to determine the relative importance of each index. The specific meanings of the scales ranges from 1 to 9 are shown in Table 1.

(2) Building the judgment matrix. A pair wise comparison was used for each index, according to the scale value of each index, yielding a $n \times n$ matrix, where diagonal elements were equal to 1. The matrix is shown in Table 2.

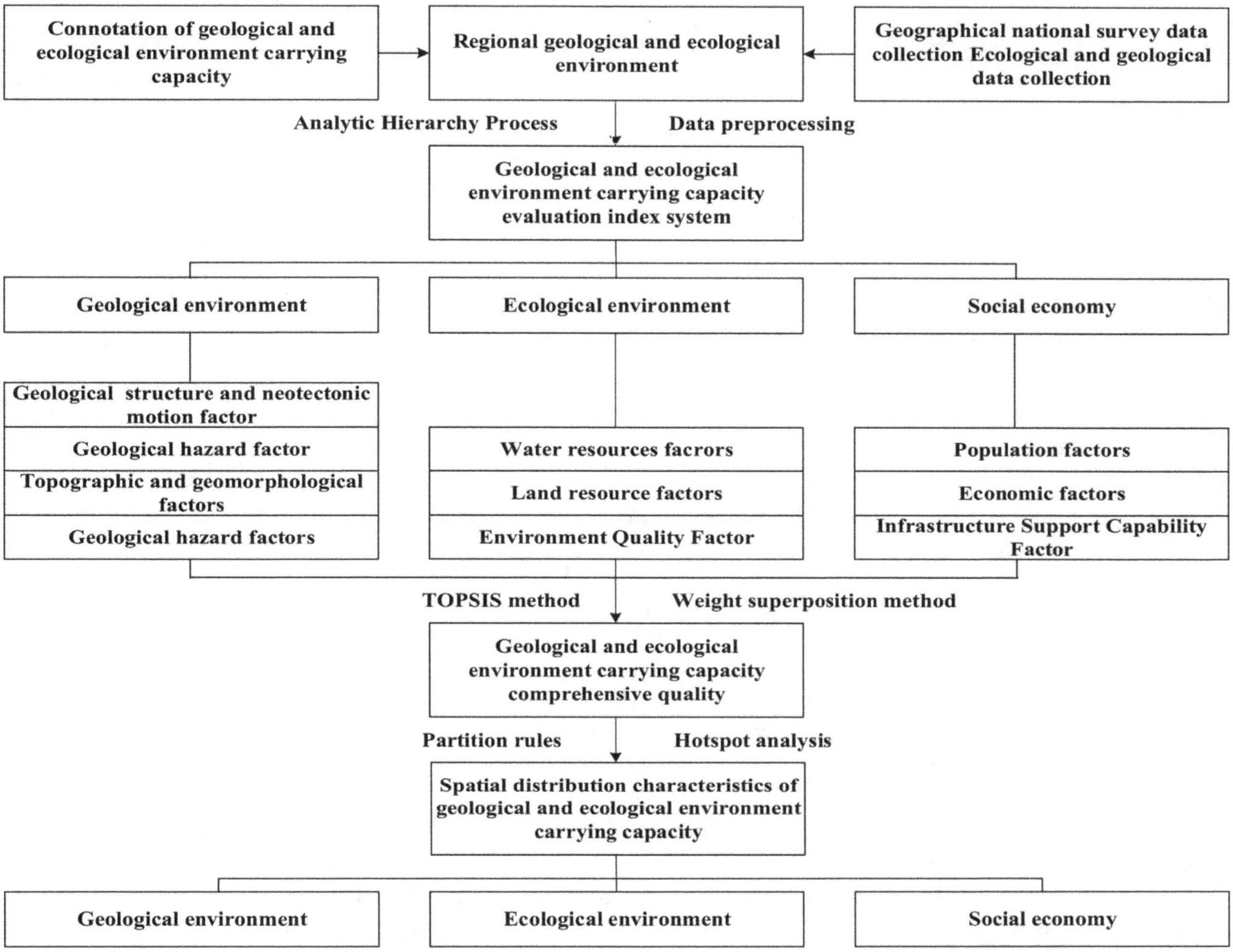

Figure 2. Overall Methodological Framework. TOPSIS = Technique for Order of Preference by Similarity to an Ideal Solution.

Table 1. Meanings of 1–9 Index Scale of the Analytic Hierarchy Process (AHP) method.

Scale $a_{i,j}$	Meaning
1	Index *i* is equally important as index *j*.
3	Index *i* is weakly more important than index *j*.
5	Index *i* is moderately more important than index *j*.
7	Index *i* is very more important than index *j*.
9	Index *i* is extremely more important than index *j*.
2,4,6,8	The comparison of the importance of index *i* and index *j* lies between two levels. For example, the importance of the index with a value of 2 is between level 1 and level 3.
Reciprocal $a_{ji}=1/a_{i,j}$	The judgment value obtained by comparing index *j* with index *i* is the reciprocal of $a_{i,j}$.

Table 2. Judgment matrix.

Judgment Index	a_1	a_2	...	a_n
a_1	1	$a_{1,2}$	...	$a_{1,n}$
a_2	$a_{2,1}$	1	...	$a_{2,n}$
...	...	...	1	...
a_n	$a_{n,1}$	$a_{n,1}$	...	1

Moreover, the judgment matrix must meet the condition ai,j = 1/aj,i.

(3) The maximum eigenvalue of the judgment matrix and its corresponding eigenvectors were calculated and a consistency check was performed. If the Consistency Ratio (C_R) was less than 0.1, it meant that the attributed weights were consistent; otherwise, the judgment matrix was reconstructed. The consistency check equation is as follows:

$$C_i = \frac{\lambda_{max} - n}{n - 1} \tag{1}$$

$$C_R = \frac{C_i}{R_i} \tag{2}$$

In the equation, C_R is the value of the consistency ratio, C_i is value of the consistency index, λ_{max} is the maximum eigenvalue of the judgment matrix, n is the number of indices and RI is the average random consistency index. The Ri values are shown in Table 3.

Table 3. Average random consistency index.

n	1	2	3	4	5	6	7
R_i	0	0	0.58	0.90	1.12	1.24	1.32

2.3.3. TOPSIS Model and Hot Spot Analysis Model

(1) TOPSIS Model

The TOPSIS model is also called "approaching the ideal solution sorting method". It is an effective means to solve the multi-objective decision analysis problem of limited schemes in system engineering. It is a comprehensive evaluation method that uses distance as the evaluation standard. By defining a measure in the workspace, the degree to which the target is close to the positive ideal and away from the negative ideal solution is calculated.

$$Q_i^+ = \sqrt{\sum_{j=1}^{n} \left[W_j\left(k_{ij} - k_j^{max}\right)\right]^2} \quad (i = 1, 2, \cdots, n) \tag{3}$$

$$Q_i^- = \sqrt{\sum_{j=1}^{n} \left[W_j\left(k_{ij} - k_j^{min}\right)\right]^2} \quad (i = 1, 2, \cdots, n) \tag{4}$$

where Q_i^+ —— the distance between the i-th evaluation area and the optimal unit

Q_i^-—— the distance between the i-th evaluation area and the worst unit

W_j—— the weight of the j-th decision indicator

k_{ij}—— the j-th evaluation index score of the i-th evaluation area

k_j^{max}—— the maximum score of the j-th evaluation index

k_j^{min}—— the j-th evaluation index minimum score

In this paper, the geological and ecological environment bearing capacity is indicated by the closeness degree of the work. According to the degree of the closeness, the geological and ecological environment bearing capacity of the study area can be determined, and then the order of good and bad is determined. Let R_i be the degree to which the bearing capacity of the evaluation area is most close to the optimal area bearing capacity. The value range of R_i is between [0,1]. The bigger the value of R_i is, the closer it is to the optimal level of bearing capacity. When the value of R_i is at 1, it indicates that the bearing capacity level is the highest. The calculation formula is as shown in Formula 5.

$$R_i = \frac{Q_i^-}{Q_i^- + Q_i^+} \ (i = 1,\ 2, \cdots, n) \tag{5}$$

(2) Hot Spot Analysis Model

In the research of determining the characteristics of spatial attributes in related fields, Fan [24] used the natural breakpoint method to evaluate the bearing capacity of resources and environment for the restoration and reconstruction after the Lushan earthquake. Zhao et al. [27] used the spatial clustering method to calculate the quality of cultivated land. Li et al. [28] used a spatial autocorrelation method to analyze the content of Ni and Cr in vegetables in the high incidence area of liver cancer in the Pearl River Delta. In order to achieve the organic combination of quality, scale and spatial distribution of regional bearing capacity, this paper chooses hotspot analysis as a method to determine the characteristics of local clustering.

Hotspot analysis (Get-Ord Gi) can calculate the position of high-value and low-value elements in the space, together with the high-low clustering by ArcGIS software. The calculation result Z score represents multiple of standard deviation, which can reflect the discrete of data set degree. A high-value clustering area with a high comprehensive score is a hot spot gathering area. Conversely, a high-value clustering area generated by a region with a lower comprehensive score is a cold spot gathering area. In this paper, the local statistics of regional bearing capacity are analyzed by using Get-Ord Gi local statistics. The hotspot analysis calculation formula is shown in Equations 6, 7, and 8.

$$G_i = \frac{\sum_{i=1}^{n} z_{hi} k_i - \overline{X} \sum_{i=1}^{n} z_{hi}}{S \sqrt{\frac{n \sum_{i=1}^{n} z_{hi}{}^2 - \left(\sum_{i-1}^{n} z_{hi}\right)^2}{n-1}}} \tag{6}$$

In the formula

$$\overline{X} = \frac{\sum_{i=1}^{n} k_i}{n} \tag{7}$$

$$S = \frac{\sum_{i=1}^{n} k_i{}^2}{n} - \overline{X}^2 \tag{8}$$

where G_i —— output statistics of Z score

k_i—— the index score of the evaluation area i

z_{hi}—— evaluate the spatial weight between areas h and i

n—— the total number of evaluation areas

$\overline{X}$—— the average of the index factor scores

S—— the standard deviation of index factor score

2.3.4. Data Preprocessing

According to the established index system, we referenced some of the index classification standards adopted in relevant home and abroad studies, including GB50218-2014, HJ192-2015, GB50188-2007 [29–31]. Wang et al. [32] researched the bearing capacity of geological and ecological environment. The index layer is quantified by ArcGIS software, and the quantitative results are divided into five levels, with scores of 1, 2, 3, 4, and 5 respectively. The higher the score, the better the bearing capacity index. Otherwise, the worse the bearing capacity index will be. The data processing flow of index factors is shown in Figure 3, Figure 4, and Figure 5.

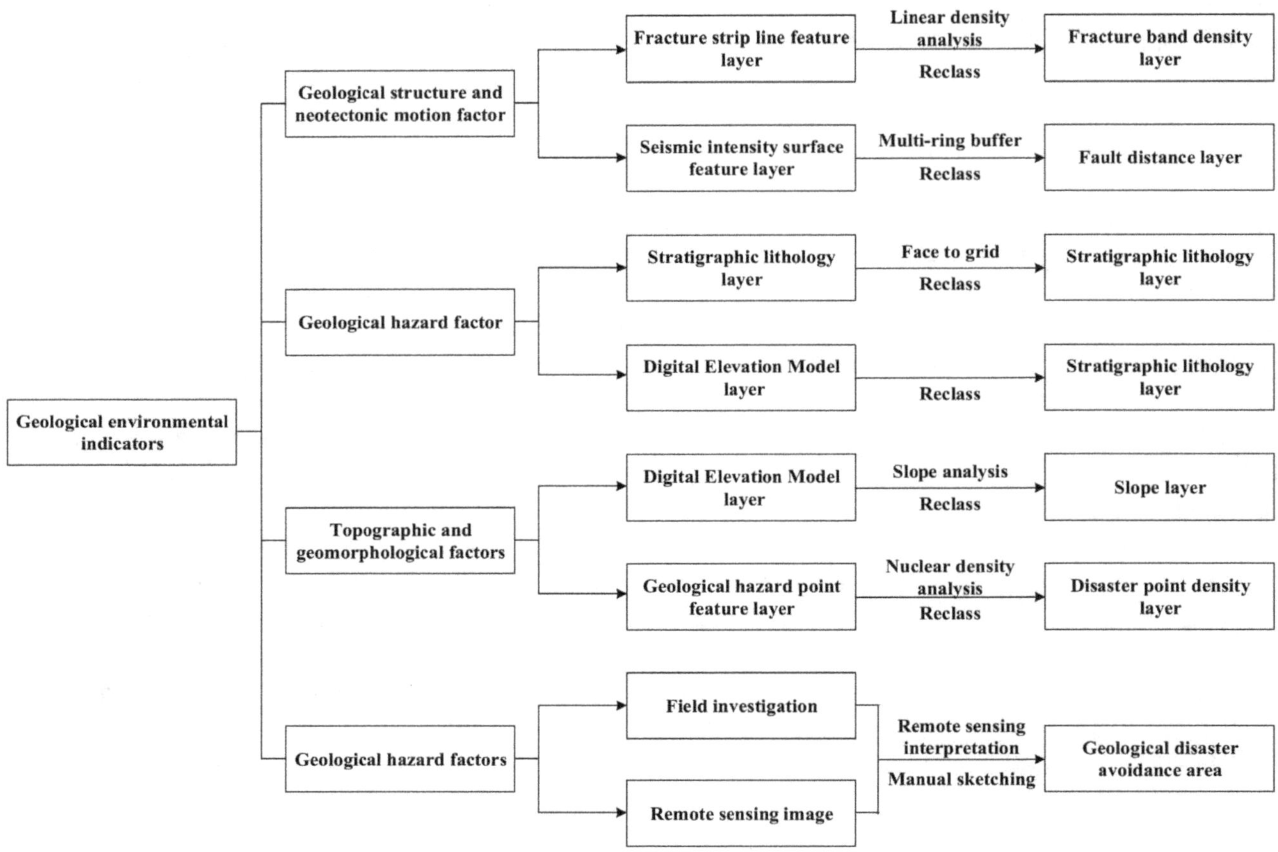

Figure 3. Processing flow chart of geological environment indicators.

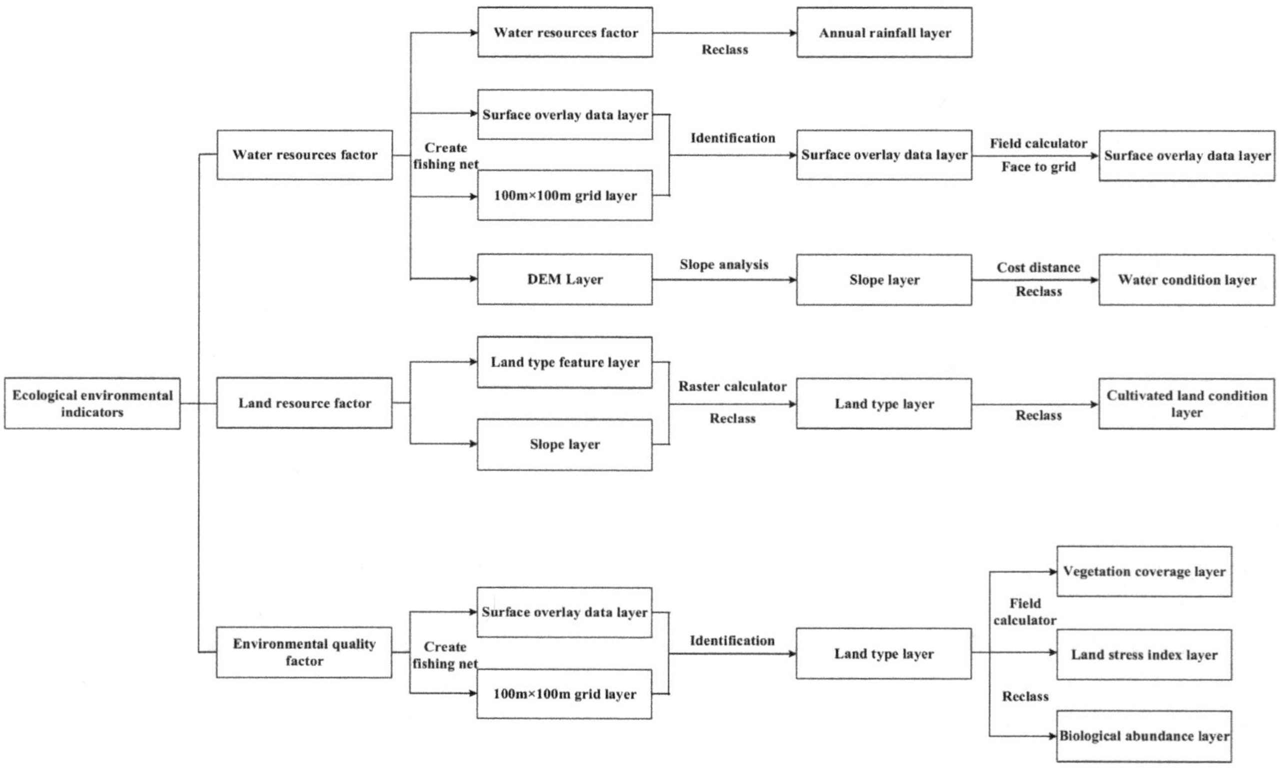

Figure 4. Processing flow chart of eco-environment indicators.

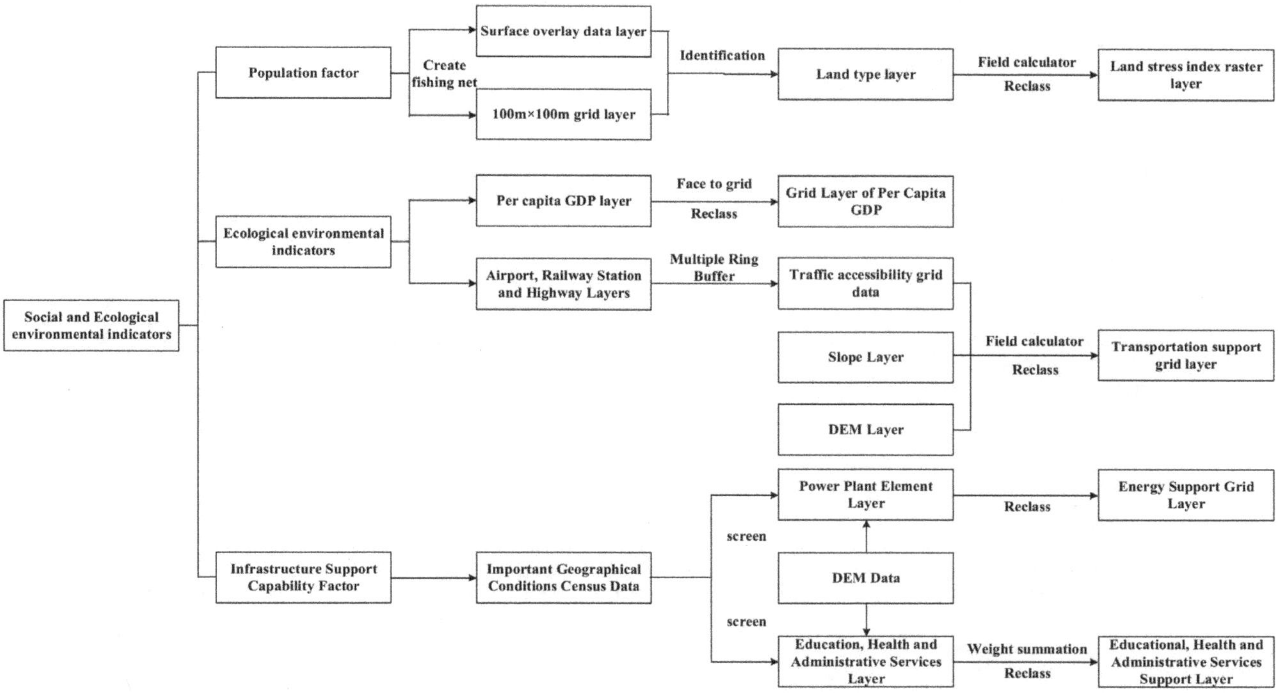

Figure 5. Processing flow chart of social and economic indicators.

3. Results and Analysis

3.1. Determination of Index Weights

In the research of bearing capacity quality evaluation, the selection and quantification of evaluation indicators is an important step in evaluating regional bearing capacity. The index system method is a widely used evaluation method. The concept of Natural-Economic-Society (NES) comes from the land ecosystem, which consists of multiple layers of subsystems. Moreover, the subsystems are coupled to form the land bearing capacity system. The evaluation index system constructed in this paper is built on top of the national geographical survey results, and it is under the principles of comprehensiveness, dominance, science, regionality, objectivity, combination of qualitative and quantitative, and the principle of inheritance. The selection should reflect the regional bearing capacity in a comprehensive, scientific, and objective manner, and build a multi-level regional bearing capacity assessment index system.

We produced an index score form and sent it to five experts. After multiple rounds of information feedback, we took the median of their scores as the final score for each index and based on the above indices and expert scores, constructed a judgment matrix, as shown in Table 4, Table 5, Table 6, Table 7, Table 8, Table 9, Table 10, Table 11.We used MATLAB to get the index weights of the judgment matrix.

Table 4. Judgment matrix of the criteria layer (layer B).

A	B1	B2	B3	B4
B1	1	5	1/2	1/2
B2	1/5	1	1/5	1/5
B3	2	5	1	1
B4	2	5	1	1
$\lambda_{max} = 4.0612, C_i = 0.0204 < 0.1, R_i = 0.9, C_R = 0.0227 < 0.1$				

Table 5. Judgment matrix of indicator layer (layer C).

B	C1	C2	C3	C5	C6	C7	C8
C1	1	7	3				
C2	1/7	1	1/5				
C2	C3	5	1				
$\lambda_{max} = 3.0655$, $C_i = 0.03275 < 0.1$, $R_i = 0.58$, $C_R = 0.05646 < 0.1$							
C5				1	2		
C6				1/2	1		
C7						1	1
C8						1	1

The consistency check ratio of the hierarchy total ranking C_R was 0.0565, that is, it was less than 0.1, and so the consistency check was passed. The weight of the evaluation system of geological environment bearing capacity is shown in Table 6.

Table 6. Weight table of geological environment bearing capacity evaluation system.

Target Layer	Criteria Layer	Weights	Index Layer	Weights	Comprehensive weight
			C_1	0.6434	0.1408
	B_1	0.2188	C_2	0.0738	0.0162
			C_3	0.2828	0.0619
A	B_2	0.0623	C_4		0.0623
	B_3	0.3595	C_5	0.6667	0.2396
			C_6	0.3333	0.1198
	B_4	0.3595	C_7	0.5	0.1797
			C_8	0.5	0.1797

Table 7. Judgment matrix of the criteria layer (layer B).

A	B_1	B_2	B_3
B_1	1	3	1
B_2	1\3	1	1/3
B_3	1	3	1
$\lambda_{max} = 3$, $C_i = 0 < 0.1$, $R_i = 0.58$, $C_R = 0 < 0.1$			

Table 8. Judgment matrix of indicator layer (layer C).

B	C_1	C_2	C_3	C_5	C_6	C_7
C_1	1	1/3	1/3			
C_2	3	1	1/2			
C_3	3	2	1			
$\lambda_{max} = 3.0538$, $C_i = 0.0269 < 0.1$, $R_i = 0.58$, $C_R = 0.0464 < 0.1$						
C_5				1	2	1
C_6				1/2	1	1/2
C_7				1	2	1
				$\lambda_{max} = 3$, $C_i = 0 < 0.1$, $R_i = 0.58$, $C_R = 0 < 0.1$		

The consistency check ratio of the hierarchy total ranking C_R was 0.0232, that is, it was less than 0.1, and so the consistency check was passed. The weight of the evaluation system of ecological environment bearing capacity is shown in Table 9.

Table 9. Weight table of ecological environment bearing capacity evaluation system.

Target Layer	Criteria Layer	Weights	Index Layer	Weights	Comprehensive weight
A	B_1	0.4286	C_1	0.1416	0.0607
			C_2	0.3338	0.1431
			C_3	0.5247	0.2249
	B_2	0.1428	C_4		0.1428
	B_3	0.4286	C_5	0.4	0.1714
			C_6	0.2	0.0857
			C_7	0.4	0.1714

Table 10. Judgment matrix of the criteria layer (layer B).

A	B_1	B_2	B_3
B_1	1	1/2	1/2
B_2	2	1	1/2
B_3	2	2	1

Table 11. Judgment matrix of indicator layer (layer C).

B	C_3	C_4	C_5
C_3	1	2	1
C_4	1/2	1	1/2
C_5	1	2	1
$\lambda_{max} = 3, C_i = 0 < 0.1, R_i = 0.58, C_R = 0 < 0.1$			

The consistency check ratio of the hierarchy total ranking C_R was 0, that is, it was less than 0.1, and so the consistency check was passed. Thus, the weight of social and economic bearing capacity evaluation system is shown in Table 12.

Table 12. Weight table of social and economic bearing capacity evaluation system.

Target Layer	Criteria Layer	Weights	Index Layer	Weights	Comprehensive weight
A	B_1	0.1976	C_1		0.1976
	B_2	0.3119	C_2		0.3119
	B_3	0.4905	C_3	0.4	0.1962
			C_4	0.2	0.0981
			C_5	0.4	0.1962

3.1.1. Weight of Geological Environment Index

In the hierarchical structure chart of geo-environmental bearing capacity evaluation (Figure 6), the geological environment bearing capacity is the target layer (layer A) of the hierarchical structure. The structural elements, geomorphic elements, and other elements are the criteria layer (layer B). The specific slope, high-rise, seismic intensity, and other eight indicators are the indicator layer (layer C) of the hierarchical level.

The consistency check ratio of the hierarchy total ranking C_R was 0.0565, that is, it was less than 0.1, and so the consistency check was passed. The weight of the evaluation system of geological environment bearing capacity is shown in Table 6.

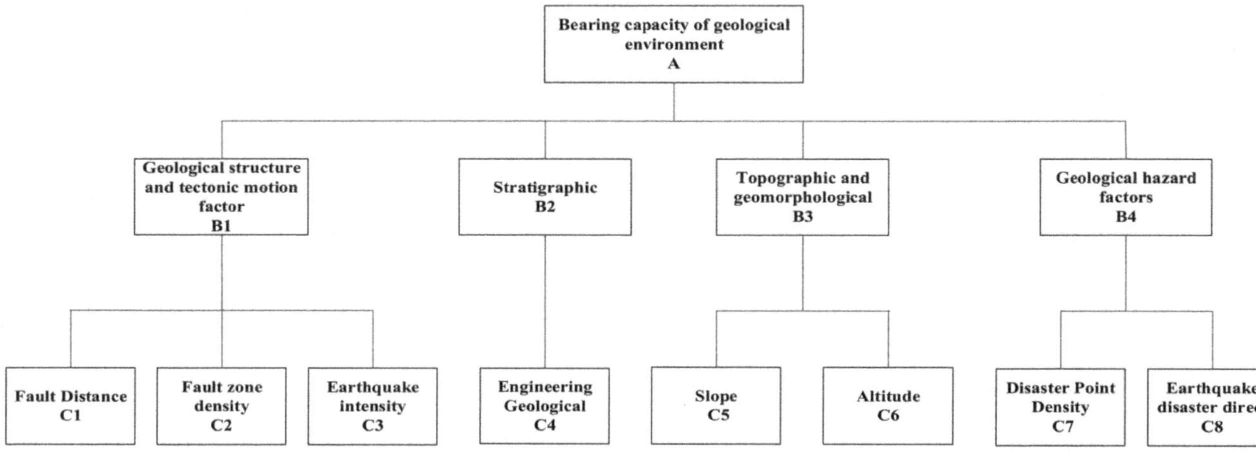

Figure 6. Hierarchical structure of geological environment bearing capacity evaluation.

3.1.2. Weight of Ecological Environment Index

In the hierarchical structure chart of eco-environmental bearing capacity evaluation (Figure 7), eco-environmental bearing capacity is the target layer (Layer A) of hierarchical structure, water resource elements, land resource elements, and other elements are the criteria layer (Layer B). Seven indicators, such as annual rainfall, water conservation, and cultivated land conditions are the indicator layer (Layer C) of hierarchical structure.

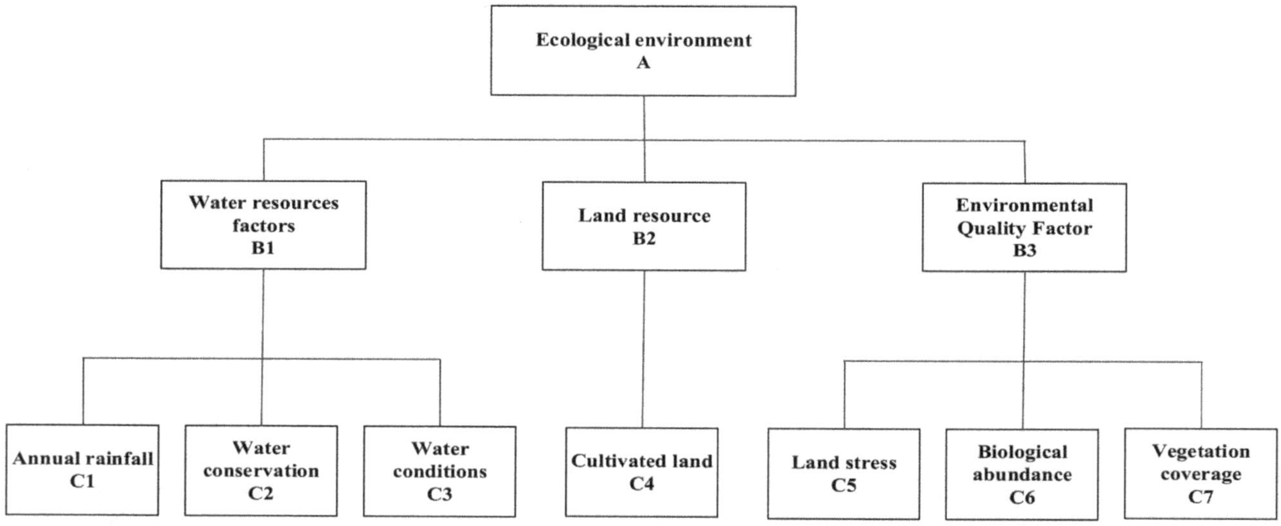

Figure 7. Hierarchical structure of ecological environment bearing capacity evaluation.

The consistency check ratio of the hierarchy total ranking C_R was 0.0232, that is, it was less than 0.1, and so the consistency check was passed. The weight of the evaluation system of ecological environment bearing capacity is shown in Table 9.

3.1.3. Weight of Social and Economic Index

In the hierarchical structure chart of social and economic bearing capacity evaluation (Figure 8), social and economic bearing capacity is the target layer of hierarchical structure (Layer A). Population factor, economic factor, and infrastructure support factor form the criteria layer (Layer B); five indicators, such as residential area distribution, GDP per capita, and transportation facility support are the indicator layers (Layer C).

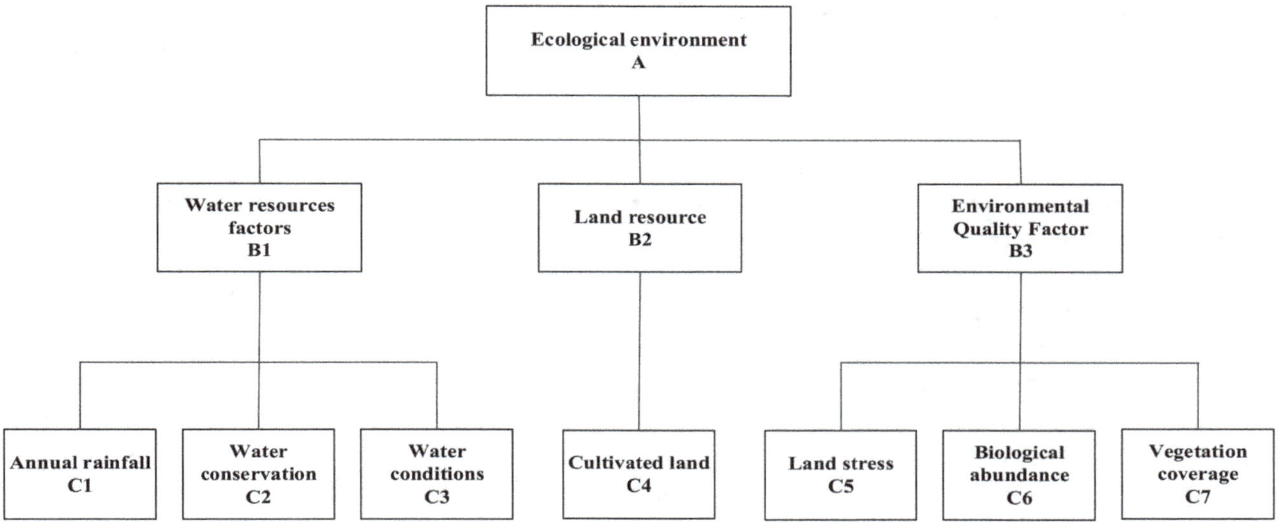

Figure 8. Hierarchical structure of social and economic bearing capacity evaluation.

The consistency check ratio of the hierarchy total ranking C_R was 0, that is, it was less than 0.1, and so the consistency check was passed. Thus, the weight of social and economic bearing capacity evaluation system is shown in Table 12.

3.2. Comprehensive Quality Evaluation of Bearing Capacity

First, the distance between each index layer grid unit and the optimal unit and the worst unit is calculated by using the grid calculator. The positive and negative ideal solutions of geological environment subsystem, ecological environment subsystem, social economy subsystem, and comprehensive bearing capacity are obtained by weighted superposition. Secondly, the grid calculator is used to calculate the corresponding closeness degree, and the natural breakpoint method is used to divide the load-bearing status closeness degree score into five grades I, II, III, IV, and V, which are converted into vector data for spatial distribution clustering research. The closeness degree score is shown in Figure 9.

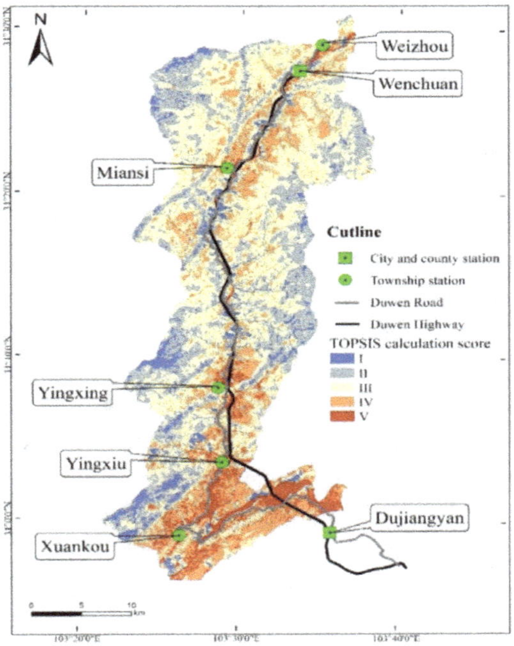

Figure 9. Closeness degree score of comprehensive assessment of geological and ecological environment.

Among them, the areas of Grade I, II, III, IV, and V are 61.93 km^2, 233.74 km^2, 345.82 km^2, 222.82 km^2, and 61.3 km^2 respectively, accounting for 6.69%, 25.25%, 37.36%, 24.07%, and 6.62% of the total area each.

3.3. Local Clustering Result

The hot spot analysis is used to calculate the closeness scores of the geological environment, ecological environment, socioeconomic, geological, and ecological environment comprehensive bearing capacity evaluation. The results of how high-valued and low-valued areas clustered in the calculation are divided into 7 levels: 99% confidence hotspot clustering (high and high neighboring), 95% confidence hotspot clustering (high and middle neighboring), 90% confidence hotspot clustering (middle and middle neighboring), which is not significant, 99% confidence cold spot clustering area (low and low neighboring, 95% confidence cold spot clustering zone (low and middle neighboring), 90% confidence cold spot clustering zone (high and low neighboring). Red zone represents high value clustering zone and blue zone represents the low. The value aggregation area is shown in Figure 10.

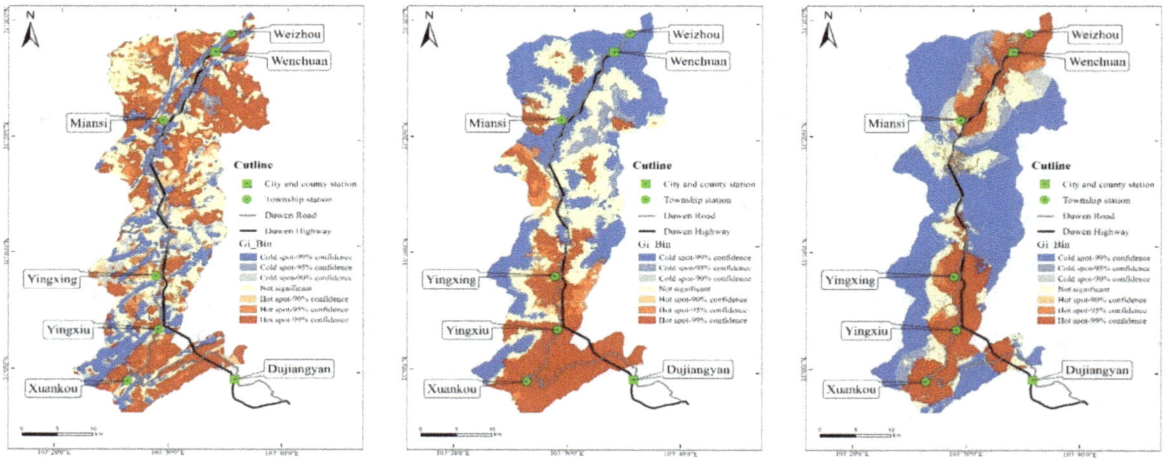

(a) (b) (c)

Figure 10. (**a**) Results of hotspot analysis: for hot spot analysis of geological environment subsystem; (**b**) hot spot analysis of ecological environment subsystem; (**c**) hotspot analysis of social economic subsystem.

As for the subsystem of geological environment, the regional geological bearing capacity of Xuankou Town, Zipingpu Town and the main stream of Minjiang River in the study area are clustered in high value. In the northwest direction of Xuankou Town, the debris flow channel and its outburst areas, such as Qipan Ditch and Taoguan Ditch, as well as the geological environment bearing capacity near Maowen fault zone (Longmen Mountain back the mountain fault) and Yingxiu fault zone, are low-value aggregated. The geological hazards in the region include debris flow, landslides, and earthquakes. The hot spot area is 430.55 km^2 and the cold point area is 234.95 km^2, as shown in Figure 10a.

As for the subsystem of ecological environment, the triangle area composed of Dujiangyan City, Yingxiu and Xuankou towns in the southern part of the study area shows high value aggregation, while the high mountain and forest areas in the northern part of the study area also show local high value aggregation. The high-altitude snow cover area in the northwest corner of Miansi Town, the high-drop vegetation-free growth area in the southwest corner of Ginkgo Township, and the area around Wenchuan County are low-value aggregation areas. The hot spot area is 317.16 km^2 and the cold point area is 373.02 km^2, as shown in Figure 10b.

As for social and economic subsystems, because of the relatively concentrated population in cities and towns, and the relatively complete construction of transportation, energy, and infrastructure, public service departments, such as hospitals and schools, are basically concentrated in the town center. The social and economic bearing capacity in space takes the town as the center, and tends to weaken outward, in accordance with the principle of attenuation. The central area of the town and the area along Du-Wen Road show high value aggregation, while the areas far from the central area of the town and the mountainous areas with higher elevation show low value aggregation. The hot area is 260.17 km^2, and the cold point area is 531.95 km^2, as shown in Figure 10c.

3.4. Graded Result

According to the calculation results of hot spot analysis, the low-low neighboring and low-middle neighboring clustering regions are defined as unsuitable construction areas according to their spatial distribution characteristics of low-value aggregation. The high-low neighboring and middle-middle neighboring clustering regions and random distribution regions are designated as backup reserve areas. The high-middle neighboring and high-high neighboring regions are designated as suitable construction areas, for their spatial distribution characteristics of high-value aggregation, as shown in Figure 11.

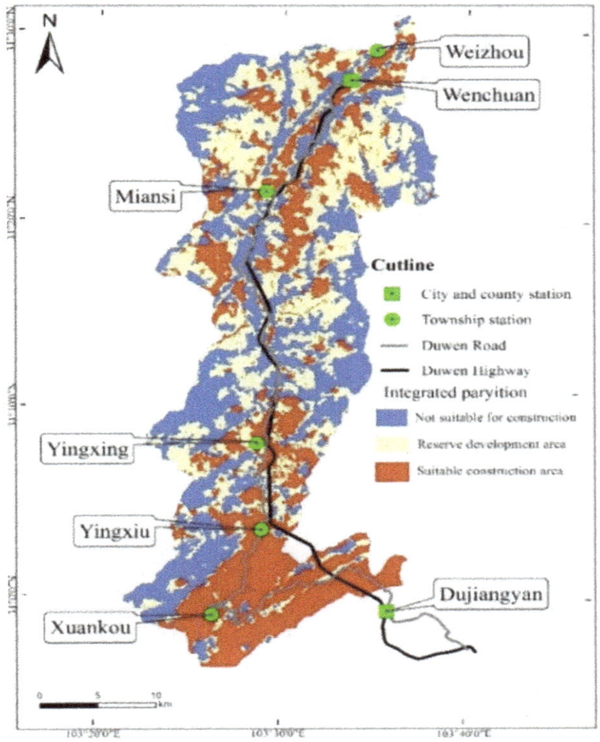

Figure 11. Comprehensive partitioning.

The bearing capacity of geological and ecological environment along Du-Wen Road has distinctive spatial clustering characteristics. In general, the bearing capacity of Xuankou Town and Yingxiu Town along the main stream of Minjiang River in the south is better than that of Weizhou Town and Miansi Town in the north. The bearing capacity is obviously low in the vicinity of faults and where debris flow, landslide, and other geological disasters have happened. Among them: the total area suitable for construction is 288.38 km^2, accounting for 31.12% of the total area of the region; the total area of the reserve development area is 296.35 km^2, accounting for 31.98% of the total area of the region; the total area unsuitable for construction area is 340.87 km^2, accounting for 36.79% of the total area of the region.

4. Conclusions

This paper selects the area along the mountain road with characteristics of geological disasters as the research area, and combines the research results of existing bearing capacity with the first national geographical survey of Sichuan Province, and adds geological environmental factors to construct a reflection research area. The evaluation index system of different aspects of geological and ecological environment bearing capacity is determined by the AHP method. The data is pre-processed by ArcGIS software. The TOPSIS method and superimposed analysis tool are used to quantitatively evaluate the geological and ecological environment bearing capacity of the study area. The following conclusions are obtained:

(1) Based on the comprehensive quality evaluation of the bearing capacity of the study area and thinking from a spatial perspective, the priority scores are calculated by hot spot analysis tools. According to the calculation results, the study area is divided into suitable construction area, backup reserve area, and unsuitable construction area. From the perspective of the spatial distribution of the construction area, the bearing capacity of Xuankou Town and Yingxiu Town along the southern Minjiang River is superior to that of Weizhou Town and Miansi Town in the north. The research provides a new idea for the zoning planning of the comprehensive evaluation of regional bearing capacity.

(2) The zoning results consider the bearing capacity relationship among the quality-scale-space distribution. The evaluation results not only consider the development suitability from the perspective of comprehensive quality, but also consider the spatial stability of sustainable development from the perspective of spatial layout. It is of great significance for optimizing the development pattern and resource allocation of land space.

(3) After the Wenchuan Earthquake in 2008 and Lushan Earthquake in 2013, the mountain environment, especially the geological environment, has become more fragile in Sichuan. Therefore, to investigate the current situation and characteristics of the geological ecological environment and analyze its bearing capacity, has become an important premise for highway management departments to formulate safe operation and management strategies, and for township construction planning departments to reasonably develop and utilize the limited environmental resources along the mountain highway, which is a work of great practical significance.

Author Contributions: Author Contributions: Conceptualization, Zhoufeng Wang and Chen Zhang; methodology, Zhoufeng Wang and Chen Zhang; software, Zhoufeng Wang; validation, Zhoufeng Wang and Jianwei Xu; formal analysis, Zhoufeng Wang and Xiangqi He; investigation, Yujun Wang; resources, Zhoufeng Wang; data curation, Zhoufeng Wang and Chen Zhang; writing—original draft preparation, Zhoufeng Wang, Chen Zhang; writing—review and editing, Zhoufeng Wang, Chen Zhang, Yujun Wang, Jianwei Xu, and Xiangqi He; visualization, Zhoufeng Wang and supervision, Zhoufeng Wang, Chen Zhang, and Xiangqi He. All authors have read and agreed to the published version of the manuscript.

Acknowledgments: We sincerely appreciate the Editor's encouragement and the anonymous reviewer's valuable support.

References

1. Ma, C.M.; Ma, Y.H. Tentative investigation of bearing capacity of geological environment for sustainable development. *Environ. Sci. Technol.* **2007**, *30*, 64–65.
2. Liu, J.J.; Dong, S.C.; Li, Z.H. Comprehensive evaluation of China's water resources carrying capacity. *J. Nat. Resour.* **2011**, *26*, 258–269.
3. Liu, M.H.; Zhao, S.; Yang, X.H. Dynamic evaluation of regional comprehensive carrying capacity in Henan Province. *Areal Res. Dev.* **2014**, *33*, 31–36.
4. Lu, J.T.; Lu, D.M.; Liu, H.P. Analysis of ecological environment resources carrying capacity of Guiyang City. *Chin. J. Agric. Resour. Reg. Plan.* **2014**, *35*, 24–28.
5. Li, Z. Land Ecological Safety Warning of Mountain Plain Transitional Area-Case Study of Dujiangyan City. Ph.D. Thesis, Chengdu University of Technology, Chengdu, China, 2012.

6. Wang, Y.Q.; Guo, J.J.; Li, K. Spatial-temporal dynamic analysis of the biocapacity pattern in the mountain area of upper Shiyang river catchment. *J. Lanzhou Univ.* **2013**, *49*, 166–172.
7. Liu, J.X.; Chen, L.J.; Huang, Q.H. Evaluation of Geo-environmental carrying capacity of Wulonggou mine based on spatial statistical analysis method. *Met. Mine* **2017**, *4*, 162–166.
8. Yang, L.; Peng, H.Y.; Zhou, M. Assessment of geological environment carrying of capacity Fengjie County based on AHP. *J. Chongqing Jiaotong Univ.* **2014**, *33*, 95–99.
9. Yang, Q.; Yu, C.P.; Zhou, F. An emergy-ecological footprint model based evaluation of ecological security at the industrial area in Northeast China, A case study of Liaoning Province. *Chin. J. Appl. Ecol.* **2016**, *27*, 1594–1602.
10. Zheng, H.; Shi, P.J.; He, J.J. THE dynamic on ecological footprint and ecological capacity of Gansu Province. *J. Arid Land Resour. Environ.* **2013**, *27*, 13–17.
11. Gu, K.K. The concept of ecological carrying capacity and its research methods. *Ecol. Environ. Sci.* **2012**, *21*, 389–395.
12. Xu, Y.Y.; LI, G.Z.; Xu, W. Dynamic evaluation of traffic environmental bearing capacity if Beijing and diagnosis of obstacle factor. *J. Highw. Transp. Res. Dev.* **2017**, *34*, 122–127.
13. Wang, Z.F.; Zhang, T.S.; Wang, C.W. Prediction model and its application for glacial lake outburst in the Himalayas area, Tibet. *J. Glaciol. Geocryol.* **2016**, *38*, 388–394.
14. Ni, Z.Y. Study on the Eco-Geological Environment Carrying Capacity of Qushi-Sangri Area in Tibet. Ph.D. Thesis, Chengdu University of Technology, Chengdu, China, 2011.
15. Tong, X.; Gan, Y.Q.; Xia, Y. Assessment and zoning of the geological environment in the city of Wuhan. *Hydrogeol. Eng. Geol.* **2015**, *42*, 149–155.
16. Wang, Z.F.; Zhang, T.S.; Wang, L. The integrated assessment system of eco-geological environmental carrying capacity for Dujiangyan-Wenchaun highway. *Sci. Surv. Mapp.* **2016**, *41*, 77–81.
17. Ma, Z.; Xie, H.; Lin, L. Analysis of geological conditions of land, resources and environment in Beijing Tianjin Hebei region. *Geol. China* **2017**, *44*, 857–873.
18. Jiang, Y.; Lin, L.; Chen, L. Resource and environmental conditions and major geological problems of the Yangtze River Economic Belt. *Geol. China* **2017**, *44*, 1045–1061.
19. Zhang, M.; Wang, Y. Theory and evaluation method of bearing capacity of geological environment based on risk. *Geol. Bull. China* **2018**, *37*, 467–473.
20. Meng, H.; Zhang, R.; Shi, J.; Li, C. Safety assessment of geological environment. *Earth Sci.* **2019**. [CrossRef]
21. Rapport, D.J.; Costanza, R.; McMichael, A. Assessing ecosystem health. *Trends Ecol. Evol.* **1998**, *13*, 397–402. [CrossRef]
22. Zhu, J.; Lu, H.; Wang, H. Ecosystem health assessment in Wenchuan earthquake disaster area during recovery period. *Acta Ecol. Sin.* **2018**, *38*, 9001–9011.
23. Tang, M.; Liu, B.; Li, S. Review and Prospect of the research on the bearing capacity of resources and environment—For the mountainous area with severe topography after the earthquake. *J. Fuzhou Univ.* **2016**, *30*, 10–13.
24. Fan, J. *Evaluation of Carrying Capacity of Resources and Environment for Restoration and Reconstruction after Lushan Earthquake*; Science Press: Beijing, China, 2014.
25. Saaty, T.L. *The Analytic Hierarchy Process: Planning, Priority Setting, Resource Allocation*; McGraw-Hill: New York, NY, USA, 1980.
26. Saaty, T.L. *Analytical Planning the Organization of Systems*; Pergamon Press: New York, NY, USA, 1985.
27. Zhao, D.L.; He, S.S.; Yang, J.Y. Quality potential calculation of arable land consolidation based on limiting factors and hot spot analysis. *Trans. Chin. Soc. Agric. Mach.* **2017**, *48*, 158–164.
28. Li, Y.; Zhou, Y.Z.; Zhang, C.B. Application of local Moran's I and GIS to identify hotspots of Ni, Cr of vegetable soils in high-incidence area of liver cancer from the pearl river delta, South China. *Environ. Sci.* **2010**, *31*, 1617–1623.
29. Ministry of Housing and Urban-Rural Development of People's Republic of China. *General Administration of Quality Supervision Inspection and Quarantine of the People's Republic of China. Standard of engineering Classification of Rock Mass, (GB50218-2014)*; Standard Press of China: Beijing, China, 2014.
30. Ministry of Environment Protection of the People's Republic of China. *Technical Criterion for Eco-Environmental Status Evaluation, (HJ192-2015)*; Environment Press of China: Beijing, China, 2015.

31. Ministry of Housing and Urban-Rural Development of People's Republic of China. *General Administration of Quality Supervision Inspection and Quarantine of the People's Republic of China. Standards for Planning of Town, (GB50188-2007)*; Standard Press of China: Beijing, China, 2007.
32. Wang, Z.F.; Wang, Y.J.; Wang, L. Research on the comprehensive evaluation system of eco-geological environmental carrying capacity based on the analytic hierarchy process. *Clust. Comput.* **2017**, *10*, 1–10. [CrossRef]

5

Spatial Relationship between Natural Wetlands Changes and Associated Influencing Factors in Mainland China

Ting Zhou [1], Anyi Niu [1], Zhanpeng Huang [2], Jiaojiao Ma [1] and Songjun Xu [1,*]

[1] School of Geography, South China Normal University, Guangzhou 510000, China; zhouting@m.scnu.edu.cn (T.Z.); niuanyi@m.scnu.edu.cn (A.N.); jjma@m.scnu.edu.cn (J.M.)

[2] International Envirotech Limited, Hong Kong 999077, China; hzp@internationalenvirotech.com

* Correspondence: xusj@scnu.edu.cn.

Abstract: Many studies have explored the dynamic change of wetlands distribution which play an important role in wetlands conservation and its sustainable management. However, given an uneven distribution of natural wetland resources in the context of global change, little is known about the spatial relationship between natural wetlands changes and associated influencing factors in mainland China. In this study, Moran-based spatial statistics are an effective methodology to examine the spatial patterns of natural wetlands and associated influencing factors at the province level, and GIS mapping is applied to help visualize spatial patterns. Results show that 1) significant spatial agglomeration and regional differences of natural wetlands distribution have been captured by Moran's *I* statistics, and the agglomeration level has increased over the past ten years; 2) Seven of the eight factors show significantly strong and positive spatial autocorrelation except for water consumption, and spatial patterns of them show significant spatial clusters or spatial outliers; 3) Spatial coordination between natural wetlands distribution and the associated influencing factors is higher in the western region than in east China and northeast China. Moreover, spatial coordination between a cultivated area or water consumption and natural wetlands distribution is weaker than that of other factors. Finally, the influences generated by neighboring provinces should not be neglected in the implementation of wetlands conservation. This study could provide a scientific basis for the policy making of wetlands conservation and sustainable management systems.

Keywords: spatial autocorrelation; spatial pattern; spatial relationship; natural wetlands changes; associated influencing factors; mainland China

---

1. Introduction

Wetlands have value owing to their useful functions to humans [1]. They not only provide goods and ecosystem services for humanity but also habitats for flora and fauna [2,3]. With their fragility and sensitivity to climatic change and anthropogenic activities, wetlands are among the most vulnerable ecosystems around the world [4]. With the increasing awareness of the important role wetlands play in wastewater storage and purification, flood regulation, food supply, wildlife habitat, etc., studies of wetlands have received considerable attention in recent years. Moreover, understanding the dynamics of wetlands is significant for the conservation of wetland biodiversity and its sustainable management [5].

Wetlands changes and their causes have been paid more attention by global environmental change research [6,7]. The dynamic change of wetlands directly reflects on the change of wetland area and landscape pattern and indirectly on the wetland structure and functions [8,9]. Most studies indicate that the area of artificial wetlands is increasing, while natural wetlands are decreasing in the form of wetland degradation and loss [10–16]. China is rich in wetland resources with the total wetland area ranked the fourth largest in the world [17]. Natural wetlands will occupy 87.37% of the overall wetland area by the end of 2013 [18]. However, wetlands in China have been severely destroyed and degraded since 1950 due to climate change and anthropogenic disturbance [16,19–22], especially natural wetlands, which has sparked considerable attention from scholars. On one hand, many studies analyzed the status and challenges of wetlands in China and put forward some recommendations for wetland protection and management [21,23–26]. On the other hand, many studies mapped and quantified wetland changes from the perspective of time and space, based on 3S technologies (remote sensing, geography information systems and global positioning systems). Landscape metrics or models at national, regional or local scales can make the long-term monitoring of wetland changes feasible and build a significant foundation for wetland resources protection and management [15,16,20,27–31].

In sum, in the context of global change, wetlands in China have suffered serious degradation and loss. Research into wetland degradation and loss in China mainly focuses on the dynamic changes of wetlands landscape patterns and their driving forces. The prevalent method is to obtain information on dynamic changes on specific wetlands and to carry out quantitative analysis by using 3S technology on the scale of nation, region or case-by-case. These research results clearly describe the dynamic process of wetlands and their causes on a large or a small scale. However, few researches have been conducted to discover the spatial autocorrelation of wetlands distribution and associated influencing factors, or to explore the spatial relationships between them at the province level. In China, policy formulation and implementation of wetlands management are carried out at the province level. Given an uneven distribution of natural wetlands in China, questions such as how targeted wetlands conservation policy should be conducted based on the local conditions and which influencing factors should be paid great attention by provincial governments remain to be resolved. However, the study of the spatial relationship between natural wetlands distribution and associated influencing factors can discover the spatial uncoordinated provinces and influencing factors that can inform the provincial government of decision-making of targeted and effective wetlands conservation. Therefore, this study explored the spatial pattern of natural wetlands dynamics and the associated influencing factors based on data of the natural wetlands area in 31 provinces, by applying Moran's I statistics. Plus, the spatial relationship between them was further examined. The objectives of this paper are to: a) explore the spatial pattern and spatial law of natural wetlands distribution; b) explore the spatial patterns of associated influencing factors and the difference in impacts of these factors on natural wetlands dynamics; and c) explore the spatial relationship between natural wetlands changes and associated influencing factors and to identify the uncoordinated provinces.

2. Methods and Data

2.1. Study Area

China's wetland is the fourth largest in area in the world. There are 26 types of natural wetlands and 9 types of artificial wetlands in China according to the delineation in the Ramsar Convention [23]. Furthermore, some Chinese researchers put forward their own wetland classification for different case areas [6,15,20,25,29]. According to national wetland resources survey reports, 5 types are included in China's wetlands, covering lake wetlands, riverine wetlands, coastal wetlands, inland marshes/swamp and artificial wetlands, all of which are natural wetlands except for the artificial wetland [18]. Natural wetlands have dominated the overall areas of China's wetlands with a rate of 3.8% of China's terrestrial area. The wetlands in China are unevenly distributed among seven regions, including the Northeast China wetland region, the Xinjiang-Inner Mongolia wetland region, the Qinghai-Tibet Plateau wetland region, the Yunnan-Guizhou Plateau wetland region, the Yellow River Mid-down Stream wetland region, the Yangtze River Mid-down Stream wetland region, and the Southeast-South China wetland region [24].

The Chinese government has conducted two national surveys of wetland resources in 1998–2003 and 2009–2013 to understand the status of wetland resources, familiarize themselves with wetland dynamics and strengthen wetland protection policies. While the statistical calibers of wetland resources in the two surveys were different from the area of an independent patch, over 100 ha were included in the first survey and over 8 ha were included in the second survey. According to the two national wetland resources survey reports, the total area of wetlands is 38.49×10^6 ha in the first survey and 53.42×10^6 ha in the second survey [18]. Compared with the first survey, based on the same statistical caliber, wetlands in the second survey have decreased by 3.40×10^6 ha with a loss rate of 8.82%. Among them, natural wetlands have decreased by 3.38×10^6 ha with a loss rate of 9.33% [18]. It is very clear that natural wetlands still suffered a great loss and degradation from 2003 to 2013.

2.2. Data Source and Data Processing

According to the two national wetland resources survey reports which were published in 2004 and 2014, the areas of natural wetlands for 31 provinces in mainland China during two periods are obtained. To eliminate the influence of different statistical calibers on data, the area occupation rate (AOR) for each province between the area of different data sources and the total area of similar data sources is adopted to explore the characteristics of the spatial distribution of natural wetlands. The raw data are subjected to statistical and spatial analysis. Then the AOR of natural wetlands is chosen to describe the dynamic change of natural wetlands. The corresponding calculation formula is as follows:

$$AOR_{ij} = \frac{S_{ij}}{S_j},$$

where AOR_{ij} is the area occupation rate in province i in the year j; s_{ij} is a natural wetlands area in province i in the year j; s_j is the total area of natural wetlands in the year j; and j represents the time at the end of 2003 and 2013 respectively.

2.3. Moran-Based Spatial Autocorrelation Statistics

Tobler's first law of geography pointed out that everything is related to everything else but near things are more related than distant things [32]. Spatial autocorrelation exists when significant similarity or dissimilarity formed between a value observed at a location and its neighboring locations. Although global spatial autocorrelation describes the global characteristics of the variables over the study area, these measures of global spatial association often hide the interesting local patterns in variables. Generally, these local patterns exist in the form of small spatial clusters or outliers [33]. Thus, local identifications of spatial association (LISA) was suggested to complete the global analysis through the detection of the existence and location with a spatial affinity at the local scale. LISA has been widely applied in the fields of economy, environment, geochemistry, epidemiology, etc. and has been proved be an effective exploratory spatial data analysis approach [34–36]. Generally, global Moran's I is commonly used to examine the spatial autocorrelation of objects under study and local Moran's I is employed to examine spatial clustering of the objects with high and low value. Formally, global and local Moran's I are expressed as Equation (1) and Equation (2), respectively [37]:

$$I = n\sum_{i=1}^{n}\sum_{j=1}^{n} w_{ij}(x_i - \overline{x})(x_j - \overline{x}) \Big/ \left\{\left(\sum_{i=1}^{n}\sum_{j=1}^{n} w_{ij}\right)\sum_{i=1}^{n}(x_i - \overline{x})^2\right\} \tag{1}$$

$$I_i = \frac{x_i - \overline{x}}{\sum_i (x_i - \overline{x})^2}\sum_j w_{ij}(x_j - \overline{x}), \tag{2}$$

where n is the number of observations; x_i is the value of the variable x at location i and x_j is the value of the variable x at all other locations (where $i \neq j$); $\overline{x}$ is the average value of x with the observation number of n; w_{ij} is the weight between locations i and j; w_{ij} is a weight which can be defined as the contiguity between location i and j. A first-order queen contiguity is adopted to calculate the weight w_{ij}. Locations with common edges to the targeted location i are given the same weight, while those without common edges are given the weight of 0. The statistical significance of spatial clusters is tested through the comparison of reference distributions which are generated by the replicated statistic 9999 times [34]. The range of Moran's I value is $[-1,1]$. A positive or negative I-value implies the existence of positive or negative spatial autocorrelation over the study area respectively. However, when the I-value is 0, it implies spatial randomness.

LISA is a decomposition of the global statistics of spatial association. It has four types of spatial association: high (low) values associated with high (low) neighboring values, high (low) values associated with low (high) neighboring values. High-high (Low-low) association with a high positive Moran's I value implies a spatial clustering of similarly high (low) values under study which are generally above (below) the mean. On the contrary, the High-low (Low-high) association is given a high negative Moran's I value with the value of the object above (below) the mean and the values of its neighbors below (above) the mean. It implies the examples of spatial outliers which are obviously different from the value of their neighbors [36]. In other words, it is a clustering of dissimilar values [37]. Additionally, significant High-low and Low-high associations are regarded as unstable spatial patterns. Moreover, significant High-high spatial clusters are regarded as "regional hotspots" and Low-low clusters are "cool spots" [35].

2.4. *Data Treatment with Computer Software*

All the data were treated using different software packages. The descriptive statistical parameters were calculated with Microsoft Excel® and SPSS® (version 23.0). Spatial autocorrelation calculations were performed using the software GeoDa (version 1.12, Spatial Analysis Laboratory, 2019); GeoDa: An introduction to spatial data analyses: website https://geodacenter.github.io/ (Last accessed: June 2019). Maps were produced with the GIS software ArcMap®(version 10.5).

3. Results and Discussion

3.1. *Statistical Characteristic of Natural Wetlands Area*

Table 1 summarizes the descriptive statistics for AOR of natural wetlands of 31 provinces in 2003 and 2013. It is clear that all the statistics values in 2013 have increased compared with those values in 2003, except the mean and the median. AOR of natural wetlands displayed strong variability in 31 provinces in 2003 and 2013, ranging from 0.0001 to 0.1445 and from 0.0005 to 0.1714, respectively. On the other hand, the medians are much lower than the means and the mean value in 2013 was more than twice its median value, indicating a more positively skewed distribution of natural wetlands than that in 2003, which is following the skewness values and suggests the existence of some high values, namely the outliers [38].

Table 1. Descriptive statistics of area occupation rate (AOR) of natural wetlands in 31 provinces in 2003 and 2013.

	Mean	Min	Q1	Median	Q3	Max	Std-Dev	Skewness
AOR in 2003	0.0323	0.0001	0.0077	0.0202	0.0346	0.1445	0.0376	1.925
AOR in 2013	0.0323	0.0005	0.0059	0.0153	0.0352	0.1714	0.0442	2.058

3.2. *Data Transformation*

In this study, AOR data is used to explore the spatial pattern of natural wetlands distribution changes. Normal Q-Q plots for raw data of AOR are displayed in Figure 1. These graphs plot the accumulative distribution function of the raw data against the empirical accumulative distribution frequency of Gaussian data sets [38]. Normal Q-Q plots show that the AOR does not follow a normal distribution, and these distributions were positively skewed. Besides, there were likely some outliers in the datasets. Therefore, normal score transformation was applied to normalize the data, as well as to weaken the negative effects of potential outliers. Q-Q plots for the normalized data are shown in Figure 2. It indicates that data transformation is helpful to further the distribution to approximate a Gaussian distribution.

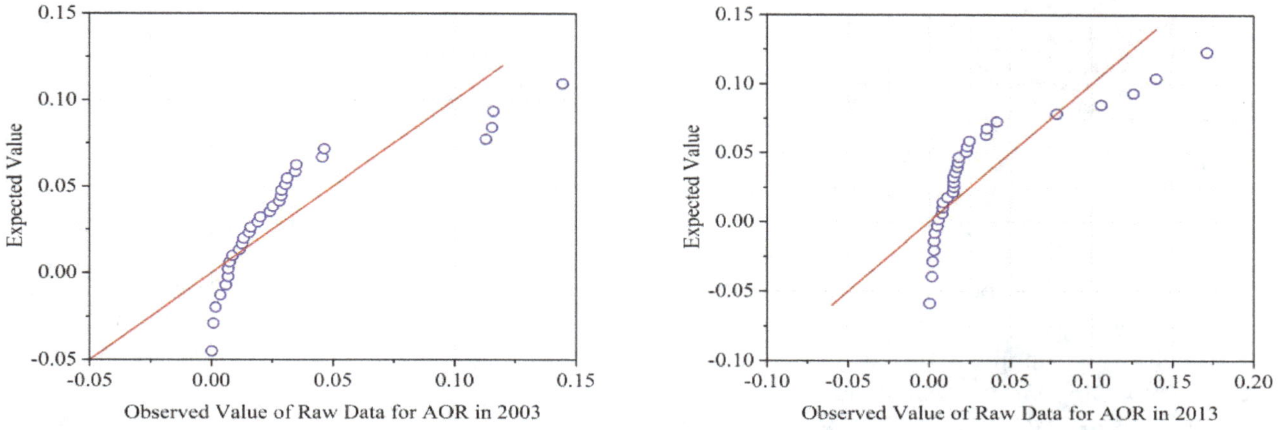

(a) (b)

Figure 1. Normal Q-Q plots: (**a**) raw data for AOR of natural wetlands in 2003; (**b**) raw data for AOR of natural wetlands in 2013.

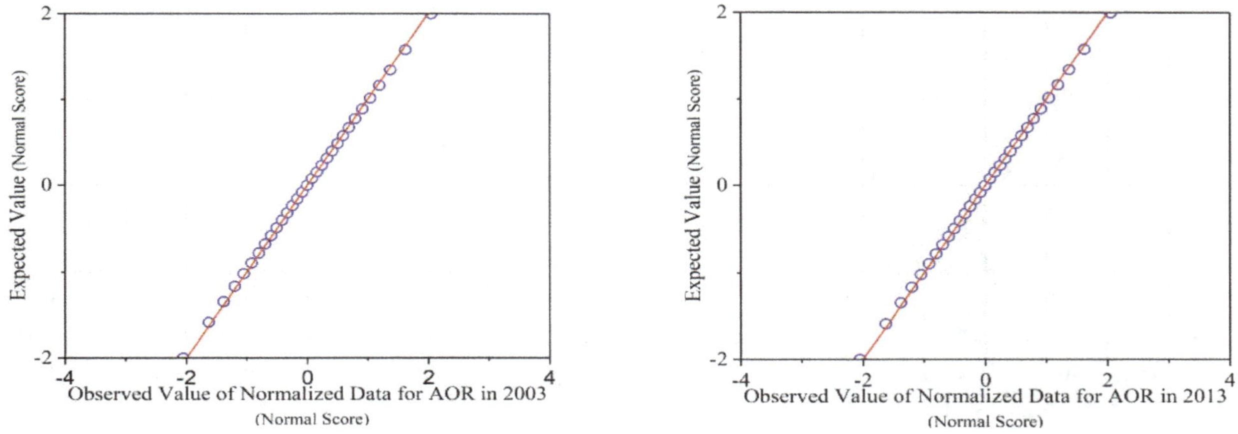

(a) (b)

Figure 2. Normal Q-Q plots: (**a**) AOR of natural wetlands in 2003 after normal score transformation; (**b**) AOR of natural wetlands in 2013 after normal score transformation.

3.3. Spatial Autocorrelation of Natural Wetlands Distribution

The choropleths spatial distribution of AOR of natural wetlands from 2003 to 2013 are illustrated in the quartile maps in Figure 3, with the darkest shade corresponding to the highest quartile. The suggestion of spatial clustering of similar values that follows from a visual inspection of these maps is confirmed by the significant and positive Moran's I values in Table 2, which imply the existence of positive spatial autocorrelation of natural wetlands distribution. The I-value in 2013 is bigger than that in 2003, which implies the spatial association of natural wetlands distribution tends to be more remarkable over the past decade.

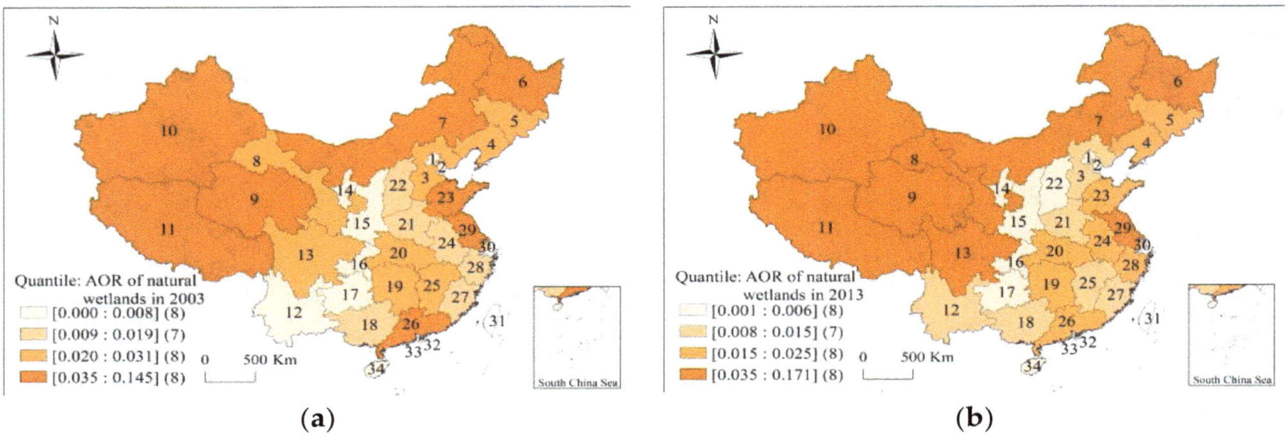

Figure 3. Quantile maps: (**a**) spatial distribution of AOR of natural wetlands in 2003; (**b**) spatial distribution of AOR of natural wetlands in 2013. Note: there are 34 provinces in China, including 1 = Beijing, 2 = Tianjin, 3 = Hebei, 4 = Liaoning, 5 = Jilin, 6 = Heilongjiang, 7 = Inner Mongolia, 8 = Gansu, 9 = Qinghai, 10 = Xinjiang, 11 = Tibet, 12 = Yunnan, 13 = Sichuan, 14 = Ningxia, 15 = Shaanxi, 16 = Chongqing, 17 = Guizhou, 18 = Guangxi, 19 = Hunan, 20 = Hubei, 21 = Henan, 22 = Shanxi, 23 = Shandong, 24 = Anhui, 25 = Jiangxi, 26 = Guangdong, 27 = Fujian, 28 = Zhejiang, 29 = Jiangsu, 30 = Shanghai, 31 = Taiwan, 32 = Hong Kong, 33 = Macau, 34 = Hainan.

Table 2. Moran's I test for global associations.

	I	z	p
AOR in 2003	0.248	2.383	0.015
AOR in 2013	0.356	3.300	0.001

3.4. Spatial Pattern of Natural Wetlands Distribution Based on LISA

In Figure 4, provinces with significant local Moran statistics for the AOR of natural wetlands distribution in 2003 and 2013 are mapped. LISA cluster map highlighted provinces are significantly clustered, which could strongly control the spatial association of natural wetlands distribution. First, LISA cluster maps in 2003 and 2013 suggest that the overall spatial pattern of natural wetlands distribution is stable. The LISA maps highlight one high-high cluster in the west of China and one isolated high value persisting in Jilin within the northeast of China. The region hotspots of natural wetlands distribution are mainly formed by Xinjiang, Qinghai, Gansu, and Tibet. Meanwhile, only one province (Tianjin) is significant for a low value. Although the overall aggregated pattern of natural wetlands distribution is stable, the scope of local hotspots has expanded significantly, with a trend of expanding southward to Tibet and northward to Gansu.

To sum up, the expansion of hotspots and the development of cool spots inform that the spatial association and spatial heterogeneity of natural wetlands distribution tend to be stronger. In addition, natural wetlands distribution in many provinces are still nonsignificant, which means that spatial aggregation of natural wetlands resources mainly occurred in the Xinjiang-Inner Mongolia wetland region, the Qinghai-Tibet Plateau wetland region and the Northeast China wetland region at the local scale, which has a strong influence on the spatial association of natural wetlands distribution in China. Therefore, the development of hotspots and cool spots indicates the existence of spatial dependence and spatial heterogeneity for the natural wetlands distribution, and the spatial spillover effects of these identified significant spatial clusters should be paid great attention when making the policies of wetlands conservation and management in China.

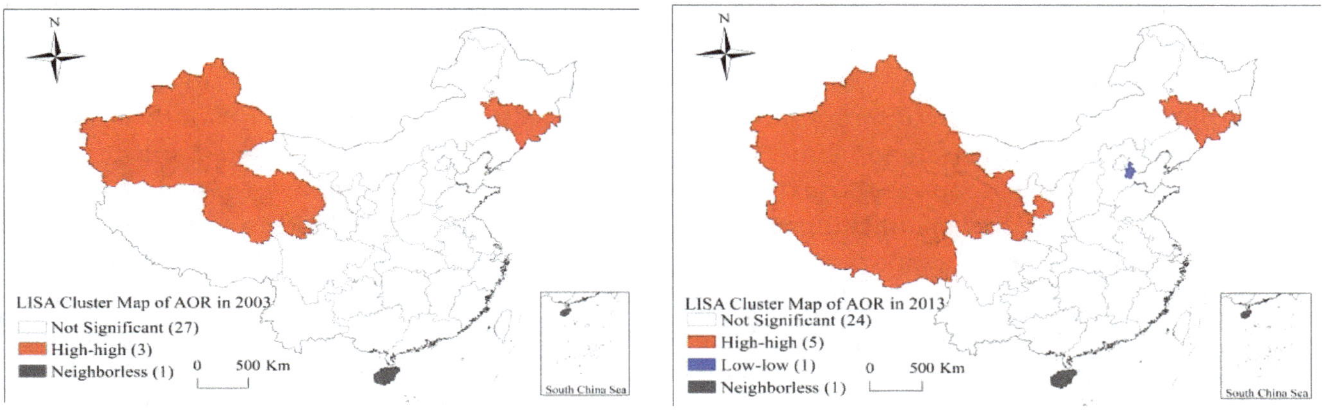

(**a**) (**b**)

Figure 4. Local identifications of spatial association (LISA) cluster map: (**a**) AOR of natural wetlands in 2003; (**b**) AOR of natural wetlands in 2013.

3.5. Spatial Pattern of the Associated Influencing Factors Based on LISA

Spatial distribution and dynamics of natural wetlands are a response to natural and anthropogenic factors. Guo et al. pointed out that natural factors provided the intrinsic drivers for the change of wetlands and human activities intensified the changes, resulting in great wetlands loss and degradation [39]. The formation, development, and evolution of natural wetlands depend on the natural geography and climate, which leads to unevenly distributed spatial patterns. On the other hand, the change of natural wetlands is partly caused by the influence of human activities. Thus, it is worth exploring what spatial pattern of the associated natural and anthropogenic factors will present when they respond to the spatial association of natural wetlands distribution. Here four factors are chosen as a representative of natural factors, including temperature, precipitation, humidity and water storage. The other four factors are chosen as representative of anthropogenic factors, including cultivated area, water consumption, per capita GDP and urbanization rate. Generally, increasing temperature and intensifying anthropogenic factors will result in the degradation of wetlands, even the loss of wetlands. However, the increase in precipitation and humidity will be beneficial to the maintenance of wetlands. Data of these indexes are collected from the China Statistical Yearbooks from 2004 to 2014 and the mean values are calculated.

In order to explore the spatial association of the influencing factors and to reveal the response spatial relationship between the change of natural wetlands distribution and the associated influencing factors, global and local Moran's I statistics were carried out. Results of the global Moran's I-values and their associated z-values and p-values are listed in Table 3. Seven of the eight factors show significant strong and positive spatial autocorrelation except for water consumption, and with temperature and precipitation and water storage, extremely significant positive autocorrelation. In addition, the spatial association of natural factors was stronger than anthropogenic factors. To further examine the spatial clustering of natural wetlands distribution at the local scale, the local Moran's I index of the eight influencing factors is calculated and the results are highlighted through LISA cluster maps in Figures 5 and 6.

Table 3. Moran's I test for global association.

Natural Factors	I	z	p	Anthropogenic Factors	I	z	p
Temperature	0.709	6.324	0.000	Cultivated area	0.232	2.278	0.018
Precipitation	0.730	6.540	0.000	Water consumption	0.105	1.186	0.123
Humidity	0.490	4.443	0.000	Per Capita GDP	0.464	4.251	0.000
Water storage	0.543	4.898	0.000	Urbanization rate	0.434	3.999	0.000

Note that LISA maps highlight the existence of a strong local spatial clustering for the associated influencing factors in Figures 5 and 6. For the associated natural factors, significant hotspots and cool spots are mainly distributed in the south and north of China, respectively. Only one outlier is identified in the local spatial association of temperature with a low value of temperature in southwest China. Others are highly spatially homogeneous, whereas spatial patterns of the associated anthropogenic factors are not so homogeneous as those of natural factors. As pointed out above, water consumption exhibits no significant global spatial autocorrelation but shows one spatial cluster and three outliers at the local scale. The regional hotspots are clustering in part of east China, including Shandong, Anhui, and Jiangxi, and the other three outliers with randomness distribution are regarded as an isolated individual hotspot. Spatial patterns of other factors are delineated as follows: the cultivated area shows one regional hotspot, including Henan and Shandong, one isolated hotspot in Jilin, and two outliers in Shanxi and Liaoning (Figure 6a). Both urbanization and per capita GDP show one spatial low-low cluster or so-called cool spot in the west of China, which indicates little human disturbance. Per capita GDP also shows two isolated individual hotspots, one in Hebei and the other one in Shanghai.

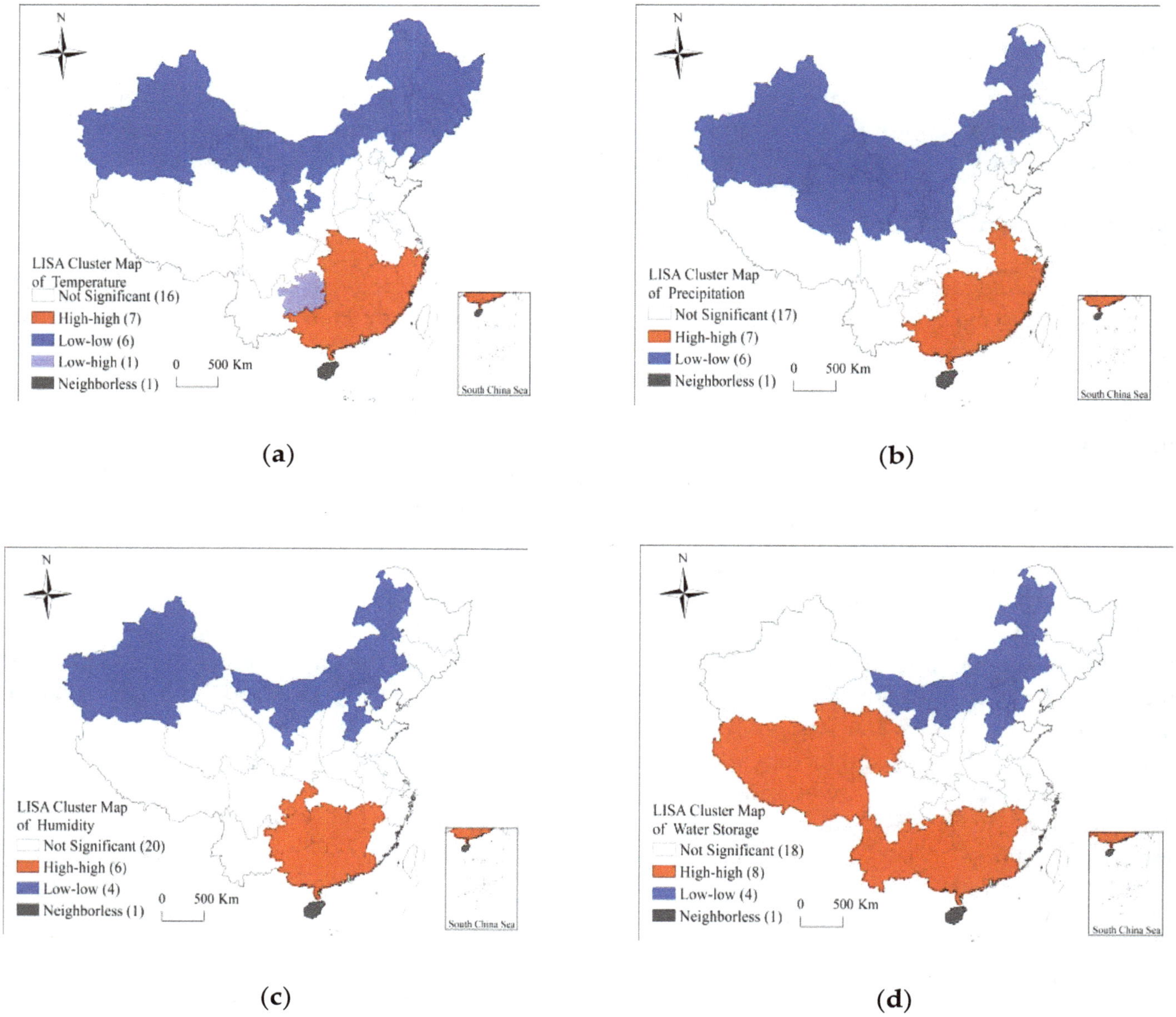

Figure 5. T LISA cluster maps for natural factors: (**a**) temperature; (**b**) precipitation; (**c**) humidity; and (**d**) water storage.

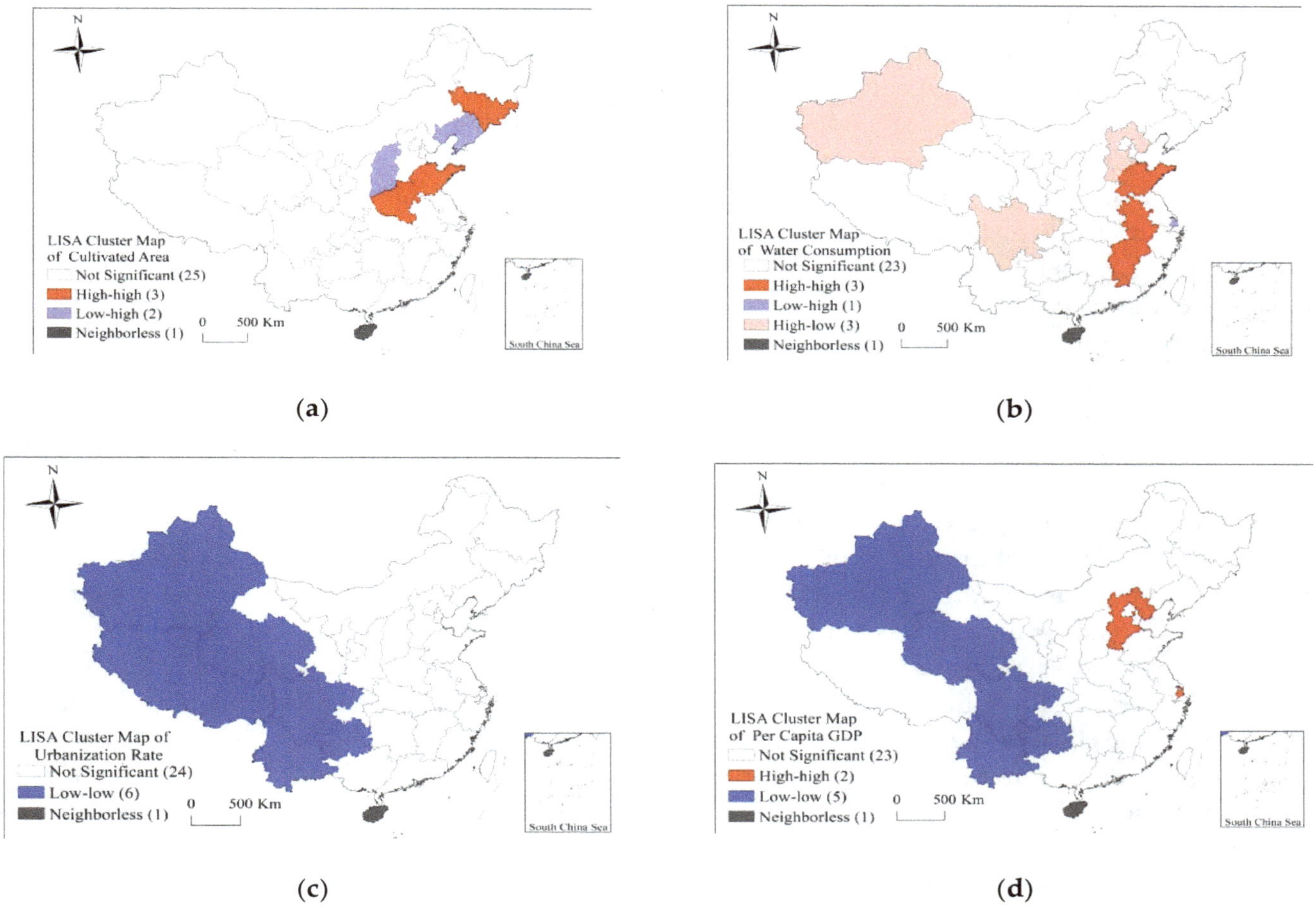

Figure 6. LISA cluster maps for anthropogenic factors: (**a**) cultivated area; (**b**) water consumption; (**c**) urbanization rate; and (**d**) per capita GDP.

3.6. Spatial Relationship Between Natural Wetlands Changes and Associated Influencing Factors

The significant spatial clusters and spatial outliers of eight influencing factors at the local scale indicate the existence of spatial dependence and spatial heterogeneity. Meanwhile, the combination of the LISA cluster maps of natural wetlands distribution (Figure 4) and the associated eight influencing factors (Figures 5 and 6) can inform one that the local spatial associations of the eight influencing factors differ from each other and have different spatial corresponding degrees in the change of natural wetlands distribution, which indicates different potential relationships between them.

Natural wetlands distribution shows two similar spatial clusters, one to the west of China and one in the northeast of China, both associated with a high area of wetland. These two clusters only partially match the spatial clusters with low temperature, high precipitation, high humidity, high water storage, low cultivated area, low water consumption, low urbanization rate, and low per capita GDP, which indicates the existence of a coordinated spatial relationship between them. Overall, spatial coordination between natural wetlands distribution and the associated influencing factors is higher in the western region than in east China and northeast China.

First, natural wetlands with high area value in Xinjiang are characterized by a statistically high value of water consumption. Wetlands in the Xinjiang-Inner Mongolia wetland region are dominated by saline lakes and swamps in a dry climate. In the context of dry climate wetlands sustainability develops a heavy dependence on water supplies. Therefore, great attention should be paid to the phenomenon why water consumption is so high in Xinjiang and how to manage water allocation in the case of a great demand for water. Natural wetlands in Qinghai and Tibet belong to the Qinghai-Tibet Plateau wetland region. This wetland region is known as the shelter of ecological security in Asia and plays an important role in China's wetland ecosystem, while it is a sensitive region of global climate change [40]. Although these two provinces are only characterized by a statistically high value of water storage and low value of urbanization rate, which show the existence of a coordinated spatial relationship between them, they are also associated with high values of water consumption in neighbors of Xinjiang and Sichuan. Hence based on Tobler's first law of geography, although no significantly negative influencing factors are revealed in Qinghai and Tibet, natural wetlands conservation here should protect themselves from the impacts from their neighbors with high values of water consumption. Additionally, drivers of changes in Qinghai-Tibet Plateau wetlands are dominated by natural factors, especially climate change, and anthropogenic factors are also regarded as the important influencing factors which also play a leading role in some areas [40,41]. Climate change will affect the hydrology of individual wetland ecosystems mostly through changes in precipitation and temperature regimes with great global variability [42]. Thus, the result obtained in this study supplements that impacts of human activities from the neighbors should be focused on together, which is suitable for natural wetlands conservation and management in Gansu. Furthermore, the success of wetlands conservation depends on the incorporation of climate change and wetland management [42,43].

Second, natural wetlands with high area value in Jilin are characterized by a statistically high value of cultivated area. A high area of cultivation will compete with wetlands for the use of water supplies and create pressures on wetlands, leading to a consequence of the weakened ecological function of wetlands, decreased biodiversity, increased rates and magnitudes of pollutant loading, etc. [42]. Jilin, one of the main agricultural bases, has converted part of its wetlands into arable land during 2008–2012, due to the task of increasing grain output by millions of tons [44]. Thus, under this circumstance, effective management of cultivation lands should be carried out to reduce the impacts from arable land on natural wetlands in Jilin. Similarly, Shandong and Henan, which are characterized by statistically high values of cultivated area, should also pay attention to this kind of problem.

Third, natural wetlands in Tianjin are characterized by a statistically low value of water storage. Most wetlands in Tianjin are located along the northeast coastal region. Tianjin is a typical industrial city with a large number of pollution emissions; pollution is one of the major threats to coastal wetlands [24]. Moreover, maps in Figure 6b,d show that natural wetlands in Hebei, the neighbor of Tianjin, are characterized by a statistically high value of water consumption and high value of per capita GDP, respectively. Thus, the environment in Tianjin and its surroundings are not conducive to the sustainability of wetlands. Note that Tianjin not only belongs to the circle of economy around the capital, which includes Beijing, Hebei, and Tianjin, but also plays a vital role in the circle of the ecology around the capital [45]. Hence, conservation of natural wetlands in Tianjin should receive great attention to deal with the uncoordinated spatial relationship.

In summary, uncoordinated spatial relationships have been revealed within provinces; furthermore, the impacts resulting from the neighbors have been highlighted. Thus, it implies that wetland sustainable conservation and management across provinces or regions [13,24,46] should be considered, especially the neighboring provinces.

4. Conclusions and Recommendations

4.1. Conclusions

Wetland loss and degradation pose great challenges for biodiversity protection in China. This study can provide some scientific references for the decision-making of effective and targeted wetlands conservation and sustainable management at national and provincial scales. Moran's I statistics are a useful tool for effectively identifying the spatial autocorrelation and spatial hotspots of natural wetlands distribution and associated influencing factors at the province level. Regional differences of the uncoordinated spatial relationship between natural wetlands changes and associated influencing factors are significant, which should be given considerable attention. GIS mapping is helpful to visually illustrate the spatial distribution of natural wetlands and associated influencing factors and to study the spatial relationship between them. To expand and advance GIS applications in wetlands research, more detailed data or case collection is important.

4.2. Recommendations

(1) Implement cross-province or cross-region wetland protection and management. As described above, the hotspot cluster is formed by two provinces in 2003 and four provinces in 2013. Although natural wetlands in different provinces that belong to different wetland regions have different dominant characteristics [23], the expansion of spatial hotspot implies the existence of a spatial diffusion effect. The aggregated provinces have strong radiation to the neighboring provinces of their wetland protection and management. Hence strengthening protection, management and cooperation cross-provinces, cross-region, or cross-sector and cross-nation are critical and necessary for the protection of wetland and biodiversity.

(2) Wetland activities should incorporate climate change, wetland restoration, management and conservation planning. Climate factors, the major part of natural factors, play a vital role in wetland sustainability. Attention to climate change and variability could make the difference between success and failure for wetlands conservation.

(3) Reduce human activity stressors on wetland ecosystems. Spatial analysis of the associated influencing factors revealed that spatial autocorrelations of anthropogenic factors are weaker than those of natural factors, which implies stronger spatial randomness of anthropogenic activities, making them hard to be controlled and managed. Negative impacts resulting from human activities should be reduced or minimized on wetland ecosystems.

Author Contributions: Conceptualization, Ting Zhou and Anyi Niu; methodology, Ting Zhou and Songjun Xu; software, Jiaojiao Ma and Anyi Niu; data curation, Zhanpeng Huang, Jiaojiao Ma; writing—original draft preparation, Ting Zhou.; writing—review and editing, Songjun Xu and Zhanpeng Huang; supervision, Songjun Xu; funding acquisition, Songjun Xu. All authors have read and agreed to the published version of the manuscript.

Acknowledgments: The authors would like to acknowledgment the guidance from Dr. Chaosheng Zhang, who is Director of International Network for Environment and Health, and head of Research Cluster of Environmental Change at Discipline of Geography, National University of Ireland, Galway, Ireland.

References

1. Mitsch, W.J.; Gosselink, J.G. The value of wetlands: Importance of scale and landscape setting. *Ecol. Econ.* **2000**, *35*, 25–33. [CrossRef]
2. Brander, L.M.; Florax, R.J.G.M.; Vermaat, J.E. The Empirics of Wetland Valuation: A Comprehensive Summary and a Meta-Analysis of the Literature. *Environ. Res. Econ.* **2006**, *33*, 223–250. [CrossRef]
3. Dudgeon, D.; Arthington, A.H.; Mark, O.G.; Kawabata, Z.I.; Knowler, D.J.; Lévêque, C.; Naiman, R.J.; Richard, A.H.P.; Soto, D.; Stiassny, M.L.J.; et al. Freshwater biodiversity: Importance, threats, status and conservation challenges. *Biol. Rev.* **2006**, *81*, 163–182. [CrossRef] [PubMed]
4. Foti, R.; Jesus, M.D.; Rinaldo, A.; Iturbe, I.R. Signs of critical transition in the Everglades wetlands in response to climate and anthropogenic changes. *Proc. Natl. Acad. Sci. USA* **2013**, *110*, 6296–6300. [CrossRef] [PubMed]
5. Carreño, M.F.; Eateve, M.A.; Martinez, J.; Palazón, J.A.; Pardo, M.T. Habitat changes in coastal wetlands associated to hydrological changes in the watershed. *Estuar. Coast. Shelf Sci.* **2008**, *77*, 475–483. [CrossRef]
6. Liu, H.Y.; Zhang, S.K.; Li, Z.F.; Lu, X.G.; Yang, Q. Impacts on wetlands on wetlands of large-scale land-use changes by agricultural development: The small Sanjiang Plain, China. *Ambio* **2004**, *33*, 306–310. [CrossRef]
7. Liu, Y.; Zha, Y.; Gao, J.; Ni, S. Assessment of grassland degradation near Lake Qinghai, West China, using Landsat TM and in situ reflectance spectra data. *Int. J. Remote Sens.* **2004**, *25*, 4177–4189. [CrossRef]
8. Han, D.Y.; Yang, Y.; Yang, Y.; Li, K. Recent advances in wetland degradation research. *Acta Ecol. Sin.* **2012**, *32*, 1293–1307.

9. Shen, G.; Yang, X.C.; Jin, Y.X.; Xu, B.; Zhou, Q.B. Remote sensing and evaluation of the wetland ecological degradation process of the Zoige Plateau wetland in China. *Ecol. Indic.* **2019**, *104*, 48–58. [CrossRef]
10. Hu, S.J.; Niu, Z.G.; Chen, Y.F.; Li, L.F.; Zhang, H.Y. Global wetlands: Potential distribution, wetland loss, and status. *Sci. Total Environ.* **2016**, *586*, 319–327. [CrossRef]
11. Bedford, B.L.; Preston, E.M. Developing the scientific basis for assessing cumulative effects of wetland loss and degradation on landscape functions: Status, perspectives, and prospects. *Environ. Manag.* **1988**, *12e*, 751–771. [CrossRef]
12. Holland, C.C.; Honea, J.; Gwin, S.E.; Kentula, M.E. Wetland degradation and loss in the rapidly urbanizing area of Portland, Oregon. *Wetlands* **1995**, *15*, 336–345. [CrossRef]
13. Davis, J.A.; Froend, R. Loss and degradation of wetlands in southwestern Australia: Underlying causes, consequences and solutions. *Wetl. Ecol. Manag.* **1999**, *7*, 13–23. [CrossRef]
14. Gibbs, J.P. Wetland loss and biodiversity conservation. *Conserv. Biol.* **2000**, *14*, 314–317. [CrossRef]
15. Zhou, D.M.; Gong, H.L.; Wang, Y.Y.; Khan, S.; Zhao, K.Y. Driving forces for the Marsh Wetland Degradation in the Honghe National Nature Reserve in Sanjiang Plain, Northeast China. *Environ. Model. Assess.* **2009**, *14*, 101–111. [CrossRef]
16. Niu, Z.G.; Zhang, H.Y.; Wang, X.W.; Wen, B.; Yao, D.M.; Zhou, K.Y.; Zhao, H.; Li, N.N.; Huang, H.B.; Li, C.C.; et al. Mapping wetland changes in China between 1978 and 2008. *Chin. Sci. Bull.* **2012**, *57*, 2813–2823. [CrossRef]
17. Chen, Y.Y.; Lü, X.G. The wetland function and research tendency of wetland science. *Wetl. Sci.* **2003**, *1*, 7–11.
18. Wetland China. The Second National Wetland Recourses Survey. Available online: http://www.shidi.org/zt/2014xwfbh/ (accessed on 21 January 2014).
19. Sica, Y.V.; Quintana, R.D.; Radeloff, V.C.; Gavier, G.I. Wetland loss due to land use change in the Lower Paraná River Delta, Argentina. *Sci. Total Environ.* **2016**, *568*, 967–978. [CrossRef]
20. Gong, P.; Niu, Z.G.; Cheng, X.; Zhao, K.Y.; Zhou, D.M.; Guo, J.H.; Lu, L.; Wang, X.F.; Li, D.D.; Huang, H.B.; et al. China's wetland change (1990–2000) determined by remote sensing. *Sci. China Earth Sci.* **2010**, *53*, 1036. [CrossRef]
21. Meng, W.Q.; He, M.X.; Hu, B.B.; Mo, X.Q.; Liu, B.Q.; Wang, Z.L. Status of wetlands in China: A review of extent, degradation, issues and recommendations for improvement. *Ocean Coast. Manag.* **2017**, *146*, 50–59. [CrossRef]
22. Sun, R.; Yao, P.P.; Wang, W.; Yue, B.; Liu, G. Assessment of wetland ecosystem health in the Yangtze and Amazon River Basins. *ISPRS Int. J. Geo-Inf.* **2017**, *6*, 81. [CrossRef]
23. An, S.Q.; Li, H.; Guan, B.H.; Zhou, C.F.; Wang, Z.S.; Deng, Z.F.; Zhi, Y.B.; Liu, Y.H.; Xu, C.; Fang, S.B.; et al. China's natural wetlands: Past problems, current status, and future challenges. *Ambio* **2007**, *36*, 335–342. [CrossRef]
24. Sun, Z.G.; Sun, W.G.; Tong, C.; Zeng, C.S.; Yu, X.; Mou, X.J. China's coastal wetlands: Conservation history, implementation efforts, existing issues and strategies for future improvement. *Environ. Int.* **2015**, *79*, 25–41. [CrossRef] [PubMed]
25. Yang, X.X.; Wang, X.L.; Qin, F.Y. Analysis of dynamic changes and influencing factors of wetland in Dalinur Nature Reserve in recent 30 years. *J. Northwest For. Univ.* **2019**, *34*, 171–178, 222.
26. Liu, Z.W.; Li, S.N.; Wei, W.; Song, X.J. Research progress on alpine wetland changes and driving forces in Qinghai-Tibet Plateau during the last three decades. *Chin. J. Ecol.* **2019**, *38*, 856–862.
27. Niu, Z.G.; Gong, P.; Cheng, X.; Guo, J.H.; Wang, L.; Huang, S.B.; Shen, S.Q.; Wu, J.Z.; Wang, X.W.; Ying, Q.; et al. Geographical characteristics of China's wetlands derived from remotely sensed data. *Sci. China Earth Sci.* **2009**, *52*, 723–738. [CrossRef]
28. Zhang, S.Q.; Na, X.D.; Kong, B.; Wang, Z.M.; Jiang, H.X.; Yu, H.; Zhao, Z.C.; Li, X.F.; Liu, C.Y.; Dale, P. Identifying wetland change in China's Sanjiang Plain using remote sensing. *Wetlands* **2009**, *29*, 302–313. [CrossRef]
29. Liu, G.L.; Zhang, L.C.; Zhang, Q.; Musyimi, Z.; Jiang, Q.H. Spatio-temporal dynamics of wetland landscape patterns based on remote sensing in Yellow River Delta, China. *Wetlands* **2014**, *34*, 787–801. [CrossRef]
30. Xue, Z.S.; Lü, X.G.; Chen, Z.K.; Zhang, Z.S.; Jiang, M.; Zhang, K.; Lü, Y.L. Spatial and temporal changes of wetlands on the Qinghai-Tibetan Plateau from the 1970s to 2010s. *Chin. Geogr. Sci.* **2018**, *28*, 935–945. [CrossRef]

31. Zhao, Y. Analysis of dynamic change and driving force of wetland landscape pattern in Dalian. *Environ. Sci. Surv.* **2019**, *38*, 49–52, 69.
32. Tobler, W. On the first law of geography: A replay. *Ann. Assoc. Am. Geogr* **2004**, *94*, 304–310. [CrossRef]
33. Talen, E.; Anselin, L. Assessing spatial equity: An evaluation of measures of accessibility to public playgrounds. *Environ. Plan. A* **1998**, *30*, 595–613. [CrossRef]
34. Jia, P.; Joyner, A. Human brucellosis occurrences in Inner Mongolia, China: A spatio-temporal distribution and ecological niche modeling approach. *BMC Infect. Dis.* **2015**, *15*, 1–16. [CrossRef] [PubMed]
35. Zhang, C.S.; Luo, L.; Xu, W.L.; Ledwith, V. Use of local Moran's I and GIS to identify pollution hotspots of Pb in urban soils of Galway, Ireland. *Sci. Total Environ.* **2008**, *398*, 212–221. [CrossRef]
36. Lalor, G.; Zhang, C.S. Multivariate outlier detection and remediation in geochemical databases. *Sci. Total Environ.* **2001**, *281*, 99–109. [CrossRef]
37. Anselin, L. Local indicators of spatial association—LISA. *Geogr. Anal.* **1995**, *27*, 93–114. [CrossRef]
38. Silva, E.F.; Zhang, C.S.; Pinto, L.S.; Reis, P. Hazard assessment on arsenic and lead in soils of Castromil gold mining area, Portugal. *Appl. Geochem.* **2004**, *19*, 887–898. [CrossRef]
39. Guo, Y.D.; He, Y.F. The dynamics of wetland landscape and its driving forces in Songnen Plain. *Wetl. Sci.* **2005**, *3*, 54–59.
40. Zhao, Z.L.; Zhang, Y.L.; Liu, L.S.; Liu, F.G.; Zhang, H.F. Advances in research on wetlands of the Tibetan Plateau. *Prog. Geogr.* **2014**, *33*, 1218–1230.
41. Lang, Q.; Niu, Z.G.; Hong, F.Q.; Yang, X.Y. Remote Sensing Monitoring and Change Analysis of the Tibet Plateau Wetlands. Geomatics and Information Science of Wuhan University. Available online: http://kns.cnki.net.scnu.vpn358.com/KCMS/detail/42.1676.TN.20190912.1340.001.html (accessed on 12 September 2019).
42. Erwin, K.L. Wetlands and global climate change: The role of wetland restoration in a changing world. *Wetlands Ecol. Manag.* **2009**, *17*, 71–84. [CrossRef]
43. Finlayson, C.M.; Davis, J.A.; Gell, P.A.; Kingsford, R.T.; Parton, K.A. The status of wetlands and the predicted effects of global climate change: The situation in Australia. *Aquat. Sci.* **2013**, *75*, 73–93. [CrossRef]
44. Lu, C.Y.; Wang, Z.M.; Liu, M.Y.; Ou, Y.L.; Jia, M.M.; Mao, D.H. Analysis of conservation effectiveness of wetland protected areas based on remote sensing in West Songnen Plain. *Chin. Environ. Sci.* **2015**, *35*, 599–609.
45. Min, Q.W.; Liu, W.W.; Xie, G.D.; Sun, X.P.; Li, N. Capital eco-sphere and its natural ecological conditions. *Resour. Sci.* **2015**, *37*, 1504–1512.
46. Bawa, K.S.; Koh, L.P.; Lee, T.M.; Liu, J.G.; Ramakrishnan, P.S.; Yu, D.W.; Zhang, Y.P.; Raven, P.H. China, India, and the environment. *Science* **2010**, *327*, 1457–1459. [CrossRef] [PubMed]

6

Evaluation of the Accuracy of the Field Quadrat Survey of Alpine Grassland Fractional Vegetation Cover Based on the Satellite Remote Sensing Pixel Scale

Jianjun Chen [1,2], Xuning Zhao [1], Huizi Zhang [1], Yu Qin [3] and Shuhua Yi [4,5,*]

[1] College of Geomatics and Geoinformation, Guilin University of Technology, 12 Jiangan Road, Guilin 541004, China; chenjj@lzb.ac.cn (J.C.); zxn@glut.edu.cn (X.Z.); gjian@glut.edu.cn (H.Z.)
[2] Guangxi Key Laboratory of Spatial Information and Geomatics, 12 Jiangan Road, Guilin 541004, China
[3] State Key Laboratory of Cryospheric Sciences, Northwest Institute of Eco-Environment and Resources, Chinese Academy of Sciences, 320 Donggang West Road, Lanzhou 730000, China; qiny@lzb.ac.cn
[4] Institute of Fragile Ecosystem and Environment, Nantong University, 999 Tongjing Road, Nantong 226007, China
[5] School of Geographic Sciences, Nantong University, 999 Tongjing Road, Nantong 226007, China
* Correspondence: yis@lzb.ac.cn

Abstract: The fractional vegetation cover (FVC) data measured on the ground is the main source for the calibration and verification of FVC remote sensing inversion, and its accuracy directly affects the accuracy of remote sensing inversion results. However, the existing research on the evaluation of the accuracy of the field quadrat survey of FVC based on the satellite remote sensing pixel scale is inadequate, especially in the alpine grassland of the Qinghai-Tibet Plateau. In this paper, five different alpine grasslands were examined, the accuracy of the FVC obtained by the photography method was analyzed, and the influence of the number of samples on the field survey results was studied. First, the results show that the threshold method could accurately extract the vegetation information in the photos and obtain the FVC with high accuracy and little subjective interference. Second, the number of samples measured on the ground was logarithmically related to the accuracy of the FVC of the sample plot ($p < 0.001$). When the number of samples was larger, the accuracy of the FVC of the sample plot was higher and closer to the real value, and the stability of data also increased with the increase of the number of samples. Third, the average FVC of the measured quadrats on the ground was able to represent the FVC of the sample plot, but on the basis that there were enough measured quadrats. Finally, the results revealed that the degree of fragmentation reflecting the state of ground vegetation affects the acquisition accuracy of FVC. When the degree of fragmentation of the sample plot is higher, the number of samples needed to achieve the accuracy index is higher. Our results suggest that when obtaining the FVC on the satellite remote sensing pixel scale, the number of samples measured on the ground is an important factor affecting the accuracy, which cannot be ignored.

Keywords: alpine grassland; fractional vegetation cover; ground survey; precision evaluation

1. Introduction

Fractional vegetation cover (FVC) refers to the percentage of the vertical projection of green vegetation in a total analyzed area [1]. It is an important parameter used to describe vegetation status, and reflects changes in the ecosystem. It is also the main factor affecting the interaction between surface, atmospheric and hydrological processes [2].

FVC is often used to monitor the process of ecological change and evaluate the ecological environment, which is an important parameter in the field of ecological environmental monitoring [3]. Therefore, accurately estimating the FVC of a region not only reveals the current ecological environment status and its changing trends, but also provides accurate data for numerous areas of simulation studies, such as ecology, hydrology and meteorology.

The acquisition methods of FVC mainly include conventional ground measurement and inversion estimation based on remote sensing data [4]. Conventional ground measurement methods include the visual estimation method, the sampling method, and the photographic method [5]. Among them, the visual estimation method directly estimates the FVC based upon experience, but the results are prone to subjectivity. The FVC calculated by the sampling method on the ground is relatively high in accuracy, but is time-consuming and laborious with low efficiency. The photographic method uses a digital camera to shoot the ground from above and then uses image processing software to interpret the image to obtain the FVC. Due to its high precision, efficiency and economical simplicity, it is one of the main methods of ground measurement. Although the ground measurement method is the most direct means to obtain FVC, these methods can only ensure the accuracy of FVC monitoring in small areas, and it is impossible to carry out long-term positioning monitoring on a wide range of vegetation conditions. With the continuous development of remote sensing technology, compared with the traditional ground measurement methods, the wide field of view, real-time and low cost of remote sensing technology are widely used in FVC estimation. For example, Yan et al. [6] use the random forest regression model utilizing FVC monitoring data and the vegetation index to better predict grassland FVC. Based on sample data, Li et al. [1] use the pixel dichotomy model to evaluate the inversion accuracy of grassland FVC at different resolutions. Jia et al. [7] use the MODIS surface reflectance data to generate a global FVC data set based on the general regression neural network inversion model. At present, research has continuously improved the inversion accuracy of FVC that is based on remote sensing inversion algorithms. However, it ignores that all satellite quantitative remote sensing inversion methods require accurate ground measurement data, in order to be calibrated and verified [8,9].

Quadrat photography is the main technique used to obtain FVC in the field. Although this method can accurately estimate the FVC at the scale of the sample quadrat, the scale of the sample quadrat is difficult to match with the scale of the remote sensing pixel. Accurate field measurements data, which can be matched with the scale of the satellite remote sensing pixels, can provide accurate calibration and verification of data for satellite remote sensing inversion, and further enhance its reliability. At present, some studies are aiming to set a certain number of quadrats in the sample plot which can match with the scale of satellite remote sensing pixels, and the mean value of FVC obtained by the quadrats would represent the "true" FVC of the sample plot. However, the heterogeneity of different areas and different underlying surfaces vary substantially.

The average value of the fixed number of quadrats represents the measured value of the observed sample plot, and its accuracy changes with the variance of the area and the underlying surface. Moreover, few studies have been conducted to investigate the influence of issue on remote sensing inversion results. Based on the above problems, this study observed five typical alpine grasslands on the Qinghai-Tibet Plateau (where the heterogeneity of the underlying surfaces of the alpine grasslands varies greatly) and obtained the field-measured FVC by the quadrat photographic method (the quadrats were evenly distributed in the sample plot, which matched the scale of the satellite remote sensing pixels). The main objectives were to: 1) Analyze the accuracy of FVC obtained by the photographic method, 2) evaluate the influence of the number of quadrats on the field survey of the satellite remote sensing pixel scale, and 3) explore the relationship between the heterogeneity of underlying surfaces and the accuracy of field investigation.

2. Materials and methods

2.1. *Study Area*

The area studied is located near the upper reaches of Shule River (Figure 1), one of the three inland rivers of the Hexi Corridor in China. It is located at 96.6° to 99.0°E and 38.2° to 40.0°N, with an elevation of 2078–5763 meters and a drainage area of 11,400 square kilometers [10]. This region has a continental, arid desert climate, which is dry, cold and windy. The annual average temperature and precipitation are -2.7 °C and 349.2 mm, respectively [11]. Permafrost is widely distributed in the region; the vegetation types are mainly alpine meadow and alpine steppe, and the overall coverage is low [12]. The main soil types found in this region are alpine cold desert soil, alpine meadow steppe soil, chestnut soil, light chestnut soil and mountain limestone soil.

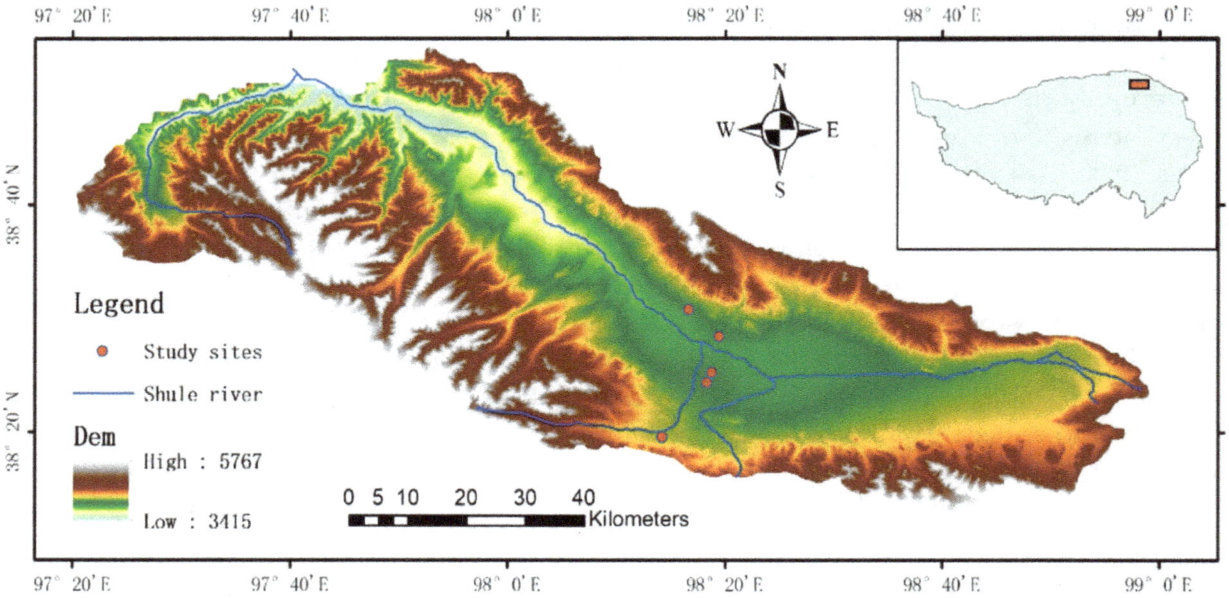

Figure 1. Spatial distribution of study sites in the upper reaches of Shule River.

2.2. *Field Data Collection*

In the summer of 2016 we laid a sample plot of 10 m × 10 m (consistent with the highest spatial resolution of Sentinel 2 images) for each of the five grassland types (swamp meadow, alpine meadow, degraded meadow, steppe meadow and alpine steppe) in the study area, and the boundary of each sample plot was fixed with white cloth strips (Figure 2a). The five selected grassland types covered all grassland types in the study area, and are also the main grassland types in the Qinghai-Tibet Plateau. In each sample plot, a quadrocopter (DJI Phantom 3 Professional, DJI Inc. Shenzhen, China) with an integrated camera (12 megapixels) was used to shoot from above the sample plot at a height of 20 m, and the aerial image of each sample plot was acquired for analyzing the underlying surface heterogeneity of the sample plot. In order to analyze the impact of the number of quadrats on the survey results, we divided each sample plot evenly into 400 0.5 m × 0.5 m quadrats (Figure 3). To facilitate data collection, we customized a 0.5 m × 0.5 m iron frame to determine the extent of each quadrat. In each quadrat, we used the digital camera to photograph an image from above (~1 m), in order to obtain the image of the quadrat (Figure 2b), which was used for fractional vegetation cover (FVC) information extraction. The camera mounted on the quadrocopter and used for ground photography is a three-band digital camera which receives irradiance in the visible region (red, green and blue spectral bands) and stores it as a gray value from 0 to 255 in JPEG format.

Figure 2. Field sampling methods and effect diagrams (**a**) unmanned aerial vehicle (UAV) sampling method; (**b**) ground sampling method.

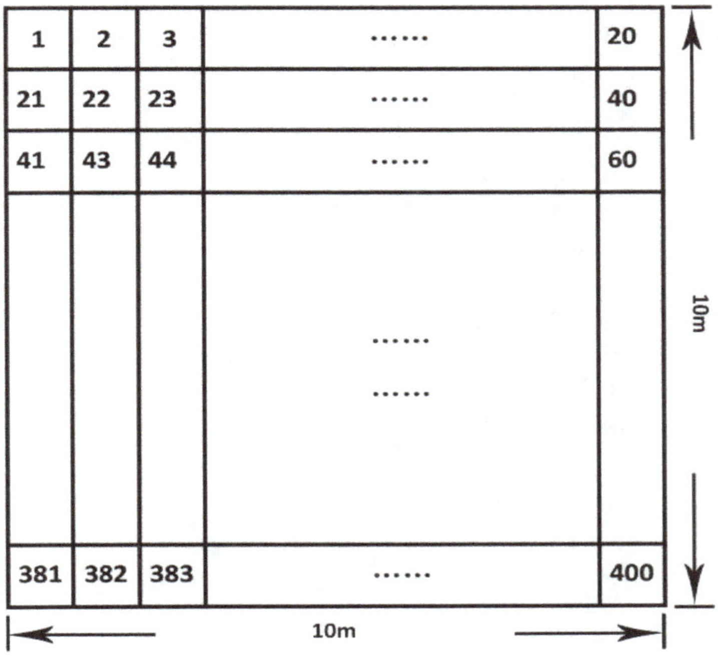

Figure 3. Diagram of each quadrat in sample plot. The sample plot is 10 m ×10 m and includes 400 sample quadrats, each measuring 0.5 m × 0.5 m, numbered from 1 to 400 and evenly distributed throughout the sample plot.

2.3. Data Processing Method

2.3.1. FVC Information Extraction

Before extracting the FVC information, the captured images were edited in Adobe Photoshop CS6 software according to the established boundary (the white cloth strips and iron frames) set by the experiment, since the coverage range of the images taken by the quadrocopter and the ground was larger than the sample plot scale (10 m×10 m) and the quadrat scale (0.5 m×0.5 m), and then the images matching the sample plot size and quadrat size were obtained. The FVC information was extracted from the clipped images by the threshold method, and the index used was the Excess Green Index (EGI = 2G-R-B, where G, R and B, respectively, represent the gray values of the green, red and blue bands in the image). Previous studies suggest that the threshold method based on the EGI had good accuracy [11–13]. More specifically, the FVC was calculated as follows. First, the EGI value of each pixel in the image was calculated. Second, an initial EGI threshold was set and compared to each pixel in the image.

If the pixel EGI value was greater than the threshold, then the pixel was regarded as vegetation, otherwise it was regarded as non-vegetation. Third, the classification result was compared with the original image. These steps were repeated to adjust the threshold value until the vegetation shapes in the classified image fit those of the original picture (Figure 4). Finally, the percentage of vegetation pixels of the total pixels was calculated as the FVC of the image. In order to evaluate the accuracy of the FVC obtained by this method, each image was analyzed by two separate people. Finally, the results of these two people were compared and analyzed.

Figure 4. Effect diagram (**a**) original image; (**b**) processing result of image (Red curves and yellow curves indicate vegetation contour, non-vegetation contour respectively).

2.3.2. FVC Estimation of Different Quadrat Combinations

After extracting the FVC information from the quadrat image by the threshold method, in order to accurately evaluate the influence of the number of quadrats on the estimation of the FVC of the sample plots, 400 quadrats of each sample plot were numbered, and the mean value of the FVC obtained from the 400 quadrats was considered to be the true value of the FVC of the corresponding sample plot. Then, according to the statistical permutation and combination theory, all permutations and combinations of different numbers of quadrats were extracted (for example, when the number of quadrats is m, there are C_{400}^{m} (from m choose 400 = $\frac{n!}{k!(n-k)!}$) combinations of quadrats), and the mean FVC obtained by each combination of quadrats was taken as the measured value of the sample plot. Finally, the influence of the number of quadrats on the field survey results was evaluated by the measured and true values of the sample plot.

2.3.3. Calculation of the Degree of Fragmentation of Sample Plot

Landscape fragmentation and FVC are correlated to a certain extent [14]. In order to analyze the correlation between landscape fragmentation and FVC, the images of the sample plot that had extracted the vegetation information by the threshold method were imported into Fragstats 4.2 software after being processed by ArcGIS 10.1 software, and the fragmentation index and the degree of fragmentation in the measurement area were calculated by selecting the patch density (PD) in the landscape pattern index.

$$PD = N/A \tag{1}$$

PD was calculated as above, where N is the number of patches and A is the total area of the region. PD is a specific quantification of the fragmentation of the area. The larger the value, the higher the degree of fragmentation [15].

2.4. Accuracy Evaluation Index

Root mean square error (RMSE), mean relative error (MRE) and variance have good statistical significance and are often used for error analyses. Therefore, this study used the above indicators to evaluate the impact of the number of quadrats on the field survey results. RMSE, MRE and variance calculation formulas are as follows:

$$RMSE = \sqrt{\sum_{i=1}^{n} (y_i - f_i)^2 / n}; \tag{2}$$

$$MRE = \sum_{i=1}^{n} \left|(f_i - y_i)\right| / n; \tag{3}$$

$$Variance = \frac{\sum_{i=1}^{n} (y_i - \overline{y})^2}{n-1} \tag{4}$$

where f_i is the predicted value, y_i is the true value, and n is the number of quadrats. Smaller RMSE and MAE values indicate smaller error and higher accuracy, which are closer to the real values. A smaller variance indicates more stable and less volatile sample data.

3. Results and Analysis

3.1. FVC Information Extraction Based on the Photographic Method

FVC information extracted from quadrat images based on the threshold method (Figure 5) show that the threshold method can distinguish between vegetation and non-vegetation well. By comparison, it is also seen that the extracted results by different people are relatively consistent in spatial distribution (Figure 5b,c), and the spatial distribution of extracted vegetation information is consistent with that of the original image (Figure 5a), indicating that the FVC information extracted by the threshold method is less affected by subjective factors. In order to further study the influence of the threshold method on the FVC estimation of the quadrat image, the FVC data extracted by the two individuals was analyzed (Figure 6). The results show that the two groups of data have very high correlation, and the correlation coefficient reached 0.936, further indicating that the threshold method for FVC information is barely affected by subjective factors and is robust.

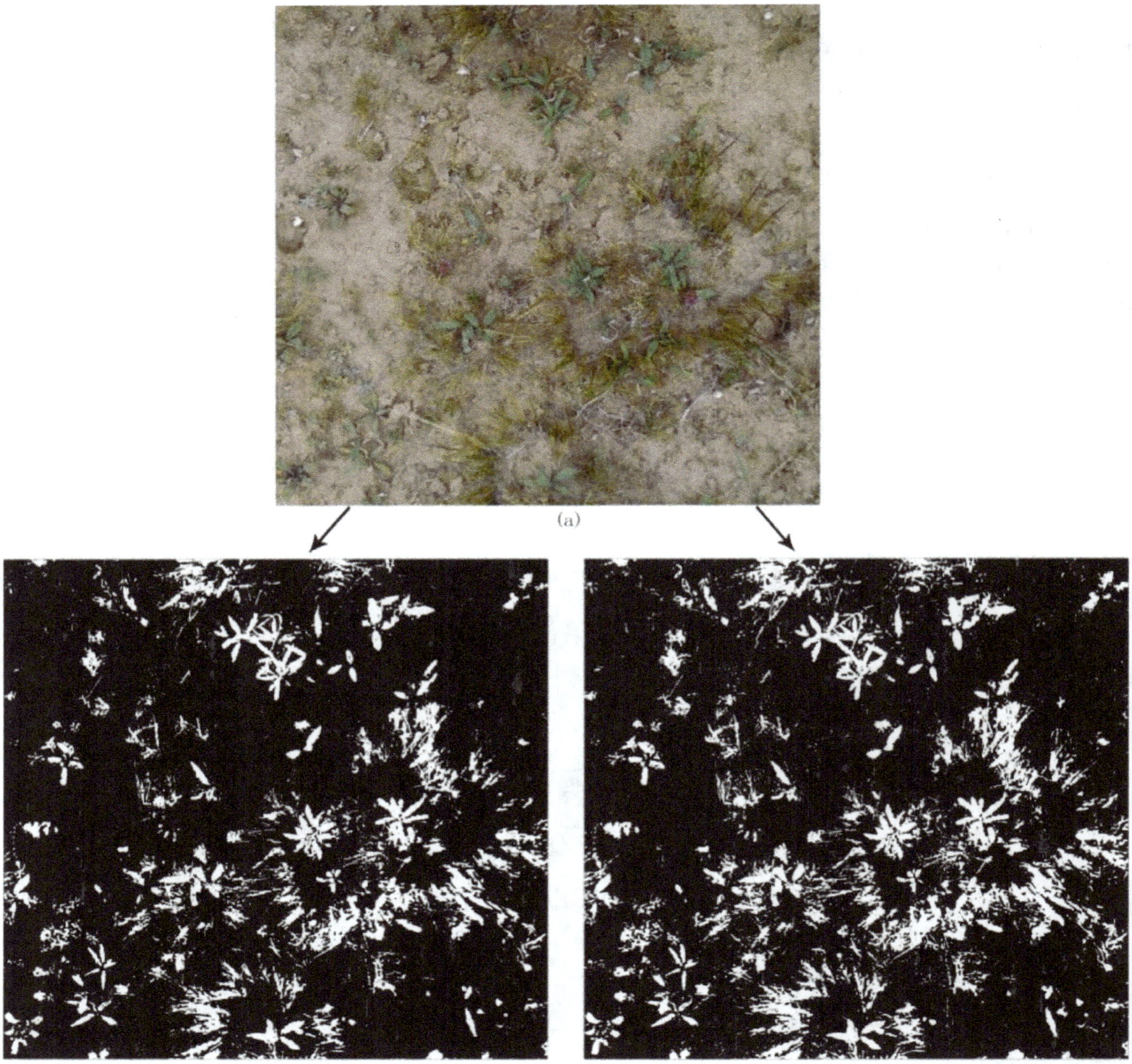

(a)

(b) (c)

Figure 5. Comparison of results of vegetation areas extracted by a different person ((**a**) is the original image, and (**b**,**c**) are the fractional vegetation cover (FVC) information extraction results of this different person, respectively, while white and black represent vegetation and non-vegetation, respectively).

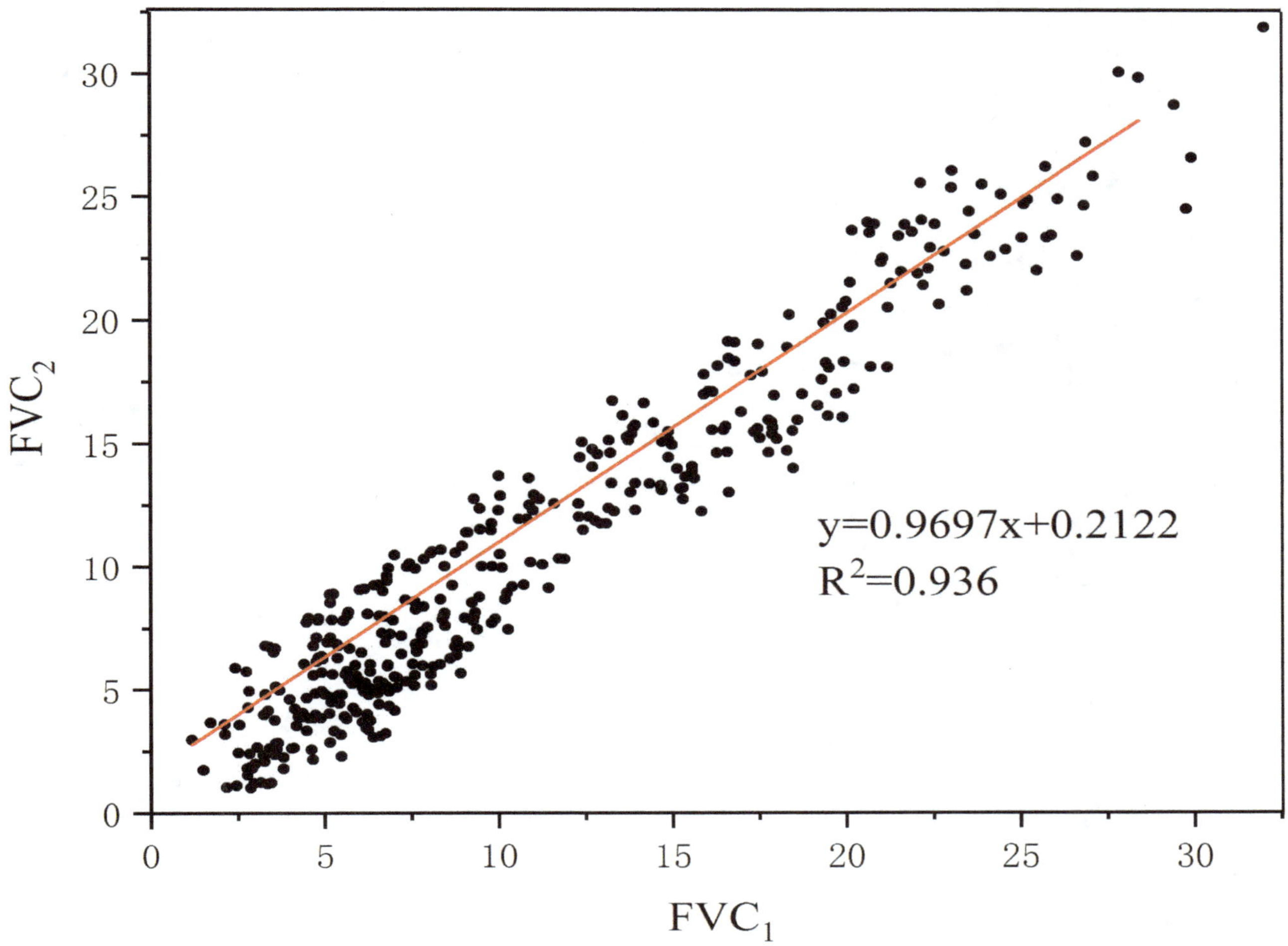

Figure 6. Comparison of the FVCs of the quadrat extracted by the different person (FVC_1 and FVC_2 represent the FVC data obtained by different people, respectively).

3.2. The Relationship between the Number of Quadrats and the Error of Survey Results

For the five grassland types in the study area, there was a logarithmic relationship between the number of quadrats and the survey error. With the increase of the number of quadrats, the error of the survey results was smaller, and the obtained FVC was closer to the true value of the remote sensing pixel scale. Additionally, different grassland types had specific differences in the relationship between the number of quadrats and the error of survey results of FVC. When the number of quadrats was the same, the largest error of the survey results was observed in the alpine meadow, followed by the swamp meadow, steppe meadow, alpine steppe and degraded meadow. When the error of the survey results was held constant, the degraded meadow required the least amount of quadrats, while the alpine meadow required the most. From the relationship between the error of the survey results and the number of quadrats (Figure 7), it can be seen that with the increase of the number of quadrats, the most obvious error trend is the alpine meadow, and the error trend of the alpine steppe is the smallest.

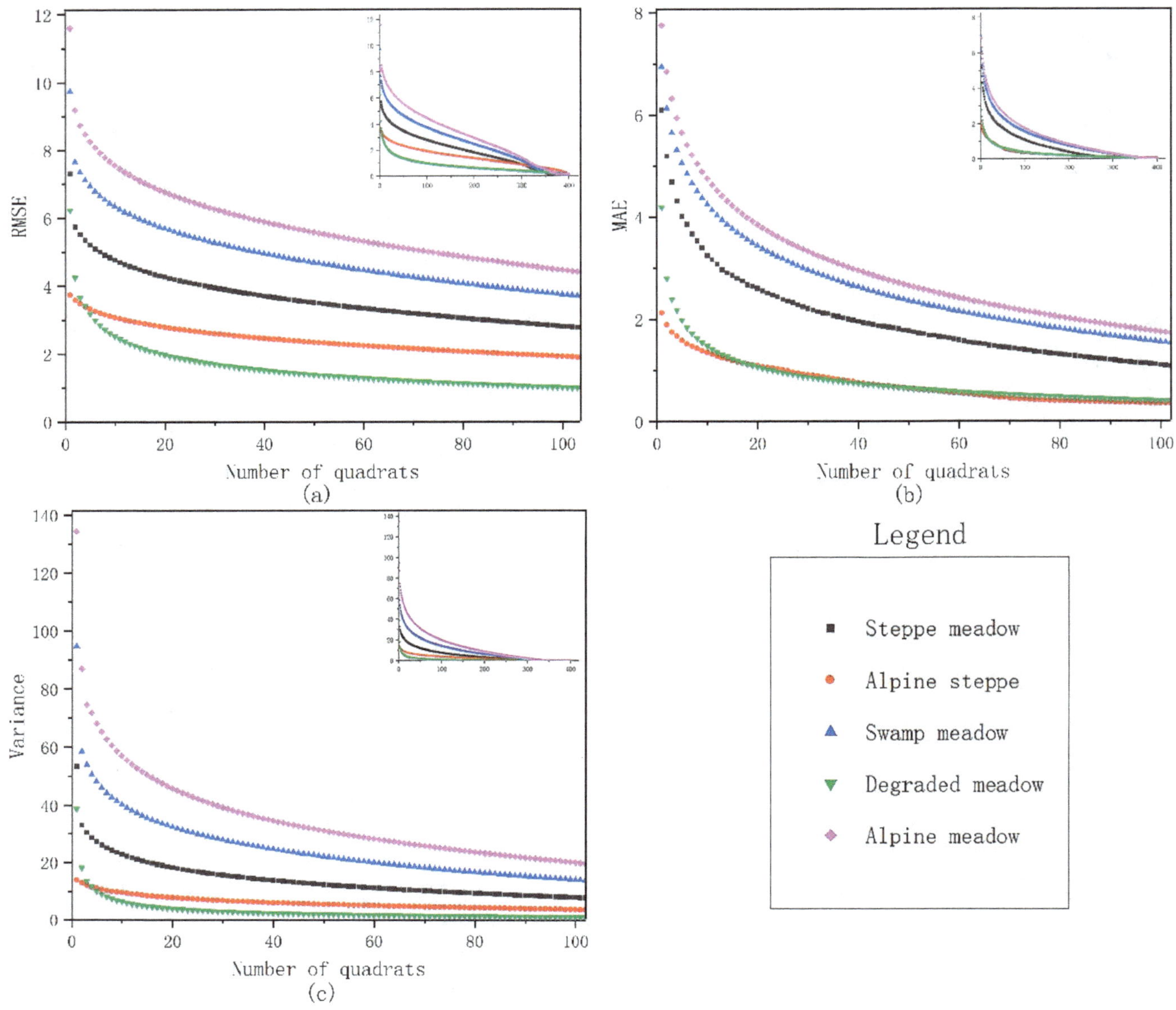

Figure 7. The relationship between the number of different quadrats and the error of survey results.

3.3. The Relationship between the Degree of Fragmentation of the Underlying Surface and the Required Number of Quadrats

In this study, the relationship between the degree of fragmentation of the landscape pattern and the survey accuracy of FVC was analyzed by calculating the patch density of five grassland types in the study area. The patch density is the ratio of the number of patches to the area, and its size reflects the degree of fragmentation and spatial heterogeneity of the surface. In order to avoid the influence of the underlying surface heterogeneity, 1 was used as the error measurement value of the three precision indicators of RMSE, MAE and variance, and the number of quadrats required for the five plots in the study area was investigated. The study found that the patch density of the alpine steppe was the smallest, indicating that the alpine steppe was dominated by large patches and the degree of fragmentation was low (Table 1); the patch density of the alpine meadow was the largest, indicating that the patches of the alpine meadow were mainly composed of small patches and the degree of fragmentation was the highest. With the increase of the degree of fragmentation of the plot, the number of ground survey quadrats required to obtain high-precision FVC also increased (Figure 8). The results show that the degree of fragmentation of the underlying surface has a high correlation with the number of quadrats required for high-precision FVC.

Table 1. The patch density of five alpine grassland types.

Grassland Type	Steppe Meadow	Alpine Steppe	Swamp Meadow	Degraded Meadow	Alpine Meadow
PD	2588.882	1906.506	3061.882	1343.480	3362.939

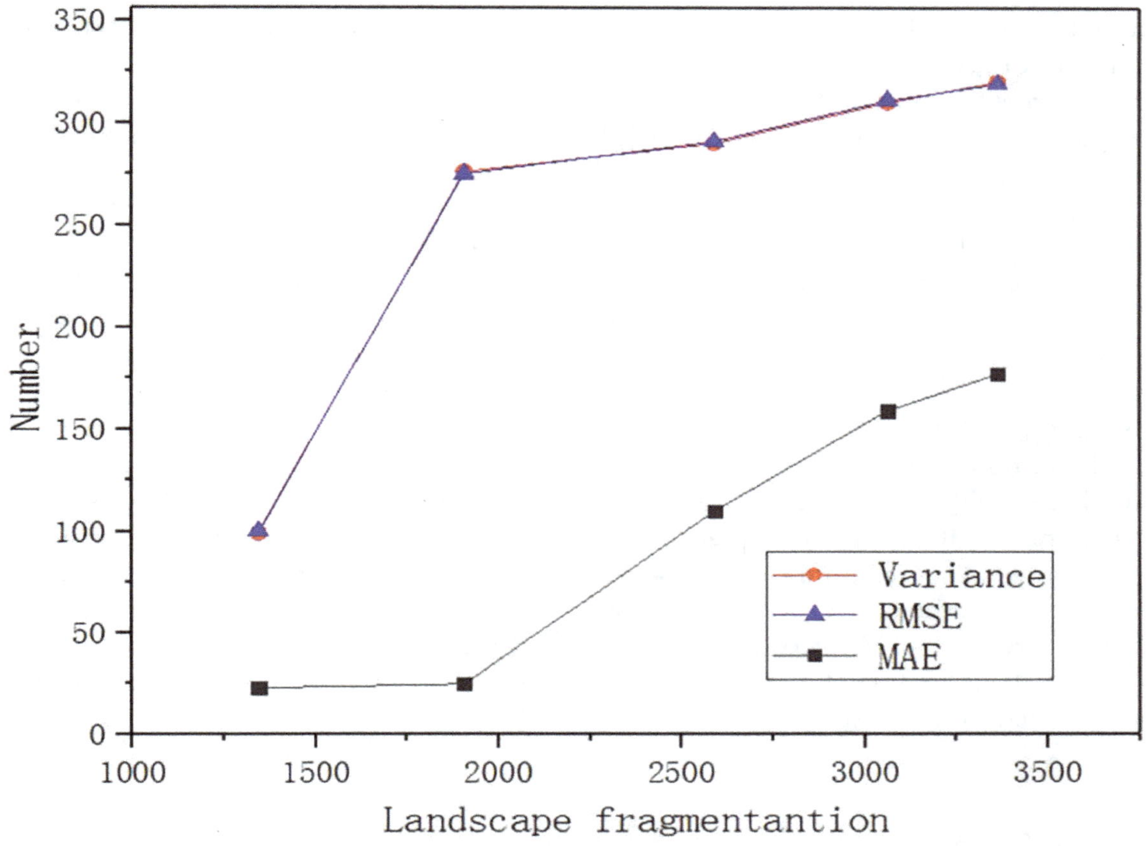

Figure 8. The relationship between the landscape fragmentation degree and the number of quadrats.

4. Discussion

4.1. FVC Extraction Effect

Most of the previous studies have reported that photography is an effective method for the field investigation of FVC, and it is also widely used in ecological environment investigation, and provides ground verification and parameters for remote sensing estimation and related models. For example, Liu et al. [16] use digital photos to estimate FVC under different background conditions. Song et al. [17] propose a new algorithm for extracting FVC from digital photos. Our study utilizes the threshold method to extract the FVC information from the quadrat image (Figure 5), and shows that this method can extract the vegetation information very well. It was also revealed that the spatial distribution results obtained by a different person are consistent, and this consistency is also observed with the spatial distribution of vegetation in the original image. This indicates that the threshold method extracts the quadrat FVC information well and with less subjectivity. In order to further study the influence of the threshold method on the FVC estimation of the quadrat image, the coefficient for the best fit line of the FVC data independently extracted by two persons reached 0.936, which exhibits very good robustness. This study verifies that the FVC information is accurate and reliable, including both the spatial distribution and the absolute value of the vegetation coverage. Therefore, this method can be used in future related research.

4.2. The Effect of the Number of Quadrats on the FVC Survey

The acquisition of ground measured data is the key to quantitative satellite remote sensing inversion, which can not only be used as data for remote sensing inversion, but is also significant for an accurate evaluation and verification of FVC remote sensing monitoring. However, acquiring large-scale measured data that matches the satellite remote sensing pixel scale is a major challenge in the study of FVC [18]. At present, the method most commonly used is to arrange a certain number of plots that match the satellite remote sensing pixel scale in the field, and then arrange several sample quadrats in the sample plot, and take the mean value of the sample survey results as the measured values of the corresponding plots. For example, Yi et al. [19] used the mean of nine quadrats in each plot as a measured value. In addition, Li et al. [1] used the mean of just five quadrats in each plot as a measured value. However, the accuracy of this method has not yet been reported. Most reports take the field survey data as the true value, but few studies have investigated the representativeness of the quadrats in regard to how the various methods and number of quadrats impact the survey results on the satellite remote sensing pixel scale. This study analyzes the spatial correlation between the number of quadrats and the FVC survey results in the satellite remote sensing pixel scale for the five grassland types. We find that there exists an exponential function relationship between the number of quadrats and the survey error for the five grassland types in the study area. With the increase of the number of quadrats, the error of the survey results is smaller, and the obtained FVC is closer to the real data of the satellite remote sensing pixel scale; however, different grassland types have differences in the relationship between the number of quadrats and the error of FVC survey results. Our study suggests that in future studies, the number of quadrats ought to be considered according to the heterogeneity of the underlying surface, as it influences the accuracy of the results when conducting field surveys with the sampling method. This study improves the understanding of quadrat error and helps to better estimate large-scale FVC. However, we have only studied five common grassland types on the Qinghai-Tibet Plateau, and whether other vegetation types also have acquaintance conclusions, needs to be demonstrated in future research.

4.3. Influence of the Degree of Fragmentation on FVC Estimation

Vegetation fragmentation causes the vegetation elements to be broken into numerous small patches, which affects their distribution structure. Therefore, FVC and fragmentation are closely related to the landscape pattern. For example, Kamusoko et al. [20] analyze landscape fragmentation by calculating different levels of landscape indicators and patch number; Saikia et al. [21] use Landsat images to evaluate the degree of fragmentation of land use. Amarnath et al. [22] evaluate the degree of fragmentation using the regression parameter slope and linear regression model based on the measured data. In addition, the fragmentation of underlying surfaces will affect the spectral signal of the remote sensing image and increase the complexity of the pixel component, which will bring uncertainty to satellite remote sensing inversion. For example, trees and grass co-exist in the savannahs, forming a landscape with vegetation fragmentation, which brings certain difficulties to the monitoring of dynamic changes of grass and tree cover in these savannahs [23,24]. In this study, fragmentation analysis was carried out on the plots. It is found (Figure 8) that there are some differences in the degree of fragmentation of the underlying surfaces of different grassland types, and these differences affect the accuracy of the estimation of FVC. For plots with a low degree of fragmentation, the number of field survey samples required to verify the accuracy of FVC is small. This analysis of the degree of fragmentation of the underlying surface provides insight for improving the accuracy of FVC estimation in future studies.

5. Conclusions

In this study, the image data of the ground sample in the source area of the Shule River was obtained by the photography method, and the FVC information was extracted by different people using the threshold method. This study finds that extracting the FVC from the original image based on the threshold method of the pixel is highly reliable, as the data was analyzed by two different people independently and was robust. During the analysis of the relationship between the number of sample quadrats and the accuracy of our FVC survey in the satellite remote sensing pixel scale, it is found that there are logarithmic relationships between the number of quadrats on the ground and the survey error for the five grassland types in the study area. Different grassland types have specific differences in the relationship between the number of sample quadrats and the error of FVC survey results. This study finds that the higher the degree of fragmentation, the more sample quadrats are needed to improve the accuracy of the FVC estimation. Therefore, when obtaining the measured FVC in the field by the sample method, the heterogeneity of the underlying surface should be considered.

Author Contributions: Jianjun Chen designed experiments; Xuning Zhao and Jianjun Chen wrote manuscripts; Xuning Zhao, Jianjun Chen, and Huizi Zhang processed the experimental data; Yu Qin and Shuhua Yi contributed in field surveying and data collection; All authors contributed to the interpretation of the results and the writing of the paper.

Acknowledgments: The authors would like to thank the editor and three anonymous reviewers for very thoughtful reviews on the previous version of this paper.

References

1. Li, F.; Chen, W.; Zeng, Y.; Zhao, Q.J.; Wu, B.F. Improving estimates of grassland fractional vegetation cover based on a pixel dichotomy model: A case study in Inner Mongolia, China. *Remote Sens.* **2014**, *6*, 4705–4722. [CrossRef]
2. Lehnert, L.W.; Meyer, H.; Wang, Y.; Miehe, G.; Thies, B.; Reudenbach, C.; Bendix, J. Retrieval of grassland plant coverage on the Tibetan Plateau based on a multi-scale, multi-sensor and multi-method approach. *Remote Sens. Environ.* **2015**, *164*, 197–207. [CrossRef]
3. Song, W.J.; Mu, X.H.; Ruan, G.Y.; Gao, Z.; Li, L.Y.; Yan, G.J. Estimating fractional vegetation cover and the vegetation index of bare soil and highly dense vegetation with a physically based method. *Int. J. Appl. Earth Obs.* **2017**, *58*, 168–176. [CrossRef]
4. Chen, W.; Sakai, T.; Moriya, K.; Koyama, L.; Cao, C.X. Estimation of vegetation coverage in Semi-arid sandy land based on multivariate statistical modeling using remote sensing data. *Environ. Model. Assess.* **2013**, *18*, 547–558. [CrossRef]
5. Okin, G.S.; Clarke, K.D.; Lewis, M.M. Comparison of methods for estimation of absolute vegetation and soil fractional cover using MODIS normalized BRDF-adjusted reflectance data. *Remote Sens. Environ.* **2013**, *130*, 266–279. [CrossRef]
6. Chen, Y.; Song, Y.Q.; Wang, W. Grassland vegetation cover inversion model based on random forest regression: A case study in Burqin County, Altay, Xinjiang Uygur Autonomous Region. *Acta Ecologica Sinica* **2018**, *38*, 2384–2394.
7. Jia, K.; Liang, S.L.; Liu, S.; Li, Y.; Xiao, Z.; Yao, Y.; Jiang, B.; Zhao, X.; Wang, X.; Xu, S.; et al. Global land surface fractional vegetation cover estimation using general regression neural networks from MODIS surface reflectance. *IEEE Trans. Geosci. Remote* **2015**, *53*, 4787–4796. [CrossRef]
8. Delamater, P.L.; Messina, J.P.; Qi, J.G.; Cochrane, M.A. A hybrid visual estimation method for the collection of ground truth fractional coverage data in a humid tropical environment. *Int. J. Appl. Earth Obs.* **2012**, *18*, 504–514. [CrossRef]
9. Jia, K.; Liang, S.L.; Gu, X.F.; Baret, F.; Wei, X.Q.; Wang, X.X.; Yao, Y.J.; Yang, L.Q.; Li, Y.W. Fractional vegetation cover estimation algorithm for Chinese GF-1 wide field view data. *Remote Sens. Environ.* **2016**, *177*, 184–191. [CrossRef]
10. Qin, Y.; Yi, S.; Ren, S.; Li, N.; Chen, J. Responses of typical grasslands in a semi-arid basin on the Qinghai-Tibetan Plateau to climate change and disturbances. *Environ. Earth Sci.* **2014**, *71*, 1421–1431. [CrossRef]
11. Chen, J.J.; Yi, S.H.; Qin, Y.; Wang, X.Y. Improving estimates of fractional vegetation cover based on UAV in alpine grassland on the Qinghai-Tibetan Plateau. *Int. J. Remote Sens.* **2016**, *37*, 1922–1936. [CrossRef]

12. Yi, S.H.; Chen, J.J.; Qin, Y.; Xu, G.W. The burying and grazing effects of plateau pika on alpine grassland are small: A pilot study in a semiarid basin on the Qinghai-Tibet Plateau. *Biogeosciences* **2016**, *13*, 6273–6284. [CrossRef]
13. Chen, J.J.; Yi, S.H.; Qin, Y. The contribution of plateau pika disturbance and erosion on patchy alpine grassland soil on the Qinghai-Tibetan Plateau: Implications for grassland restoration. *Geoderma* **2017**, *297*, 1–9. [CrossRef]
14. Baldi, G.; Guerschman, J.P.; Paruelo, J.M. Charactering fragmentation in temperate South America grasslands. *Agric. Ecosyst. Environ.* **2006**, *116*, 197–208. [CrossRef]
15. Dewan, A.M.; Yamaguchi, Y.; Rahman, M.Z. Dynamics of land use/cover changes and the analysis of landscape fragmentation in Dhaka Metropolitan, Bangladesh. *GeoJournal* **2012**, *77*, 315–330. [CrossRef]
16. Liu, Y.K.; Mu, X.H.; Wang, H.X.; Yan, G.J. A novel method for extracting green fractional vegetation cover from digital images. *J. Veg. Sci.* **2012**, *23*, 406–418. [CrossRef]
17. Song, W.J.; Mu, X.H.; Yan, G.J.; Huang, S. Extracting the green fractional vegetation cover from digital images using a shadow-resistant algorithm (SHAR-LABFVC). *Remote Sens.* **2015**, *7*, 10425–10443. [CrossRef]
18. Chen, X.X.; Vierling, L.; Rowell, E.; DeFelice, T. Using lidar and effective LAI data to evaluate IKONOS and Landsat 7 ETM+ vegetation cover estimates in a ponderosa pine forest. *Remote Sens. Environ.* **2004**, *91*, 14–26. [CrossRef]
19. Yi, S.H.; Zhou, Z.Y.; Ren, S.L.; Xu, M.; Qin, Y.; Chen, S.Y.; Ye, B.S. Effects of permafrost degradation on alpine grassland in a semi-arid basin on the Qinghai-Tibetan Plateau. *Environ. Res. Lett.* **2011**, *6*, 045403. [CrossRef]
20. Kamusoko, C.; Aniya, M. Land use/cover change and landscape fragmentation analysis in the Bindura District, Zimbabwe. *Land Degrad. Dev.* **2007**, *18*, 221–233. [CrossRef]
21. Saikia, A.; Hazarika, R.; Sahariah, D. Land-use/land-cover change and fragmentation in the Nameri Tiger Reserve, India. *Geogr. Tidsskrift-Dan. J. Geogr.* **2013**, *113*, 1–10. [CrossRef]
22. Amarnath, G.; Babar, S.; Murthy, M.S.R. Evaluating MODIS-vegetation continuous field products to assess tree cover change and forest fragmentation in India-A multi-scale satellite remote sensing approach. *Egypt. J. Remote Sens. Space Sci.* **2017**, *20*, 157–168. [CrossRef]
23. Ibrahim, S.; Balzter, H.; Tansey, K.; Tsutsumida, N.; Mathieu, R. Estimating fractional cover of plant of functional types in African savanna from harmonic analysis of MODIS time-series. *Int. J. Remote Sens.* **2018**, *39*, 2718–2745. [CrossRef]
24. Ibrahim, S.; Balzter, H.; Tansey, K.; Mathieu, R.; Tsutsumida, N. Impact of Soil Reflectance Variation Correction on Woody Cover Estimation in Kruger National Park Using MODIS Data. *Remote Sens.* **2019**, *11*, 898. [CrossRef]

7

Modeling Major Rural Land-Use Changes using the GIS-Based Cellular Automata Metronamica Model: The Case of Andalusia (Southern Spain)

Rafael M. Navarro Cerrillo [1], Guillermo Palacios Rodríguez [1,*], Inmaculada Clavero Rumbao [1], Miguel Ángel Lara [1], Francisco Javier Bonet [2] and Francisco-Javier Mesas-Carrascosa [3]

[1] Department of Forestry Engineering, Laboratory of Silviculture, Dendrochronology and Climate Change, DendrodatLab-ERSAF, Instituto Interuniversitario de Investigación del Sistema Tierra de Andalucía (IISTA), University of Cordoba, Campus de Rabanales, Crta.IV, km. 396, E-14071 Córdoba, Spain; rmnavarro@uco.es (R.M.N.C.); iclavero@idaf.es (I.C.R.); mlara@idaf.es (M.Á.L.)

[2] Department of Ecology, University of Cordoba, Campus de Rabanales, Crta.IV, km. 396, E-14071 Córdoba, Spain; bv2bogaf@uco.es

[3] Department of Graphic Engineering and Geomatics, University of Cordoba, Campus de Rabanales, Crta.IV, km. 396, E-14071 Córdoba, Spain; ig2mecaf@uco.es

* Correspondence: gpalacios@uco.es

Abstract: The effective and efficient planning of rural land-use changes and their impact on the environment is critical for land-use managers. Many land-use growth models have been proposed for forecasting growth patterns in the last few years. In this work; a cellular automata (CA)-based land-use model (Metronamica) was tested to simulate (1999–2007) and predict (2007–2035) land-use dynamics and land-use changes in Andalucía (Spain). The model was calibrated using temporal changes in land-use covers and was evaluated by the Kappa index. GIS-based maps were generated to study major rural land-use changes (agriculture and forests). The change matrix for 1999–2007 showed an overall area change of 674971 ha. The dominant land uses in 2007 were shrubs (30.7%), woody crops on dry land (17.3%), and herbaceous crops on dry land (12.7%). The comparison between the reference and the simulated land-use maps of 2007 showed a Kappa index of 0.91. The land-cover map for the projected PRELUDE scenarios provided the land-cover characteristics of 2035 in Andalusia; developed within the Metronamica model scenarios (Great Escape; Evolved Society; Clustered Network; Lettuce Surprise U; and Big Crisis). The greatest differences were found between Great Escape and Clustered Network and Lettuce Surprise U. The observed trend (1999–2007–2035) showed the greatest similarity with the Big Crisis scenario. Land-use projections facilitate the understanding of the future dynamics of land-use change in rural areas; and hence the development of more appropriate plans and policies

Keywords: future scenarios; prelude; dynamic of land use; Spatial Decision Support System, CORINE Land Cover

---

1. Introduction

Continuous land-use changes, both urban and rural, are caused mainly by anthropogenic activities [1]. The use of tools, such as models of land-use change, supports the analysis of the causes and consequences of such changes in order to understand the functioning of the land-use system and to support land-use planning and policy [2]. Over the past century, changes in demographic development, urbanization, and industrialization have constantly induced land use and land cover (LULC) changes in many regions around the world, producing new biophysical and socio-economic conditions [3]. Therefore, LULC changes are associated with socioeconomic factors, changes in industrial development, agricultural production conversion [4,5], immigration [6,7], protected area status [8], and climate change [9]. As a result of these drivers, the current land-use patterns are a consequence of a historical series of previous and incremental land-use changes, which makes the present land-use pattern highly path-dependent [10]. Many researchers have focused on understanding the factors that lead to these transitions [11,12]. However, more research is needed to explain land-use changes at different spatial and temporal scales [13].

Policy and decision makers demand land-use change scenarios to enable them to develop sustainable strategies that anticipate future trends [14]. Over recent decades, abundant research has been conducted on the spatial modeling of land use, based on the improved accessibility of spatial information, increased computational capacities, and the demand for more accurate planning tools for decision support [15–18]. Several spatially explicit approaches for modeling land use have been proposed [19,20]. Among them, cellular automata (CA) models are have been frequently used for modeling land-use change due to their ability to simulate dynamic spatial processes [21,22], creating complex patterns [23]. CA models include integrated CA models such as the Research Institute for Knowledge Systems (RIKS) model [24], fuzzy CA model [25], artificial neural networks CA model [26], and multi-CA model [27]. More recently, CA land-use models have been generalized to support land-use planning and policy analysis [22,28] as well as to predict future changes and the impacts of economic, development, and climate-change scenarios [29].

The CA models of land-use dynamics are generated by a set of cell states, neighborhoods, and transition rules and time [30]. The cell state can be made to represent any attribute of the rural or urban environment (e.g., land use, population density, land cover, terrain factors, or road networks) [22,31] and has been used in many land-use change studies oriented to policy and land-use decision support [14,32,33]. Metronamica is a spatial decision support system (SDSS), based on the constrained cellular automata (CCA) model [24,34], increasingly used to assist the simulation of the spatial implications of future land-use scenarios. This approach is very useful in long-term planning decisions. Metronamica has been used by public administrations and researchers for the simulation of rural areas dynamics, including agricultural and forest uses, under different change scenarios [35–39]. The susceptibility cartography obtained from this analysis is a useful tool in natural and rural planning.

Andalucía is the second largest region in Spain and has a population of over 8.5 million inhabitants. During the last two decades, it has experienced rapid economic growth, focused on urban development, tourism, and agriculture. In this context, planning of the rural environment faces many challenges in the balancing of current and future land use with sustainability and limiting environmental problems. In this study, Metronamica [34] was applied to model future trends of land-use changes in Andalusia for the period 2010–2035. There were three specific objectives: (i) to describe the major rural land-use changes at a regional scale between 1999 and 2007, (ii) to test the Metronamica land-use model, and (iii) to simulate future land-use patterns in the Prospective Environmental Analysis of Land Use Development in Europe (PRELUDE) scenarios. This study tests the applicability of this model at a regional scale regarding its ability to generate realistic results about land-use changes and to support policy makers.

2. Materials and Methods

2.1. Study Area

Andalusia is in the south of the Iberian Peninsula, in south-western Europe and extends over 87,600 km^2 (Figure 1). It includes three major mountain ranges running in an east to west direction, Sierra Morena and the Baetic System, consisting of the Subbaetic and Penibaetic Mountains, separated by the Intrabaetic Basin, while Lower Andalusia is in the Baetic Depression of the valley of the river Guadalquivir. A Mediterranean climate, with two rainfall peaks in spring and autumn and a dry season during summer, is the most frequent. The average annual precipitation is in the range of 300 to 2000 mm, and the mean annual temperature is 17°C, with maximum temperatures averaging above

36°C in summer. Topographic variations mean the climate ranges from arid (including the driest zone of the European Mediterranean) to humid. The dominant rural land uses are extensive and intensive agriculture, including non-irrigated and irrigated crops, olive orchards, and greenhouse horticulture. During the second half of the 20th century and more recent decades, progressive land abandonment has led to a significant increase in extensive rangeland and wood landscapes (e.g., shrubs and forests). Most of the natural vegetation is Mediterranean forests, mainly evergreen trees, such as Holm and cork oaks and pines, with large areas covered by Mediterranean shrub land. According to the data released by the National Statistics Institute of Spain, in Andalusia the population increased by approximately 750,000 inhabitants during the period 1999–2007, mainly due to the concentration of the rural population in some areas.

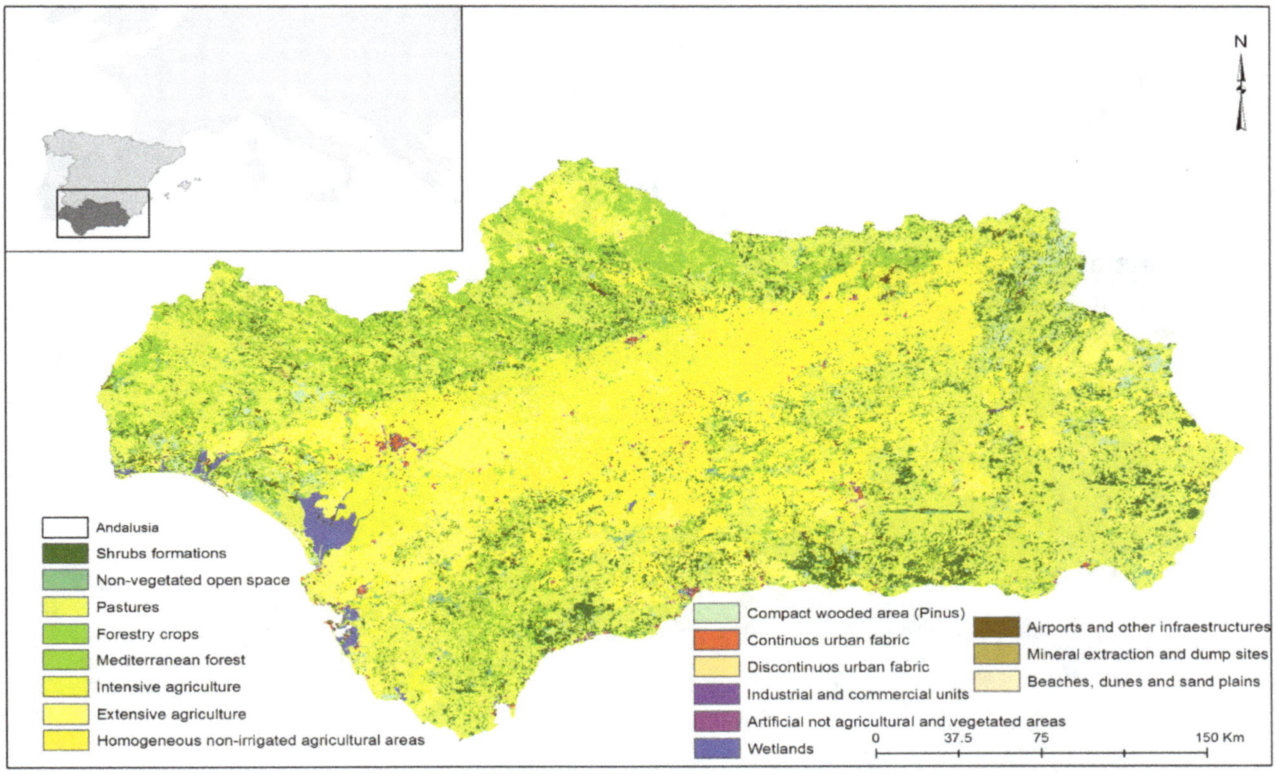

Figure 1. Distribution of the most representative rural land uses in Andalusia region (Southern Spain) in 2007.

2.2. Land-Use Data Sets

Land-use data were obtained from CORINE Land Cover (CLC) for the years 1999 to 2007 (henceforward CLC1999, CLC2007), from the Andalusia Regional Government (http://www.juntadeandalucia.es/medioambiente/site/rediam). Although the problems related to the thematic accuracy and reliability of the CORINE Land Cover datasets are known (e.g., [40]), the CLC for 1999–2007 can be regarded as generally of high quality because of the correction applied by the Andalusia Regional Government. For the study area investigated here, raster data layers were represented on a grid with a 100-m spatial resolution, which was appropriate for use at a regional scale. The classification procedure resulted in 17 LULC classes, listed in Table S1 (Supplementary Material), which were grouped in vacant land uses (e.g., land-use changes as a consequence of the dynamic of other land uses; e.g., shrub cover), functional uses (land uses with a high change trend; e.g., urban use), and features uses (land uses with a low change trend; e.g., bare soils).

2.3. Methodological Framework

The methodological framework for the application of the Metronamica model implies two main processes: calibration and simulation (Figure 2). The calibration was designed to identify the key elements of the CA model based on the change in the dynamic behavior of a given land-use cell as a response to the changes observed between 1999 and 2007: its current suitability (step 1), zoning (step 2), and accessibility parameters (step 3) were calculated. The influence of the three key components and the land-use rules (step 4) generates a transition potential (step 5), which determines whether, in a particular year, the cell will shift to a different land use or will persist in its current state. Finally, the future scenarios were created with updated suitability parameters according to the PRELUDE scenarios. Further details on Metronamica and the different elements of the calibration process are available in the Metronamica model documentation [34].

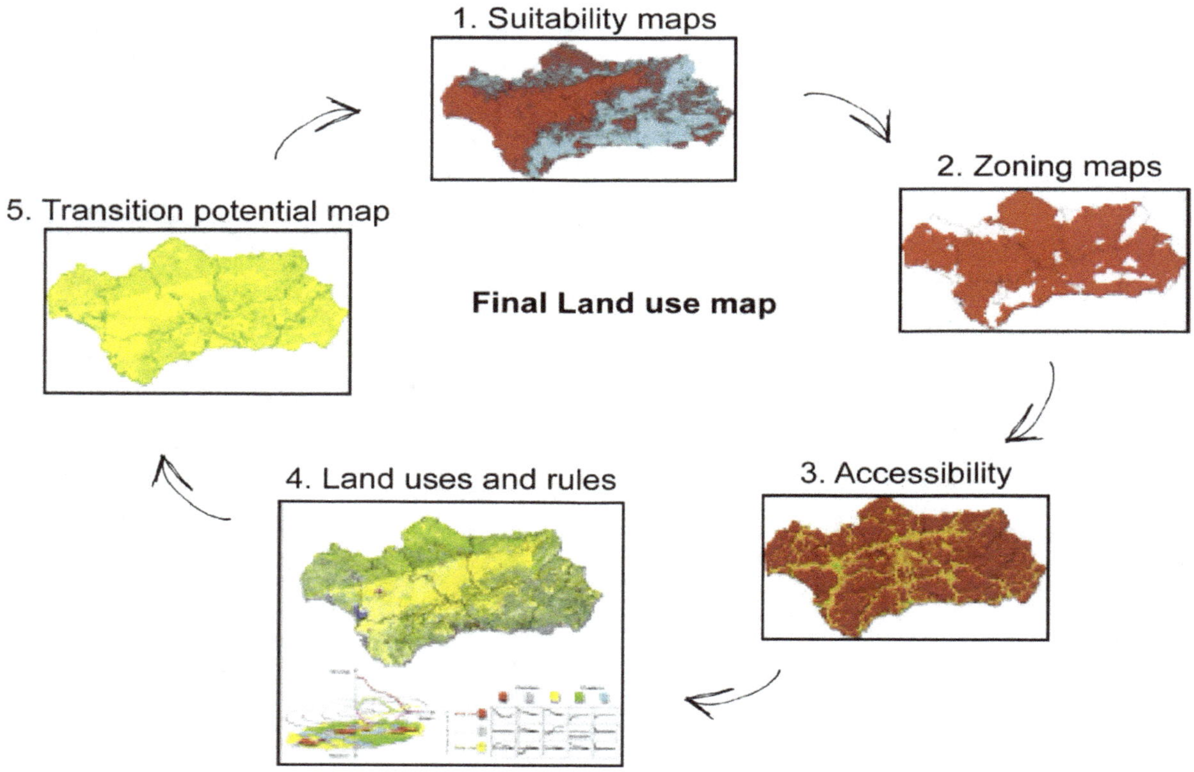

Figure 2. Metronamica system diagram of the land use-based model implemented for Andalusia (see text for explanation).

2.4. Changes in Land-Luse Calibration

Four groups of factors driving changes in land use at the regional level were selected (Figure 2 and Table S2 Supplementary Material) [34]: (i) the physical suitability of land uses, based on physical or geographical determinants (e.g., elevation, slope, soil quality, and climatic variables), (ii) zoning or institutional suitability model constraints on the allocation of land uses that reflect current and future decision making (e.g., transformation of agricultural land to forest uses), (iii) accessibility as an expression of the strength of the influence of transport networks (e.g., roads and rail networks) and other infrastructure, and (iv) land-use interactions in the area, represented by the neighborhood potential land-use dynamics. Physical suitability and zoning maps were generated using the Metronamica Overlay Tool, and included (i) average annual temperature, average maximum and minimum temperature and annual precipitation, at a resolution of 100 m, obtained from the environmental repository of Andalusia (http://www.juntadeandalucia.es/medioambiente/site/rediam) and (ii) a digital elevation model (1998, 20 m resolution; 2006, 5 m resolution) [41]. All drivers were resampled at a resolution of 100 m × 100 m and re-projected according to the coordinate reference system UTM zone 30 European Datum 1950. Land-use change was simulated for this period, using time steps of one year. The local land-use model for Andalusia included 15 land-use classes, of which nine were dynamically modeled and then grouped in three rural dominant uses: extensive and intensive agriculture and natural vegetation (Figure 1, Table S1 Supplementary Material). The remaining classes represented non-modeled classes (e.g., water bodies, infrastructures, dump sites, and beaches) and we assumed that they would remain unchanged during the simulation period.

For the calibration, based on the four driver groups, the model advanced through a series of stages representing time steps to simulate the transition potential for each cell and the land use based on the Metronamica land-use model. The land-use model was calibrated to simulate land-use changes in Andalusia between 1999 and 2007. The neighborhood rules to establish the transition potential were defined using a multiple linear regression analysis of the available spatial-temporal land-use data for Andalusia (1999–2007). The transition potential reflects the pressures exerted on each land use and constitutes the information needed for modeling land-use changes. The three key components of the model described above collectively determine the land transition rules.

Visual comparison was used to assess if the simulated land-use patterns were realistic and if the locations of simulated land-use changes coincided with those of observed land-use changes. To rigorously test the model calibration, the predictive accuracy of the calibrated land-use model was assessed by means of the Kappa index [35]. The Kappa index expresses the agreement (a value of 1 indicates a maximum level of agreement among land-use types, and values close to 0 indicate a near random spatial arrangement) between the final year modeled (2007) and the data from CLC2007. Under conditions of statistical significance, a Kappa value of 1 indicates a maximum level of clustering of a land-use type, while values close to 0 indicate a near random spatial arrangement and a value of −1 indicates a maximum level of dispersion. The Kappa test was implemented with the Map Comparison Kit (MCK) [42].

2.5. PRELUDE Scenarios

After completing the calibration, Metronamica was used to simultaneously model future potential land-cover changes associated with the PRELUDE scenarios [43]. We selected the PRELUDE scenarios which are based on those developed by the European Union [44] for analysis of the environmental implications of different scenarios for the future of Europe. They represent five global socio-economic development pathways: Great Escape (Andalusia of Contrasts) shows the highest population and economic growth, Evolved Society (Andalusia of Harmony) is characterized by a strong increase in extensive agriculture, Clustered Network (Andalusia of Structure) expects high growth in the European population combined with the impact of climate change, increasing the demand for land for agriculture, Lettuce Surprise U (Andalusia of Innovation) features high technological innovation, leading to economic growth and internal migration, and Big Crisis (Andalusia of Cohesion), in which economic and planning intervention increase (Table 1; Figure S1 Supplementary Material).

Subsequently, the four global-change models were calibrated to create the rural land-use system in the initial year of the simulation. The parameters included in the four global-change models were mainly those for climatic and socio-economic generation development, but also those related to social preferences. These parameters were derived from climatic-change models for Andalucía [45], and from other studies related to rural land uses and socio-economic variables (Table S2 Supplementary Material). These were fixed and not calibrated further. The other settings used to calibrate the Metronamica model (neighborhood rules, accessibility settings, suitability data, and zoning restrictions) were left untouched. In this phase, the initial land-use map (CLC2007) was the input into the model to simulate the global-change scenario for the reference time interval (2007–2035). Finally, an assessment was then made of the scenario-based land-cover changes to be introduced using the Metronamica model interface to create a visual representation of rural land uses in 2035 under each scenario.

Table 1. Influence of the driving forces of the PRELUDE scenarios [32]. The "+" symbol indicates that this force increases, "–" indicates that this force decreases, and "o" indicates that this force has no effect. The presence of two "+" indicates that this force has more influence (see also Figure S2 Supplementary Material).

Driving Forces	E1 Great Escape	E2 Evolved Society	E3 Clustered Networks	E4 Lettuce Surprise	E5 Big Crisis
Policy intervention	-	+	+	-	++
Population growth	++	+	++	+	+
Internal migration	+	+	+	++	+
Social equity	–	+	O	o	o
Economic growth	++	+	+	++	+
Self-sufficiency	o	++	O	+	o
Technological growth	+	+	+	++	+
Climate change	+	++	++	o	+
Renewable energy	+	o	+	++	o

2.6. *Comparison of the Land-Use and Global-Change Interaction Models*

The results of the four global-change models (for the year 2035) were assessed by visual comparison of the models among themselves, as well as with the results from the reference land-use map (CLC2007). A change matrix was used to compare the land-use pattern and measure the similarity among the simulated land-use patterns [46]. The Kappa index, implemented with the MCK [42], was also used to quantify the mismatch between simulated land-use maps [35].

3. Results

3.1. *Recent Changes in Land Cover Across Andalusia*

Using the Metronamica interface, a change matrix between 1999 and 2007 was obtained that displayed data for the observed changes in rural land cover across the region for the period 1999 to 2007 (Figure 3, Table S3 Supplementary Material). This information provides a valuable insight into how Andalusian land uses have evolved over recent years. The overall area in which land use changed was 674,971 ha, representing 8.7% of the total area in 2007. In 1999, the three land uses that occupied the largest area were shrubs (31.1%), woody crops on dry land (16.1%), and herbaceous crops on dry land (16%). Similarly, the land uses that occupied the largest area in 2007 were shrubs (30.7%), woody crops on dry land (17.3%), and herbaceous crops on dry land (12.7%). However, the largest change in land cover was for non-irrigated tree crops, which decreased by 3.8%. The most important changes were due to the transition from Mediterranean forest to shrubs and from agricultural to urban and industrial areas, together with the increase in non-irrigated agricultural areas, mainly derived from changes in the extensive agriculture areas (Figure 3). As a general view, the aggregated land-use changes between 1999 and 2007 (Table 2) indicate that the principal land-use change occurred in areas of natural vegetation.

Table 2. Transition matrix of major land-use categories (ha) in Andalusia (southern Spain) for different rural land-cover classes, between 1999 and 2007.

	Land Uses 2007			
Land Uses 1999	**Urban Areas**	**Natural Areas**	**Agricultural Land**	**Overall**
Urban areas	191,987.16	2401.62	2151.35	196,540.13
Natural areas	29,782.97	4,528,079.76	65,945.05	4,623,807.77
Agricultural land	42,452.25	55,480.42	3,842,557.86	3,940,490.53
Overall	264,222.38	4,585,961.80	3,910,654.26	8,760,838.44

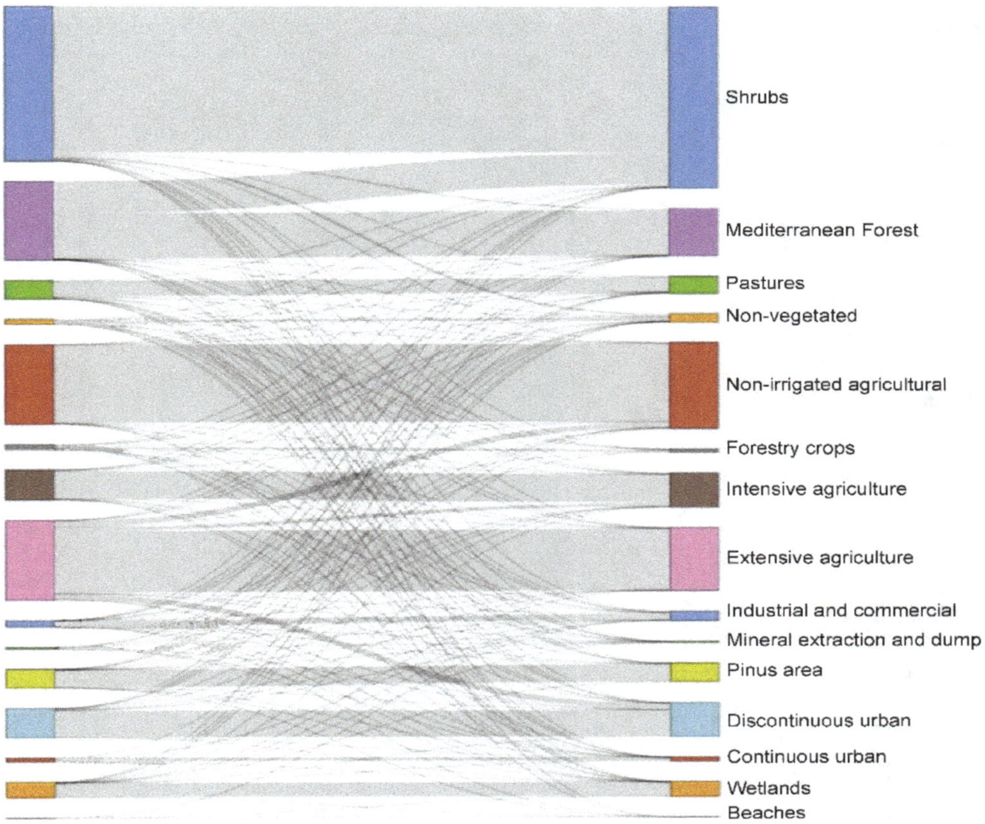

Figure 3. Transition matrix of major land-use categories in Andalusia (southern Spain) for different rural land-cover classes, between 1999 and 2007.

3.2. Metronamica Projection of Land Cover across Andalusia between 1999 and 2007

A projection of the land uses in 2007 was obtained by applying the Metronamica algorithm across the region. The comparison of the reference and the simulated land-use map of 2007 showed a Kappa index of 0.91. These results show that the model was valid and, therefore, it could be used to perform the simulations under the different hypotheses established in the scenarios.

The land-cover map for the projected PRELUDE scenarios is displayed in Figure 4. Table 3 provides the land-cover characteristics of 2035 in Andalusia developed within the Metronamica model scenarios.

For the Great Escape (Andalusia of Contrast) scenario, among the potential rural land uses over the period 2007–2035, the clearest change is the increase in intensive agriculture (Figure 4b) at the expense of extensive agriculture (Figure 4a), mainly in the most productive areas (Guadalquivir Valley and western areas). In addition, a spatial reduction in extensive agriculture is observed in the central-eastern mountains of Andalusia (Sierra Morena, Córdoba, and Sevilla), mainly due to the increase in intensive agriculture. In addition, there is a slight reduction in shrub cover, while the semi-natural areas remain almost unchanged. In 2035, the change in shrub cover (Figure 4c) is more apparent in eastern Andalusia (Jaén, Almería, and Granada), mainly due to the transformation of those areas for intensive agriculture. The impact of climate change is not very severe in this scenario.

In the Evolved Society (Andalusia of Harmony) scenario, the extensive (Figure 4d) and intensive (Figure 4e) agricultural areas increase because of the reduction in shrub cover, with a shift, in particular, to extensive agricultural uses. Self-sufficiency and climate change mark the changes in land use in this scenario. The decline in shrub cover is a result of "extensification", with a loss of semi-natural areas, and the resulting grasslands become very important and valued by the people. In addition, there is a "migration" of extensive agriculture to the mountainous areas (Sierra Morena, Sierra Nevada, Subbéticas, etc.), creating a mix of agriculture, grasslands, and semi-natural areas.

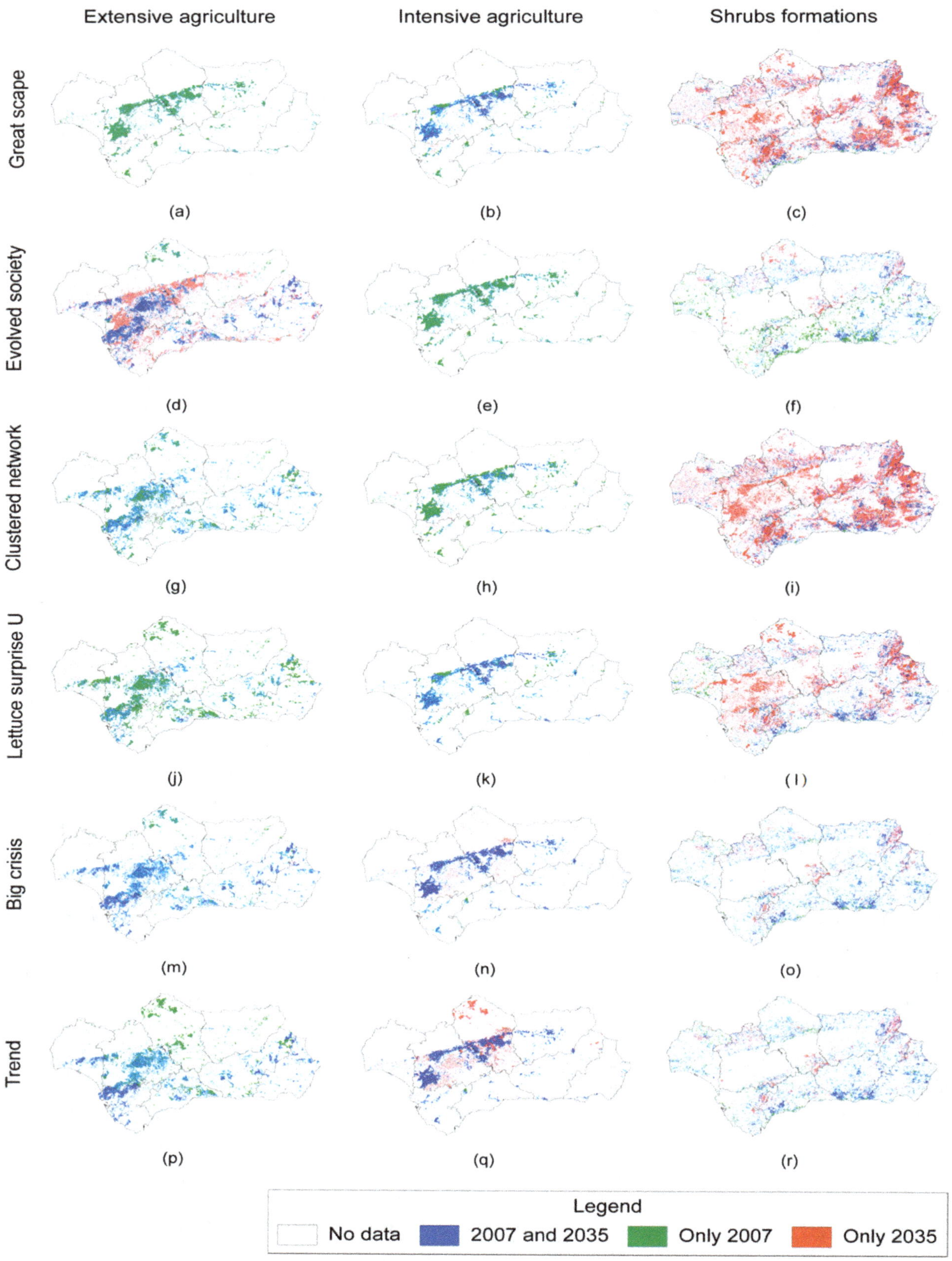

Figure 4. Simulated land-use change between 2007 and 2035 in Andalusia under the PRELUDE scenarios (**a**–**r**) (Great Escape—Andalusia of Contrasts, Evolved Society—Andalusia of Harmony, Clustered Network—Andalusia of Structure, Lettuce Surprise U—Andalusia of Innovation, and Big Crisis—Andalusia of Cohesion). Letters in parentheses express the combination of the scenarios and the main rural land uses

Projecting the trends forward to 2035, the Clustered Network (Andalusia of Structure) scenario (Figure 4c) shows a reduction in all agricultural sectors, including extensive and intensive uses, with a high increase in shrub cover. Population growth and climate change mark the change in the discontinuous rural fabric attracted to the more productive areas. There is a less marked increase in extensive and intensive agricultural uses, coupled with an increase in semi-natural areas around the entire region. The impact of climate change is very severe in this scenario.

In the Lettuce Surprise U (Andalusia of Innovation) scenario (Figure 4j,k), intensive and extensive agriculture both decrease because of their abandonment and subsequent cover by shrubs, moving to forest uses. In this scenario, a social and technological transformation leads the change in the discontinuous rural fabric, marked by the reduction in agriculture uses. The decline in extensive agriculture is a consequence of abandonment and changes in the use of semi-natural areas. The major changes in this scenario affect rural areas on the periphery of cities (Guadalquivir Valley).

Finally, in the Big Crisis (Andalusia of Cohesion) scenario, extensive (Figure 4m) and intensive (Figure 4n) agricultural areas increase significantly due to the development of small-scale intensive–extensive agriculture matrixes for self-sufficiency. This produces a mix of extensive and intensive agriculture and semi-natural areas, giving rise to a very diverse landscape. Political intervention marks the change in rural areas, with a better balance among land uses. The loss of shrub area is compensated by an increase in extensive agriculture. The value of the landscape is very important and valued by the people, so in this scenario the integration between extensive agriculture and semi-natural areas is less important. The impact of climate change is not very severe in this scenario. There is migration from cities to rural areas, with the development of a mix of agricultural and semi-natural areas.

Table 4 provides the Kappa indexes for the rural land cover of the five scenarios at the end of the period. It shows that the greatest differences are between Great Escape and both Clustered Network and Lettuce Surprise U. Although in all three scenarios, large areas of shrubs remain, a severe decrease in agricultural uses can be seen in the Lettuce Surprise U scenario, and yet, in the Clustered Network scenario, the shrub area is greater in the year 2035.

Table 3. Predicted distributions (ha) of major land uses in Andalusia for the forecasted year (2035) (PRELUDE scenarios) (see Figure 4).

	2035
E1—Great Escape—Andalusia of contrasts	
Shrubs formations	2,326,599
Intensive agriculture	513,781
Extensive agriculture	517,443
E2—Evolved Society—Andalusia of harmony	
Shrubs formations	797,311
Intensive agriculture	308,168
Extensive agriculture	1,497,530
E3—Clustered Networks— Andalusia of structure	
Shrubs formations	2,546,337
Intensive agriculture	347,399
Extensive agriculture	634,098
E4—Lettuce Surprise U— Andalusia of innovation	
Shrubs formations	1,805,649
Intensive agriculture	129,956
Extensive agriculture	522,375
E5— Big Crisis— Andalusia of cohesion	
Shrubs formations	992,330
Intensive agriculture	575,298
Extensive agriculture	819,297

Comparing the observed trend (1999–2007–2035) with the rest of the scenarios in the year 2035, the greatest similarity was found with the Big Crisis scenario (Kappa index of 0.87, Table 4). This can be explained by the fact that the agricultural uses, both extensive and intensive, are more balanced in these scenarios, yielding the great similarity with the initial situation.

Table 4. Kappa index for the land-use model in the various scenarios in 2035. The values vary between 0 and 1, where 0 is totally different and 1 is identical.

SCENARIOS	Great Escape	Evolved Society	Clustered Network	Lettuce Surprise U	Big Crisis
Great Escape		0.50	0.78	0.79	0.62
Evolved Society			0.48	0.59	0.73
Clustered Network				0.69	0.57
Lettuce Surprise U					0.69
Big Crisis					
Trend	0.60	0.69	0.56	0.69	0.87

4. Discussion

In this study, Metronamica was applied in the modeling of the rural land-use changes and trends in Andalusia, to forecast the pattern for the period 2007–2035. There is clear extensification and intensification of agricultural areas over time according to the different PRELUDE scenarios. The projection of these trends forward to 2035 will help policy makers and land-use managers to plan Andalusian landscapes in the near future.

4.1. Land-Use Changes in Andalusia between 1999 and 2007

In Europe, ecosystems and landscape dynamics are modified by land-use changes, in particular in rural areas. The impact of agricultural scenarios has yet to be analyzed in detail in southern Europe. In terms of rural land uses, in Andalusia, the period 1999 to 2007 was characterized by a decline in agriculture, pasture, and forestry uses, reflecting both a continuation of the reduction in the extensive agricultural sector and the competition from intensive agriculture faced by the lands of lower agricultural potential. A significant amount of this expansion appears to have occurred on land in mountainous agricultural areas (with complex cultivation patterns and sparsely vegetated areas). The growth in these two land uses also occurred at the expense of other types of agricultural use, including non-irrigated crop land. The changes in rural land uses have been mainly due to the concentration of the rural population in some areas and migration to urban areas [47]. Thus, according to the data released by the National Statistics Institute of Spain, in Andalusia the population increased by approximately 750,000 inhabitants during the period 1999–2007. This increase in urban land use is also a consequence of the massive constructions built along the coastline of Andalusia [48]. In this regard, the growth of tourism led to the increase in the development of infrastructures. Therefore, there is a connection between the transport network and the economic development of the region, both being a main cause of landscape transformation [49].

Agricultural activities constitute an important part of the Andalusian economy. Therefore, the decrease in agricultural areas in the period 1999–2007 may be due to the abandonment of traditional land uses in marginal areas [50] and the policies that promote agricultural intensification [51]. This decrease in these areas is also an immediate consequence of changes in land uses: for example, from herbaceous crops to grassland or from dense wooded formations of eucalyptus to scrubland interspersed with grassland, among others. The transformations also occur as a result of the change from scattered scrub with grass or rock to irrigated woody crops (in particular, citrus) or from grassland to arable crops. Approximately 38,000 hectares of natural areas were lost from 1999 to 2007, but the area occupied by forests generally does not show significant variations, since its stability is covered by the Andalusian government's environmental protection policies.

The areas occupied by shrubs and agricultural uses have altered. These changes are clearly seen in the rural land-cover changes during the studied period and the land-use agenda did not seem to strongly influence the planning and policy making in rural areas. However, agricultural and forest cover changes have significant implications in a context of global change given the role that green space plays in adaptation to climate change impacts [52,53].

The generation of scenarios based on parameterized models improves the capacities for adaptation [54]. CA models have been used in several studies in Europe, showing their ability for conceptual and practical land use change studies (see, for example, [55,56]). In this work, we implemented the land-use changes and scenarios in the cell-based Metronamica model which has already been applied in Europe at regional and national scales (see for example, [57,58]). The model parameterization was done with two reference land-use maps, one for 1999 and one for 2007. According to the Kappa value (0.91), the model results indicate very high overall agreement between the observed and simulated maps, not only in quantity but also in the ability to distinguish spatial land-use patterns.

4.2. PRELUDE Land-Use Scenarios in Andalusia 2007–2035

Once the calibrated model had been satisfactorily validated, land-use maps covering Andalusia were produced, considering the effects of regional driving forces, for five PRELUDE scenarios for 2035 [43,44]. We generated five probability maps by combining the transition probability maps for the global change scenarios tested.

The Great Escape (Andalusia of contrast) scenario shows the highest population and economic growth. An expansion of agricultural land uses, in particular of intensive agricultural areas, directly reflects this situation. This increase is mostly located in the areas close to the most productive land and large urban areas (Guadalquivir Valley), whereas the shrub areas show less reduction than in the other scenarios. The growth of intensive rural uses (intensive and extensive agriculture) is mainly concentrated around already existing productive agricultural land (perimetric areas of the Guadalquivir Valley), while there is a widespread increase in shrubs in mountain areas.

Characterized by high levels of self-sufficiency and climate-change impacts, the Evolved Society (Andalusia of Harmony) scenario shows a strong increase in extensive agriculture. Indeed, a high proportion of the shrub area becomes covered by grasses and extensive livestock areas. In this scenario, measures oriented to global mitigation and adaptation to cut greenhouse gas emissions are implemented. This would partially explain conversion-naturalization in countryside areas, bringing potential adaptation benefits related to the promotion of "green" land uses. The indirect benefits include increases in organic agriculture and the provision of ecosystems services [59].

The Clustered Network (Andalusia of Structure) scenario expects high growth in the European population combined with the impact of climate change, increasing the demand for land for agriculture. Furthermore, over the period 2007–2035, a process of proportional growth of both extensive and intensive agricultural land was detected. As a result, this situation, which could be considered a positive scenario for agricultural uses, could have negative implications for other "natural" land uses such as shrubs.

The Lettuce Surprise U (Andalusia of Innovation) scenario features high technological innovation, leading to economic growth and internal migration. This is reflected by a loss of agricultural areas. In this scenario, more land is converted into urban land, providing technological opportunities and population mobility. Of the rural land uses, the shrub area shows a moderate decline.

Big Crisis (Andalusia of Cohesion) is a scenario in which economic and planning intervention increase. Unlike the others, it may be described as 'business as usual'. However, there is a greater emphasis on land-use planning to orient decisions and actions linked to optimized land-use changes and development. It is characterized by more progressive changes and the avoidance of intense development; consequently, rural uses are more balanced. In land-use terms, this is related to decisions on increases in extensive agriculture to reduce greenhouse gas emissions, and to conservation measures for traditional landscapes (e.g., "cultural" landscapes) more adapted to climate change and able to provide extensive ecosystem service functions. Although some shrub areas increase, the general trend is towards a reduction in this land cover, especially in moderately marginal areas. This reflects the land-uses policy followed in the Big Crisis scenario, and the promotion of agricultural activities closer to transport networks.

The scenarios were compared based on the principal land-use changes, to give an overview of the landscape trends. Strong similarity was observed between the Big Crisis scenario and the trend scenario. Therefore, the agricultural intensification was strongly supported; consequently, areas of extensive agriculture tend to decline, mainly in areas with productive (e.g., climatic and topographic) restrictions for agriculture. The combination of land abandonment and the focus on planning means that both scenarios showed a very high percentage of "naturalization" processes.

The use of prediction models of land use changes, such those used in this study, help policy makers meet their needs for current knowledge of spatial locations and magnitude of likely land use changes under different socio-economic and environmental scenarios to select more appropriate decisions and the consequences in the short and long term [60].

5. Conclusions

This study shows the usefulness of Metronamica land-use model to assess future trends of land-use changes in Andalusia (south Spain). Our results allow us to identify the major rural land-use changes at a regional scale and demonstrate the validity of Metronamica to evaluate these processes in other regions on a global scale. In our study, the main limitation was related to the simulation period (8 years between 1999 and 2007), which allowed the model to be calibrated and validated. However, since land use data for the CLC2018-Andalusía are available, it could have been used to confirm the trends of the projections obtained in the different scenarios for 2035. This extra phase would reinforce the interactions between observed land-use classes and projections for different plausible scenarios (e.g., PRELUDE), facilitating the transferability of results to other regions and situations. The use of the PRELUDE scenarios makes the work particularly interesting for the countries of the European Union. It would allow for comparing the situation on a regional or national scale and generate realistic results regarding land-use changes and to support European policies related to future land uses. Based on our results, further research may be developed such as the interaction between complex land-use changes with environmental services or further insights into the capacity of Metronamica for simulating land use including a more diverse landscapes (e.g., "Mediterranean cultural landscapes).

Supplementary Materials:
Table S1: Land-uses legends reclassified and used in the Metronamica model, Table S2: Drivers of the land-use model at the local level. For agriculture, suitability is calculated on a yearly basis in the physical suitability model based on climate-change scenarios and local characteristics, Table S3: Change matrix 1999–2007, Figure S1: Main driving forces of the PRELUDE scenarios [42].

Author Contributions: Conceptualization, Rafael M. Navarro Cerrillo and Inmaculada Clavero Rumbao; Methodology, Rafael M. Navarro Cerrillo, Inmaculada Clavero Rumbao, and Guillermo Palacios Rodríguez; Formal Analysis, Inmaculada Clavero Rumbao, Rafael M. Navarro Cerrillo, and Francisco-Javier Mesas-Carrascosa; Investigation, Rafael M. Navarro Cerrillo, Inmaculada Clavero Rumbao, Guillermo Palacios Rodríguez, Miguel Ángel Lara, and Francisco Javier Bonet; Writing—Original Draft Preparation, Rafael M. Navarro Cerrillo, Guillermo Palacios Rodríguez, and Francisco Javier Bonet; Writing—Review & Editing, Rafael M. Navarro Cerrillo, Guillermo Palacios Rodríguez, Francisco Javier Bonet and Francisco-Javier Mesas-Carrascosa. All authors have read and agreed on the published version of the manuscript.

Acknowledgments: The authors thank the Andalusia Department of Agriculture and Environment, which provided access to and background information on the field site (in particular, the REDIAM authorities, J.M. Moreira, F. Cáceres, F. Gímenez de Azcarate, and J.J. Guerrero-Álvarez). We thank our financial supporters: the Junta de Andalucía-Consejería de Medio Ambiente and the European Union's Horizon 2020 ECOPOTENTIAL project (grant agreement No. 641762). We also thank David Walker, for the linguistic revision of the manuscript. We acknowledge the institutional support of the University of Cordoba-IISTA and Campus de Excelencia CEIA3.

References

1. Parker, D.C.; Hessl, A.; Davis, S.C. Complexity, land-use modeling, and the human dimension: Fundamental challenges for mapping unknown outcome spaces. *Geoforum* **2008**, *39*, 789–804. [CrossRef]
2. Verburg, P.H.; Schot, P.P.; Dijst, M.J.; Veldkamp, A. Land use change modelling: current practice and research priorities. *GeoJournal* **2004**, *61*, 309–324. [CrossRef]
3. Bajocco, S.; De Angelis, A.; Perini, L.; Ferrara, A.; Salvati, L. The Impact of Land Use/Land Cover Changes on Land Degradation Dynamics: A Mediterranean Case Study. *Environ. Manag.* **2012**, *49*, 980–989. [CrossRef] [PubMed]
4. Perz, S.G.; Skole, D.L. Secondary Forest Expansion in the Brazilian Amazon and the Refinement of Forest Transition Theory. *Soc. Nat. Resour.* **2003**, *16*, 277–294. [CrossRef]
5. Rudel, T.K.; Defries, R.; Asner, G.P.; Laurance, W.F. Changing Drivers of Deforestation and New Opportunities for Conservation. *Conserv. Biol.* **2009**, *23*, 1396–1405. [CrossRef] [PubMed]
6. Deng, J.S.; Wang, K.; Hong, Y.; Qi, J.G. Spatio-temporal dynamics and evolution of land use change and landscape pattern in response to rapid urbanization. *Landsc. Urban Plan.* **2009**, *92*, 187–198. [CrossRef]
7. Wu, Q.; Li, H.; Wang, R.; Paulussen, J.; He, Y.; Wang, M.; Wang, B.; Wang, Z. Monitoring and predicting land use change in Beijing using remote sensing and GIS. *Landsc. Urban Plan.* **2006**, *78*, 322–333. [CrossRef]
8. Ellis, E.A.; Porter-Bolland, L. Is community-based forest management more effective than protected areas?: A comparison of land use/land cover change in two neighboring study areas of the Central Yucatan Peninsula, Mexico. *For. Ecol. Manage.* **2008**, *256*, 1971–1983. [CrossRef]
9. Pielke, R.A., Sr.; Pitman, A.; Niyogi, D.; Mahmood, R.; McAlpine, C.; Hossain, F.; Goldewijk, K.K.; Nair, U.; Betts, R.; Fall, S.; et al. Land use/land cover changes and climate: modeling analysis and observational evidence. *WIREs Clim. Chang.* **2011**, *2*, 828–850.
10. Brown, D.G.; Johnson, K.M.; Loveland, T.R.; Theobald, D.M. Rural land-use trends in the conterminous united states, 1950–2000. *Ecol. Appl.* **2005**, *15*, 1851–1863. [CrossRef]
11. Lambin, E.F.; Meyfroidt, P. Land use transitions: Socio-ecological feedback versus socio-economic change. *Land Use Policy* **2010**, *27*, 108–118. [CrossRef]
12. Hurtt, G.C.; Chini, L.P.; Frolking, S.; Betts, R.A.; Feddema, J.; Fischer, G.; Fisk, J.P.; Hibbard, K.; Houghton, R.A.; Janetos, A.; et al. Harmonization of land-use scenarios for the period 1500–2100: 600 years of global gridded annual land-use transitions, wood harvest, and resulting secondary lands. *Clim. Chang.* **2011**, *109*, 117. [CrossRef]
13. Rudel, T.K.; Coomes, O.T.; Moran, E.; Achard, F.; Angelsen, A.; Xu, J.; Lambin, E. Forest transitions: towards a global understanding of land use change. *Glob. Environ. Chang.* **2005**, *15*, 23–31. [CrossRef]
14. Bryan, B.A.; Nolan, M.; McKellar, L.; Connor, J.D.; Newth, D.; Harwood, T.; King, D.; Navarro, J.; Cai, Y.; Gao, L.; et al. Land-use and sustainability under intersecting global change and domestic policy scenarios: Trajectories for Australia to 2050. *Glob. Environ. Chang.* **2016**, *38*, 130–152. [CrossRef]
15. Wang, F.; Hasbani, J.-G.; Wang, X.; Marceau, D.J. Identifying dominant factors for the calibration of a land-use cellular automata model using Rough Set Theory. *Comput. Environ. Urban Syst.* **2011**, *35*, 116–125. [CrossRef]
16. Pradhan, B. Use of GIS-based fuzzy logic relations and its cross application to produce landslide susceptibility maps in three test areas in Malaysia. *Environ. Earth Sci.* **2011**, *63*, 329–349. [CrossRef]
17. Bathrellos, G.D.; Gaki-Papanastassiou, K.; Skilodimou, H.D.; Papanastassiou, D.; Chousianitis, K.G. Potential suitability for urban planning and industry development using natural hazard maps and geological–geomorphological parameters. *Environ. Earth Sci.* **2012**, *66*, 537–548. [CrossRef]
18. Thill, J.-C. *Spatial Multicriteria Decision Making and Analysis. A Geographic Information Sciences Approach*; Ashgate: London, UK, 2019.
19. Ren, Y.; Lü, Y.; Comber, A.; Fu, B.; Harris, P.; Wu, L. Spatially explicit simulation of land use/land cover changes: Current coverage and future prospects. *Earth-Sci. Rev.* **2019**, *190*, 398–415. [CrossRef]
20. Noszczyk, T. A review of approaches to land use changes modeling. *Hum. Ecol. Risk Assess.* **2019**, *25*, 1377–1405. [CrossRef]

21. Barredo, J.I.; Demicheli, L.; Lavalle, C.; Kasanko, M.; McCormick, N. Modelling Future Urban Scenarios in Developing Countries: An Application Case Study in Lagos, Nigeria. *Environ. Plan. B Plan. Des.* **2004**, *31*, 65–84. [CrossRef]
22. Aburas, M.M.; Ho, Y.M.; Ramli, M.F.; Ash'aari, Z.H. The simulation and prediction of spatio-temporal urban growth trends using cellular automata models: A review. *Int. J. Appl. Earth Obs. Geoinf.* **2016**, *52*, 380–389. [CrossRef]
23. Batty, M. Agents, Cells, and Cities: New Representational Models for Simulating Multiscale Urban Dynamics. *Environ. Plan. A Econ. Sp.* **2005**, *37*, 1373–1394. [CrossRef]
24. White, R.; Engelen, G. High-resolution integrated modelling of the spatial dynamics of urban and regional systems. *Comput. Environ. Urban Syst.* **2000**, *24*, 383–400. [CrossRef]
25. Wu, F. A linguistic cellular automata simulation approach for sustainable land development in a fast growing region. *Comput. Environ. Urban Syst.* **1996**, *20*, 367–387. [CrossRef]
26. Li, X.; Yeh, A.G.-O. Neural-network-based cellular automata for simulating multiple land use changes using GIS. *Int. J. Geogr. Inf. Sci.* **2002**, *16*, 323–343. [CrossRef]
27. Cechini, A.; Rinaldi, E. Building Urban Models with Multi Cellular Automata. In Proceedings of the 6th International Conference: Computers in Urban Planning & Urban Management, Venice, Italy, 8–11 September 1999.
28. Omrani, H.; Tayyebi, A.; Pijanowski, B. Integrating the multi-label land-use concept and cellular automata with the artificial neural network-based Land Transformation Model: an integrated ML-CA-LTM modeling framework. *GISci. Remote Sens.* **2017**, *54*, 283–304. [CrossRef]
29. Rimal, B.; Zhang, L.; Keshtkar, H.; Haack, B.N.; Rijal, S.; Zhang, P. Land use/land cover dynamics and modeling of urban land expansion by the integration of cellular automata and markov chain. *ISPRS Int. J. Geo-Inf.* **2018**, *7*, 154. [CrossRef]
30. Santé, I.; García, A.M.; Miranda, D.; Crecente, R. Cellular automata models for the simulation of real-world urban processes: A review and analysis. *Landsc. Urban Plan.* **2010**, *96*, 108–122. [CrossRef]
31. Al-shalabi, M.; Billa, L.; Pradhan, B.; Mansor, S.; Al-Sharif, A.A.A. Modelling urban growth evolution and land-use changes using GIS based cellular automata and SLEUTH models: the case of Sana'a metropolitan city, Yemen. *Environ. Earth Sci.* **2013**, *70*, 425–437. [CrossRef]
32. van Delden, H.; Luja, P.; Engelen, G. Integration of multi-scale dynamic spatial models of socio-economic and physical processes for river basin management. *Environ. Model. Softw.* **2007**, *22*, 223–238. [CrossRef]
33. Verburg, P.H.; Kok, K.; Pontius, R.G.; Veldkamp, A. Modeling Land-Use and Land-Cover Change. In *Land-Use and Land-Cover Change*; Series, G.C.-T.I., Ed.; Springer Berlin, Heidelberg: Berlin, Germany, 2006; ISBN 978-3-540-32201-6.
34. Research Institute for Knowledge Systems (RIKS BV). *Metronamica—Documentation*; Research Institute for Knowledge Systems (RIKS BV): Maastricht, The Netherlands, 2011.
35. van Vliet, J.; Bregt, A.K.; Hagen-Zanker, A. Revisiting Kappa to account for change in the accuracy assessment of land-use change models. *Ecol. Modell.* **2011**, *222*, 1367–1375. [CrossRef]
36. Kok, K.; van Delden, H. Combining Two Approaches of Integrated Scenario Development to Combat Desertification in the Guadalentín Watershed, Spain. *Environ. Plan. B Plan. Des.* **2009**, *36*, 49–66. [CrossRef]
37. Allen, J.; Lu, K. Modeling and Prediction of Future Urban Growth in the Charleston Region of South Carolina. *Conserv. Ecol.* **2003**, *8*, 2. [CrossRef]
38. Xu, X.; Du, Z.; Zhang, H. Integrating the system dynamic and cellular automata models to predict land use and land cover change. *Int. J. Appl. Earth Obs. Geoinf.* **2016**, *52*, 568–579. [CrossRef]
39. Weynants, M.; Montanarella, L.; Tóth, G.; Arnoldussen, A.; Anaya Romero, M.; Bilas, G.; Borresen, T.; Cornelis, W.; Daroussin, J.; Da Conceiçao Gonçalves, M.; et al. *European Hydropedological Data Inventory (EU-HYDI)*; Publications Office of the European Union: Ispra, Italy, 2013.
40. Catalá Mateo, R.; Bosque Sendra, J.; Plata Rocha, W. Análisis de posibles errores en la base de datos Corine Land Cover (1990–2000) en la Comunidad de Madrid. *Estud. Geogr.* **2008**, *69*, 264.
41. Red de Información Ambiental de Andalucía. Available online: http://www.juntadeandalucia.es/medioambiente/site/rediam (accessed on 14 June 2020).
42. Research Institute for Knowledge Systems (RIKS BV). *MCK Reader: Methods of Map Comparison Kit*; Research Institute for Knowledge Systems (RIKS BV): Maastricht, The Netherlands, 2011.

43. Van Delden, H.; Engelen, G.; Uljee, I.; Hagen, A.; Van der Meulen, M.; Vanhout, R. *Prelude Quantification and Spatial Modelling of Land Use/Land Cover Changes*; Research Institute for Knowledge Systems (RIKS BV): Maastricht, The Netherlands, 2005.
44. Hoogeveen, Y.; Volkery, A.; Henrichs, T.; Ribeiro, T. *Land Use Scenarios for Europe–Modelling at the European Scale*; European Environment Agency: Copenhagen, Denmark, 2005.
45. Consejería de Medio Ambiente y Ordenación del Territorio. El clima de Andalucía en el siglo XXI. Escenarios locales de cambio climático de Andalucía. Available online: http://aeclim.org/wp-content/uploads/2016/01/Climate-Change-in-Andalusia-Libro-completo.pdf (accessed on 19 July 2020).
46. Li, X.; Chen, Y.; Liu, X.; Li, D.; He, J. Concepts, methodologies, and tools of an integrated geographical simulation and optimization system. *Int. J. Geogr. Inf. Sci.* **2011**, *25*, 633–655. [CrossRef]
47. Bindereif, L.; Rentschler, T.; Bartelheim, M.; Bonilla, M.D.-Z.; Gries, P.; Schmidt, K.; Scholten, T. Analysis and mapping of spatio-temporal land use dynamics in Andalusia, Spain using the Google Earth Engine cloud computing platform and the Landsat archive. In Proceedings of the Geophysical Research Abstracts, Vienna, Austria, 7–12 April 2019; Volume 21.
48. Molina, R.; Anfuso, G.; Manno, G.; Gracia Prieto, F.J. The Mediterranean Coast of Andalusia (Spain): Medium-Term Evolution and Impacts of Coastal Structures. *Sustainability* **2019**, *11*, 3539. [CrossRef]
49. Rokicki, B.; Stępniak, M. Major transport infrastructure investment and regional economic development—An accessibility-based approach. *J. Transp. Geogr.* **2018**, *72*, 36–49. [CrossRef]
50. Romero-Díaz, A.; Ruiz-Sinoga, J.D.; Robledano-Aymerich, F.; Brevik, E.C.; Cerdà, A. Ecosystem responses to land abandonment in Western Mediterranean Mountains. *CATENA* **2017**, *149*, 824–835. [CrossRef]
51. Muñoz, M.P.M. *Sustainable Agriculture and Resource use under Climate Change: A Multi-scale and Cross-sectoral Approach with a Focus on Andalusia (Spain)*; Universidad Politécnica de Madrid: Madrid, Spain, 2019.
52. Gill, S.E.; Handley, J.F.; Ennos, A.R.; Pauleit, S. Adapting cities for climate change: the role of the green infrastructure. *Built Environ.* **2007**, *33*, 115–133. [CrossRef]
53. Dunford, R.W.; Smith, A.C.; Harrison, P.A.; Hanganu, D. Ecosystem service provision in a changing Europe: adapting to the impacts of combined climate and socio-economic change. *Landsc. Ecol.* **2015**, *30*, 443–461. [CrossRef]
54. Jeong, J.S. Design of spatial PGIS-MCDA-based land assessment planning for identifying sustainable land-use adaptation priorities for climate change impacts. *Agric. Syst.* **2018**, *167*, 61–71. [CrossRef]
55. Gounaridis, D.; Chorianopoulos, I.; Symeonakis, E.; Koukoulas, S. A Random Forest-Cellular Automata modelling approach to explore future land use/cover change in Attica (Greece), under different socio-economic realities and scales. *Sci. Total Environ.* **2019**, *646*, 320–335. [CrossRef] [PubMed]
56. Gomes, E.; Abrantes, P.; Banos, A.; Rocha, J. Modelling future land use scenarios based on farmers' intentions and a cellular automata approach. *Land Use Policy* **2019**, *85*, 142–154. [CrossRef]
57. Van Delden, H.; Vanhout, R. Territorial Scenarios and Visions for Europe. Available online: http://bsr-espon.infeurope.lu/export/sites/default/Documents/Projects/AppliedResearch/ET2050/FR/ET2050_FR-03_Volume_5_-_Land-use_Trends_and_Scenarios.pdf (accessed on 19 July 2020).
58. Van Delden, H.; Stuczynski, T.; Ciaian, P.; Paracchini, M.; Hurkens, J.; Lopatka, A.; Shi, Y. Integrated assessment of agricultural policies with dynamic land use change modelling. *Ecol. Modell.* **2010**, *18*, 2153–2166. [CrossRef]
59. Locatelli, B.; Lavorel, S.; Sloan, S.; Tappeiner, U.; Geneletti, D. Characteristic trajectories of ecosystem services in mountains. *Front. Ecol. Environ.* **2017**, *15*, 150–159. [CrossRef]
60. Holman, I.; Brown, C.; Janes, V.; Sandars, D. Can we be certain about future land use change in Europe? A multi-scenario, integrated-assessment analysis. *Agric. Syst.* **2017**, *151*, 126–135. [CrossRef]

8

Assessing the Distribution of Heavy Industrial Heat Sources in India between 2012 and 2018

Caihong Ma [1,2,3], Zheng Niu [1,2], Yan Ma [2,*], Fu Chen [2], Jin Yang [2] and Jianbo Liu [2]

1 State Key Laboratory of Remote Sensing Science, Institute of Remote Sensing and Digital Earth, Chinese Academy of Sciences, Beijing 100101, China; mach@radi.ac.cn (C.M.); niuzheng@radi.ac.cn (Z.N.)

2 Remote Sensing and Digital Earth, Chinese Academy of Sciences, Beijing 100094, China; chenfu@radi.ac.cn (F.C.); yangjin@radi.ac.cn (J.Y.); liujb@radi.ac.cn (J.L.)

3 Sanya Institute of Remote Sensing, Sanya 572029, China

* Correspondence: mayan@radi.ac.cn.

Abstract: The heavy industry in India has witnessed rapid development in the past decades. This has increased the pressures and load on the Indian environment, and has also had a great impact on the world economy. In this study, the Preparatory Project Visible Infrared Imaging Radiometer (NPP VIIRS) 375-m active fire product (VNP14IMG) and night-time light (NTL) data were used to study the spatiotemporal patterns of heavy industrial development in India. We employed an improved adaptive K-means algorithm to realize the spatial segmentation of long-term VNP14IMG data and artificial heat-source objects. Next, the initial heavy industry heat sources were distinguished from normal heat sources using a threshold recognition model. Finally, the maximum night-time light data were used to delineate the final heavy industry heat sources. The results suggest, that this modified method is a much more accurate and effective way of monitoring heavy industrial heat sources, and the accuracy of this detection model was higher than 92.7%. The number of main findings were concluded from the study: (1) the heavy industry heat sources are mainly concentrated in the north-east Assam state, east-central Jharkhand state, north Chhattisgarh and Odisha states, and the coastal areas of Gujarat and Maharashtra. Many heavy industrial heat sources were also found around a line from Kolkata on the Eastern Indian Ocean to Mumbai on the Western Indian Ocean. (2) The number of working heavy industry heat sources (NWH) and, particularly, the total number of fire hotspots for each working heavy industry heat source area (NFHWH) are continuing to increase in India. These trends mirror those for the Gross Domestic Product (GDP) and total population of India between 2012 and 2017. (3) The largest values of NWH and NFHWH were in Jharkhand, Chhattisgarh, and Odisha whereas the smallest negative values, the *Slope_NWH* in Jharkhand and Chhattisgarh were also the two largest values in the whole country. The smallest negative values of *Slope_NWH* and *Slope_NFHWH* were in Haryana. The *Slope_NFHWH* in the mainland Gujarat had the second most negative value, while the value of the *Slope_NWH* was the third-highest positive value.

Keywords: adaptive K-means algorithm; heavy industry heat sources; NPP-VIIRS; active fire data; night-time light data

---

1. Introduction

Over the few past decades, India has become one of the world's fastest-growing major economies and is now considered a newly industrialized country [1]. The amount of heavy industry, which is an important component of basic industry and provides technical equipment, power, and raw materials for all sectors of the national economy, has also soared in India [2]. This industry effectively supports the economic development of the country. However, this growth has been accompanied by a large increase in greenhouse gas emissions and other air pollutants from heavy industrial production [3].

Therefore, real-time maps of the layout of heavy industrial development are becoming important for studies of Indian economic development and air pollution issues [2,4].

Many scholars and nonprofit organizations or institutions have focused their attention on the global distribution of one or more energy types or industries. The British Petroleum (BP) company [5] and the International Energy Agency (IEA) [6] provide regular, annual reports of energy (coal, oil, gas, etc.) prospects. The Global Power Emissions Database (GPED) [7] was formed from individual power-generating units for 2010 [3]. In addition, the India Coal-Fired Power Plant Database (ICPD) [8] is also available for India. These databases include a large amount of information that can be used for mining and strategic development in India. However, traditional statistical methods usually involve a lot of human error; in addition, the real-time distribution of heavy industry in India is not available.

Satellite images, which can be considered to be objective, true data, have become the most effective way to monitor the dynamics of Land-Cover (LC) and Land-Use (LU) (also referred to as LULC) [9,10]. Heat sources, such as the combustion of fossil fuels in cement plants and steelworks and the flaring of petroleum gas in oil fields [2,11], are also vital for most heavy industries. Therefore, thermal anomaly products derived from remote sensing data provide new ways of revealing the objective and real-time distribution of heavy industry in India. Recently, it has been widely and well-used in the detection of global-scale self-ignition fire point data [12–16]. Also, the night-time thermal anomaly product from the National Polar-orbiting Operational Environmental Satellite System (NPOESS) Preparatory Project (NPP) Visible Infrared Imaging Radiometer (VIIRS) has been successfully applied in studies of volcanic activity [17] and oil exploitation [18]. NPP VIIRS night-time fire data (resolution 750 m) were used to identify industrial heat sources considering their time, space, and temperature information [11,19]. Also, better active global fire-points product named NPP VIIRS active fire product (VNP14IMG), with 375-m resolution and covering day- and night-time thermal anomaly, was provided by Schroeder et al. [20] and Giglio et al. [21]. It effectively provided an improved response for fires with small areas. Then, Ma et al. [2] proposed a heavy industry heat source detection model based on an improved adaptive K-means algorithm using long-term VNP14IMG data. This produced good results for mainland China; however, due to the complexity of the Indian geographical coverage, the precision was not so good when this was applied to India.

In addition, large and heavy equipment and facilities (such as heavy equipment, large machine tools and large buildings) are also important characteristics of heavy industry. So, the use of lighting is also common and necessary in those areas. Night-time light (NTL) data, especially the VIIRS day/night band (DNB) data, can provide the day and night distribution of lights for the whole world [22,23]. Therefore, in this study, NTL data were used to modify Ma's model [2]. The new heavy industry heat source detection model for revealing spatiotemporal patterns in and the development of heavy industry in India based on an improved adaptive K-means using VNP14IMG and NTL was then developed. As part of this study, VNP14IMG and NTL data were acquired and preprocessed. We adopted an improved adaptive K-means algorithm using long-term VNP14IMG data to construct heat-source objects. Then, many hot features, including geometric, statistical, and heat source attribute features, were extracted for each heat-source object. In addition, the initial heavy industry heat sources were discriminated from other heat-source objects using a threshold recognition model based on hot features. Finally, maximum night-time light data were used to delineate the final heavy industry heat sources.

The remainder of this article is organized as follows. Section 2 describes the study area, data sources, main data preprocessing steps, and methodology. Section 3 shows the experimental results that were obtained using the VNP14IMG and NTL data and discusses and assesses the distribution of heavy industrial heat sources in India. Conclusions are drawn in Section 4, and recommendations for future research are given.

2. Materials and Methods

2.1. Study Area

India is a country in South Asia, lying to the north of the equator between 6°44′ N and 35°30′ N and 68°7′ E and 97°25′ E. It is surrounded by the Indian Ocean, the Arabian Sea, and the Bay of Bengal. Since market-based economic reforms began in 1991, India has emerged as a global player with one of the fastest-growing major economies and is now considered a newly industrialized country [24]. It is also the world's second-most populous country (with more than 1.3 billion people) as well as being the most populous democracy in the world. India is a federal republic governed under a parliamentary system and comprises 29 states and seven union territories, giving a total of 36 entities (as shown in Figure 1). It should be noted, however, that Jammu and Kashmir state, marked by the red dashed line, lies within the disputed Kashmir region.

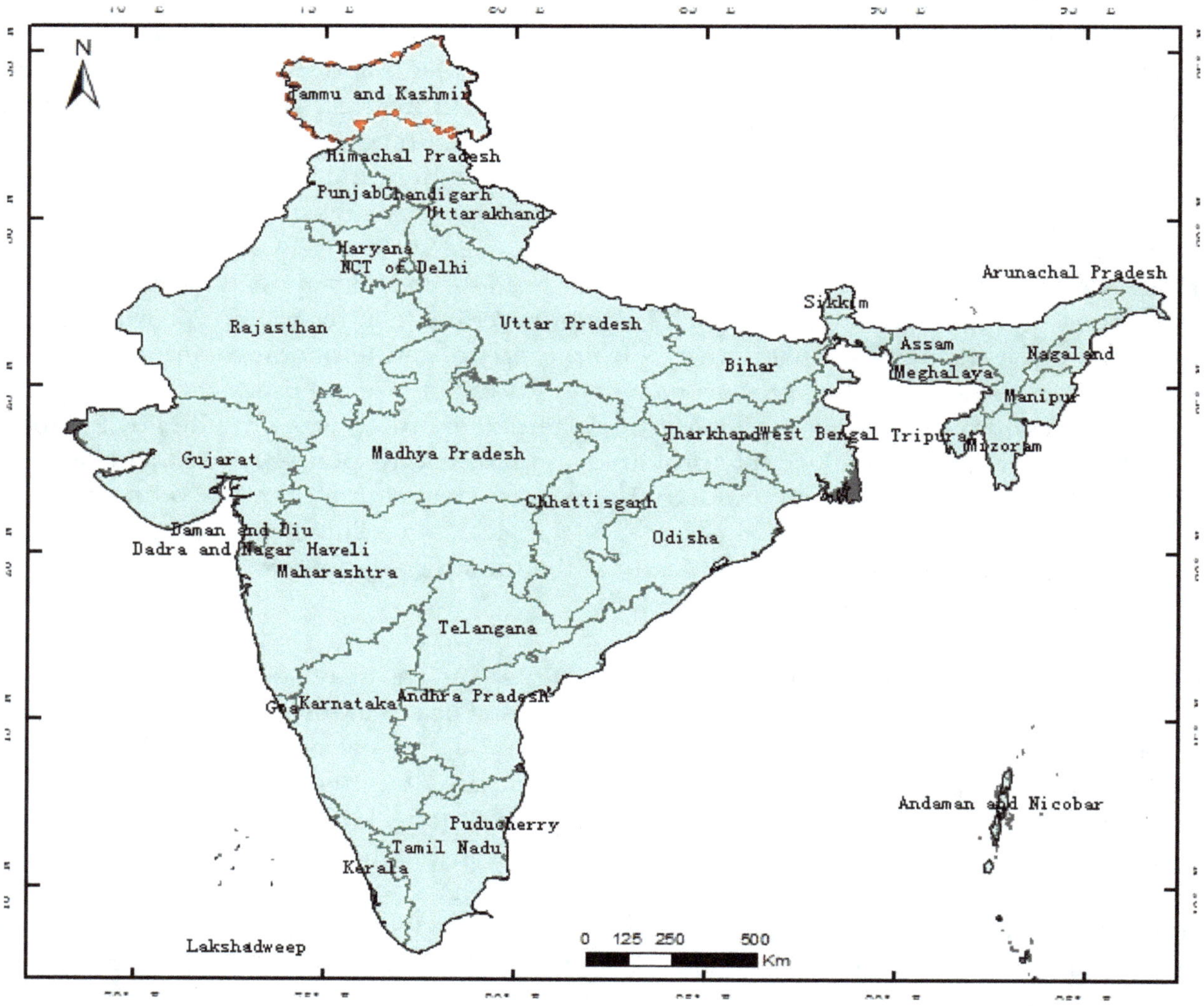

Figure 1. The 36 States and Union Territories of India.

2.2. Data Sources

2.2.1. VIIRS Active Fire/hotspot Data

In this study, the VNP14IMG data were selected as input data for the evaluation of the distribution of heavy industrial heat sources in India. This product is based on reprocessed nominal-resolution Collection 1 data from the NASA Land Science Investigator Processing System (Land-SIPS) [20]. Using the MOD14/MYD14 algorithm, several modifications were implemented to accommodate the unique characteristics associated with the VIIRS 375-m data [25]. The newly improved 375-m data, compared to the traditional coarser-resolution (≥ v1 km) fire products, provide a greater response for fires that cover relatively small areas and improved mapping of large fire perimeters. So, it is well suited to support fire management as well as to meet other scientific applications' needs. VNP14IMG data (19 January 2012 to now) can be freely obtained from the Fire Information for Resource Management System (FIRMS) [26]. Three million nine hundred ninety-eight thousand four hundred sixty-five observed Indian fire hotspots, ranging from 19 January 2012 to 31 December 2018, were used in this paper, and their spatial density is shown in Figure 2.

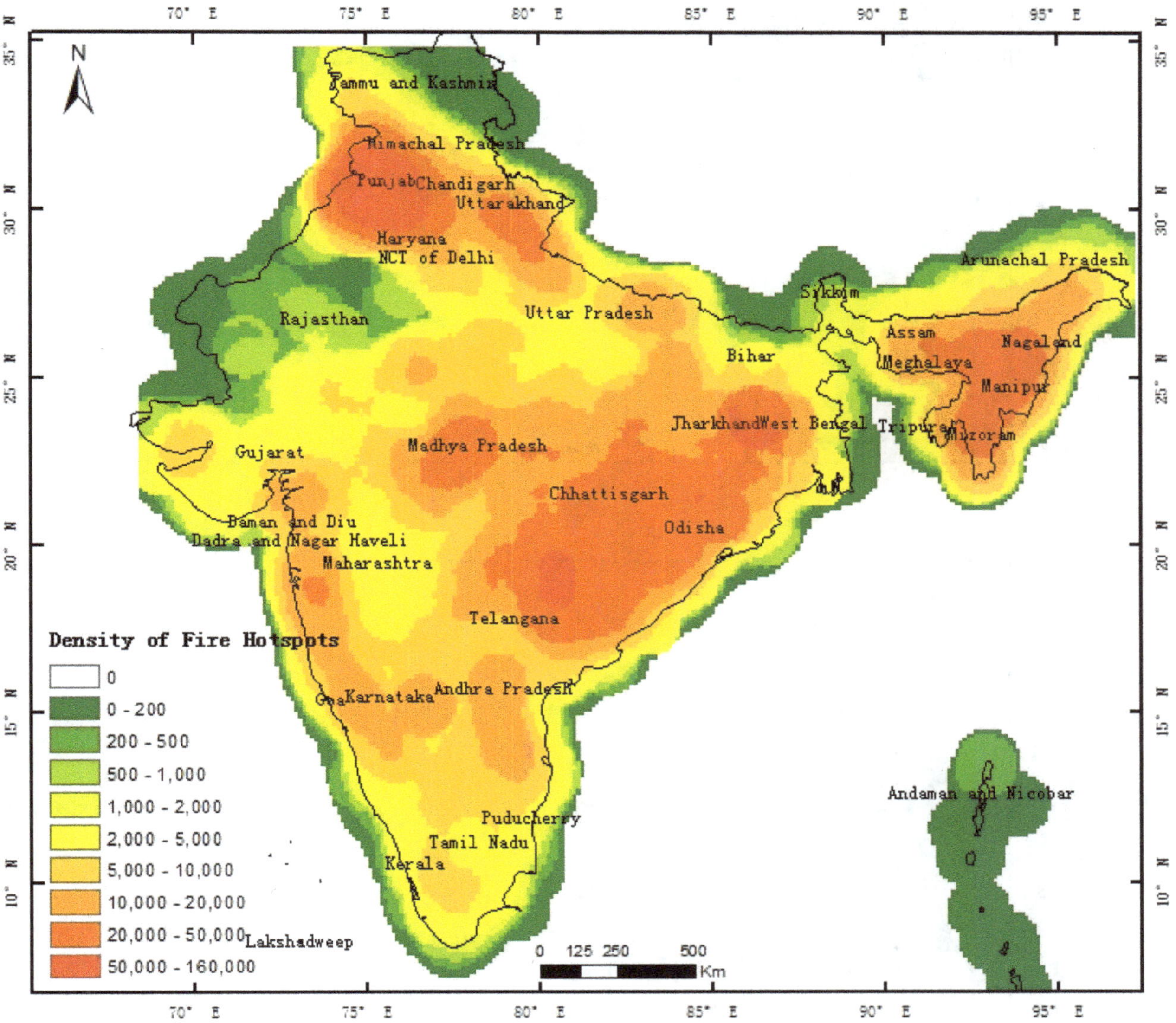

Figure 2. Spatial density of the 3,998,465 fire hotspots in Indian regions (including Jammu and Kashmir state).

VIIRS Nightfire product (VNF), using Day/Night Band (DNB), near-infrared (M7 and M8), short-wave infrared (M10), and mid-wave infrared (M12 and M13) to detect subpixel heat sources, has been used in gas [27] and industrial heat sources detection [11]. So, VNF data were downloaded from the Earth's Observation Group (EOG) [26]. Their spatial distribution maps from VNF data and VNP14IMG data were made to compare in the study area (Figure 3) on 01/01/2018. It showed that VNP14IMG data were quite abundant in India. The fire/hotspot number of VNP14IMG was more tban five times than VNF. Its spatial distribution range was also bigger than the VNF data. So, VNP14IMG data were used lastly to detect heavy industries.

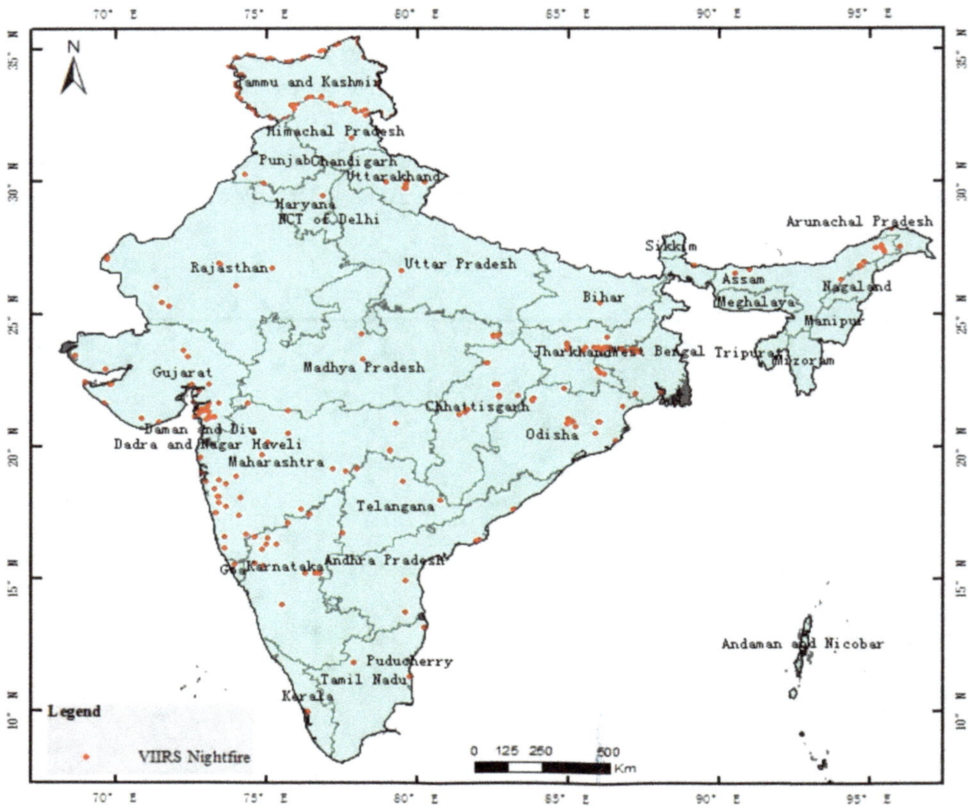

(a)

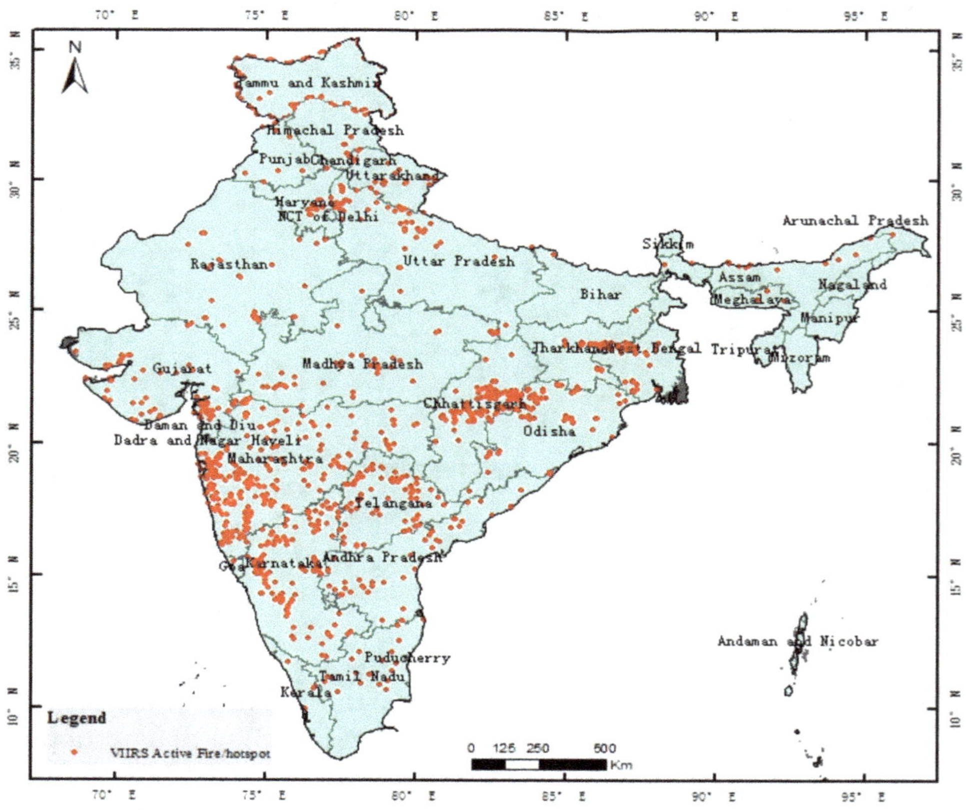

(b)

Figure 3. Spatial distribution comparison of VIIRS Nightfire product (VNF) and NPP VIIRS active fire product (VNP14IMG) on 01/01/2018. (**a**) The spatial distribution comparison of 301 VNF on 01/01/2018. (**b**) The spatial distribution comparison of the 1760 VNP14IMG hotspot.

2.2.2. NPP–VIIRS Night-time Light Data

NPP–VIIRS night-time light (NTL) data were also used in this study. Compared with the Defense Meteorological Satellite Program/Operational Linescan System (DMSP/OLS) data, the night-time light data had a higher spatial resolution (15 arc-seconds, about 750 m) and a wider radiometric detection range [22,27]. These data can be obtained from NOAA's National Centers for Environmental Information (NOAA/NCEI) website [28]. However, as it is a preliminary product, these data are not filtered to remove detected light associated with gas flares, fires, volcanoes, or aurorae, and the dataset has not been processed to remove background noise [29]. In addition, the VIIRS annual night-time light data are being discontinued by NOAA, and only annual data from 2015 and 2016 are supported [30]. Therefore, the 'Flint' annual data were also obtained from the Chinese Academy of Sciences [31]. These data are not affected by fires, volcanoes, and background noise as they have been through statistical cleaning and average noise reduction preprocessing. Therefore, these annual products can be considered as the surface light, and 'Flint' version Beta 1 [32] was used in this study. This 'Flint' imagery consists of 15 arc-second grids, spanning the range −180 to 180 degrees longitude and from −65 to 75 degrees latitude. The digital pixel numbers (DN) range from 0–255. The 'Flint' India light data for India in 2018 is shown below as Figure 4.

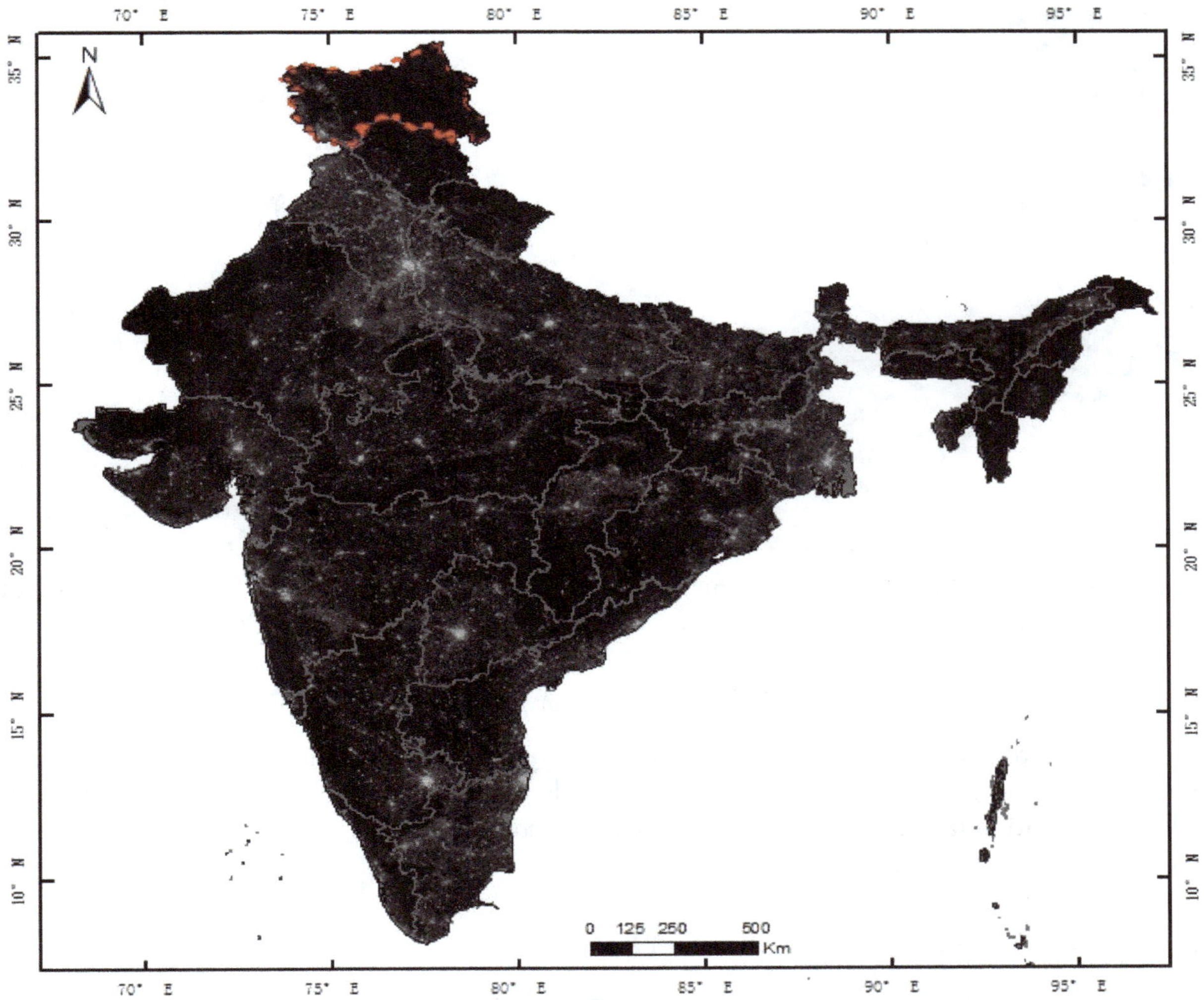

Figure 4. 'Flint' India light data for 2018 (including Jammu and Kashmir state).

2.2.3. Auxiliary Data

Indian national, state, and taluk boundaries were acquired from the Global Administrative Areas (GADM) provided by the Center for Spatial Sciences at the University of California, Davis [33].

The latest version (version 3.6, released on 6 May 2018) was used. The coordinate reference system based on the WGS84 datum was adopted for the boundary files. In order to support the verification of heavy industry heat sources in India, high-resolution images from Google Earth were also utilized in this paper.

2.3. Data Preprocessing

The size of the long-term time series of active fire/hotspot data was huge, and the 'Flint' data consisted of global data; therefore, some preprocessing work was necessary for this study. In order to obtain information about heavy industry heat sources in India, the VNP14IMG and NTL data were processed, as shown below (Figure 5). This processing consisted of two main parts: data preprocessing and a heavy industry heat source detection model.

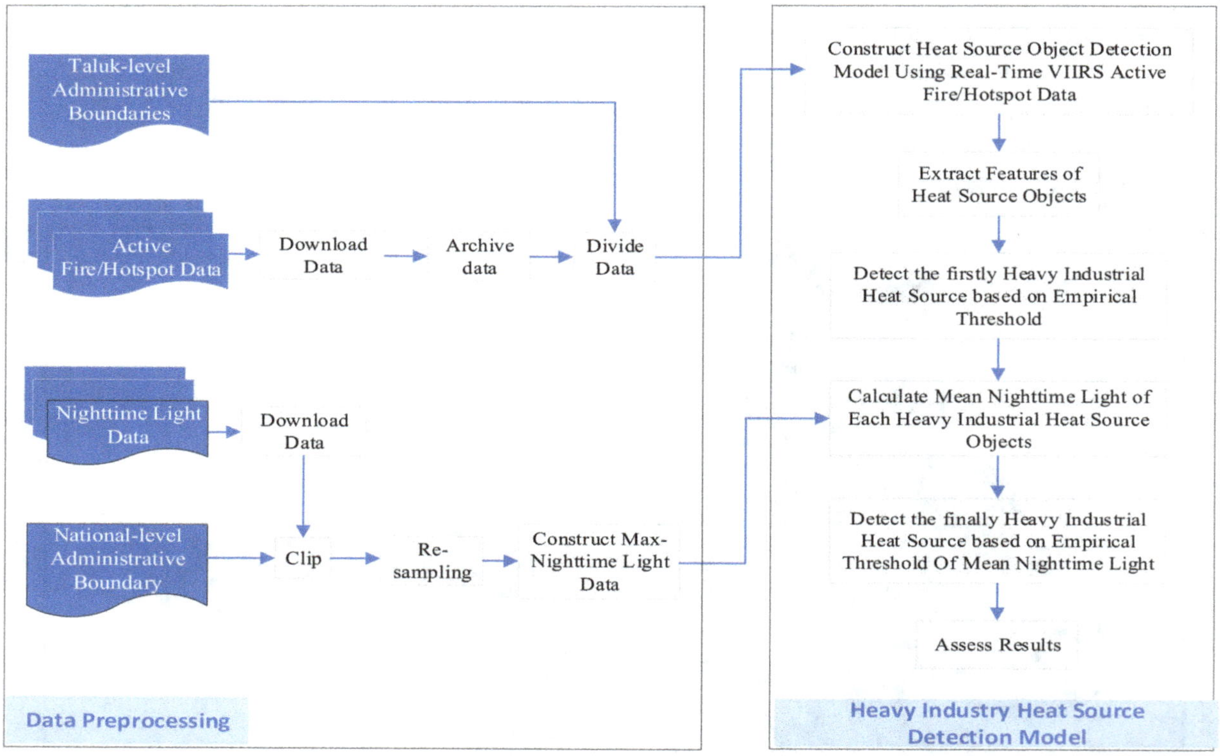

Figure 5. The architecture of the heavy industry heat source detection model using the active fire/hotspot data and night-time light data for India.

2.3.1. NPP-VIIRS Active Fire/Hotspot Data Preprocessing

For the same reason in a previous paper [2], the long-term time series of VNP14IMG products was also needed to be divided. It was almost impossible to divide one area of heavy industry into two or more administrative taluks in India. So, according to the taluk-level administrative boundaries, the 3,998,465 fire hotspots were then divided according to the taluk-level administrative boundaries.

2.3.2. Preprocessing of NPP-VIIRS Night-time Light Data

For most heavy industrial production activities, the use of lighting is also necessary. Therefore, superimposed light data can be used to verify industrial heat sources and filter out false ones. Also, due to economic problems or policy decisions, including regional plans and environmental protection policies, only a small fraction of large, heavy enterprises worked continuously between 2012 and 2018: most enterprises operated for only a few years or months. Thus, some preprocessing of the data was needed. The main processing step was as follows.

Step 1: The annual and global 'Flint' night-time light data were clipped according to the Indian national boundary to obtain annual Indian 'Flint' night-time light data.

Step 2: The annual Indian 'Flint' night-time light data were re-sampled from 750m to 375 m in order to maintain the same spatial resolution as the NP14IMG products.

Step 3: Maximum night-time light data were produced by selecting the maximum value from the annual Indian night-time light data for 2012 to 2018.

2.4. *Heavy Industry Heat Source Detection Model*

In this study, we propose an Indian heavy industry heat source detection model that uses VNP14IMG and NTL data. This model consists of six parts: constructing the heat source object detection model using real-time VNP14IMG data, extracting the hot features of the heat source objects, detecting the initial heavy industrial heat sources based on an empirical threshold, calculating the mean night-time light value for each heavy industrial heat source object, detecting the final heavy industrial heat sources based on the empirical threshold for the mean night-time light, and, finally, assessing the results. Details of the model are described in this section.

Step 1: Static and persistent industrial heat sources in the VNP14IMG time series were found to be concentrated around the hot centers due to the stability of their positions and temporal consistency. The heat source object detection model that used long-order VNP14IMG data based on an improved adaptive K-means algorithm was then implemented [2].

Step 2: Extraction of the hot features of heat source objects. In this study, geometric, statistical, and heat source attribute features were used. The central point of the heat source, as well as the width and the height of the max-circumscribed rectangle, were used as the geometric features. For the statistical feature extraction, the number of fires/hotspots, the density of fires/hotspots per unit area, the initial and final detection times of the heat source object, and the mean and variance of the time interval sorted by date were adopted. For the heat source attributes, the minimum, maximum, mean, and variance attribute information of the VIIRS I-4 band brightness temperature (bright_ti4), the I-5 band brightness temperature (bright_ti4), scan direction pixel size (scan), track direction pixel size (track), and fire point radiation Power (FRP) were extracted for each heat source object.

Step 3: Heavy industrial heat source objects are static and persistent, whereas biomass fires are usually sparsely distributed. The initial heavy industrial heat source identification was based on an empirical threshold [2]. Subsequently, the initial heavy industry heat sources were identified from heat-source objects.

Step 4: Once the initial vector data of the initial heavy industry heat sources had been registered to the raster data of the max night-time light data using the same WGS84 projection, the mean night-time light value was calculated for each initial heavy industrial heat source object.

Step 5: The final detection of the heavy industry heat sources was carried out by applying the empirical threshold algorithm to the mean night-time light data.

Step 6: Assessment of results. The number of working heavy industry heat sources (NWH), the total number of fire hotspots for each working heavy industry heat source area (NFHWH), as well as *Slope_NWH* and *Slope_NFHWH* [2], were used to analyze the distribution of the heavy industry heat sources in different statistical areas for different years.

3. Results and Discussion

3.1. *Heavy Industrial Heat Source Distribution Characteristics at the National Level*

The spatial distribution of 711 heavy industrial heat sources in Indian regions (Figure 6) revealed that heavy industrial heat sources were mainly concentrated in north-east Assam, east-central Jharkhand, the north of Chhattisgarh and Odisha, and coastal areas of Gujarat and Maharashtra. Another interesting phenomenon was that a large number of heavy industrial heat was found lying close to a line between Kolkata on the Eastern Indian Ocean and Mumbai on the Western Indian Ocean. The spatial distribution of the 711 heavy industrial heat sources across India was not the same as that shown by the spatial density distribution image for the 3,998,465 fire hotspots (Figure 2), especially in the case of Punjab and Madhya Pradesh. Further investigation revealed that most of the fire hotspots in Punjab and Madhya Pradesh were due to burning straw, especially in May, October, and November. Additionally, heavy industrial heat sources founded in regions 1, 2, and 3 were mainly connected to petroleum development, whereas in region 4, they were linked to coal mining and steel production. And, each heavy industrial heat source detected were verified using Google Earth Map one by one. Six hundred fifty-nine heat sources can be easily confirmed as heavy industrial factories by Google Earth images. The type of the other 52 results cannot be curtained due to the lack of more field measured data. So, the accuracy of this detection model was higher than 92.7%. As the database of real heavy industrial heat sources has not been obtained, the recall ration can be calculated.

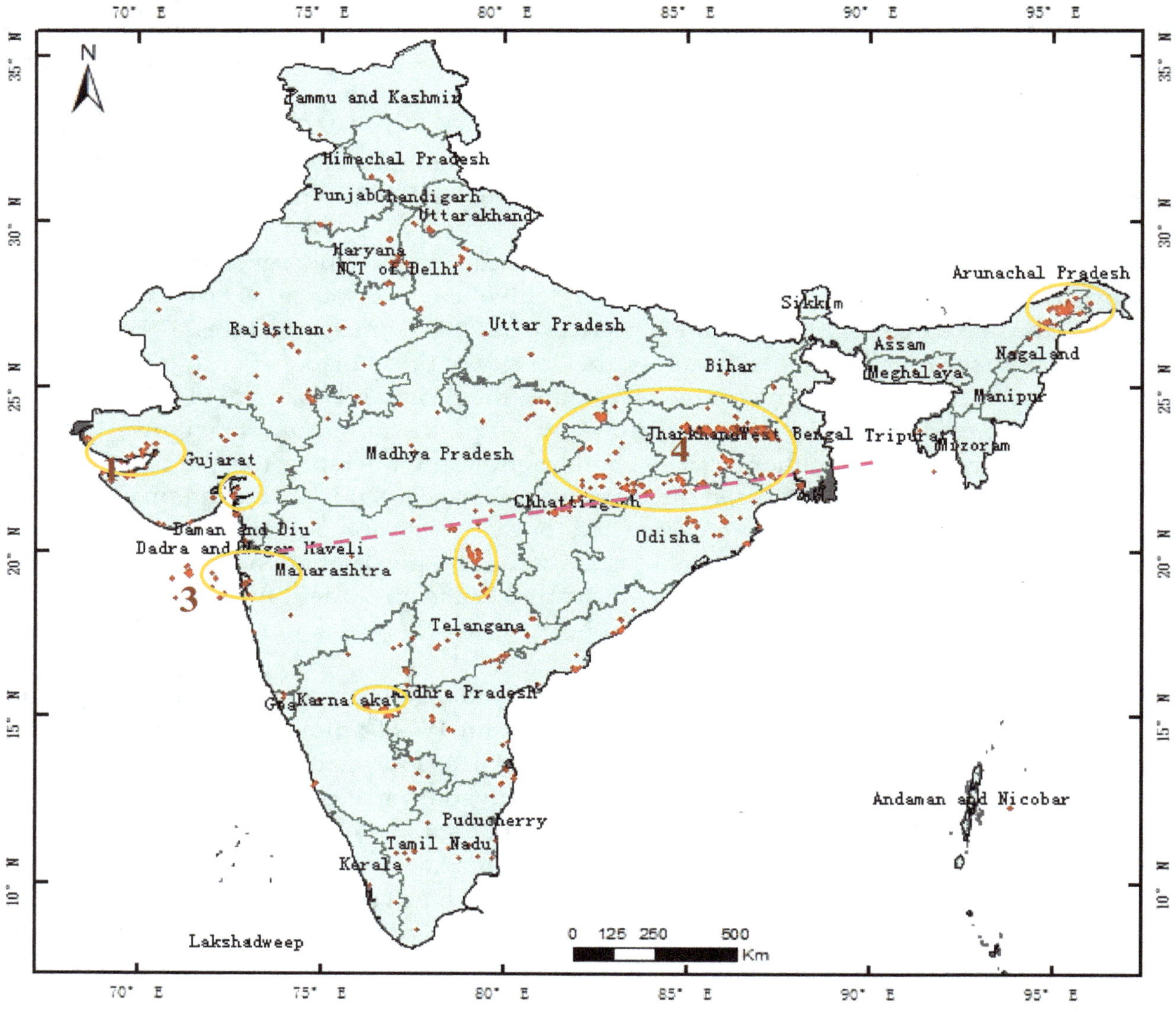

Figure 6. Spatial distribution of 711 heavy industrial heat sources in Indian regions (including Jammu and Kashmir state).

Recent changes in working heavy industry heat sources were compared, and the values of the NWH and NFHWH for each year during the period 2012 to 2018 were calculated (Figure 7). The values of NWH, and in particular, the NFHWH increased during this period. The trends in GDP and total population in India (Figure 8) between 2012 and 2017 were similar [24], demonstrating that heavy industries developed along with the development of the Indian economy as a whole.

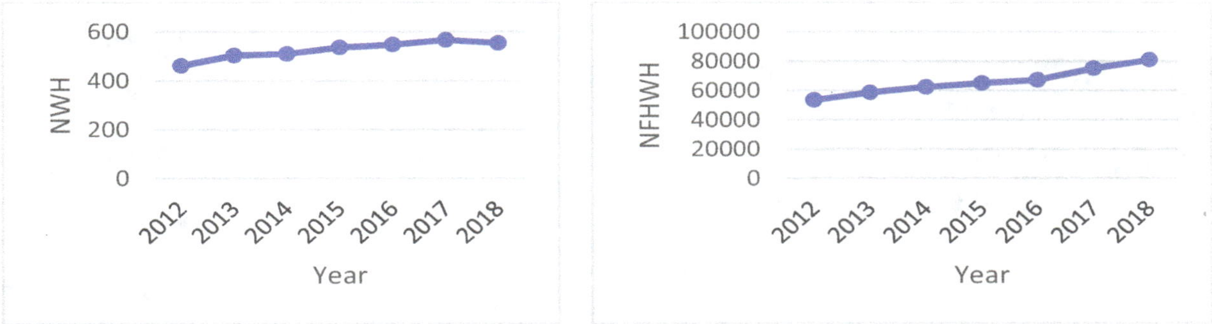

(**a**) (**b**)

Figure 7. Changes for heavy industrial heat sources in India. (**a**) The number of working heavy industrial heat sources (NWH) during the period 2012 to 2018. (**b**) The number of fire hotspots in working heavy industrial heat sources areas (NFHWH) during the period 2012 to 2018.

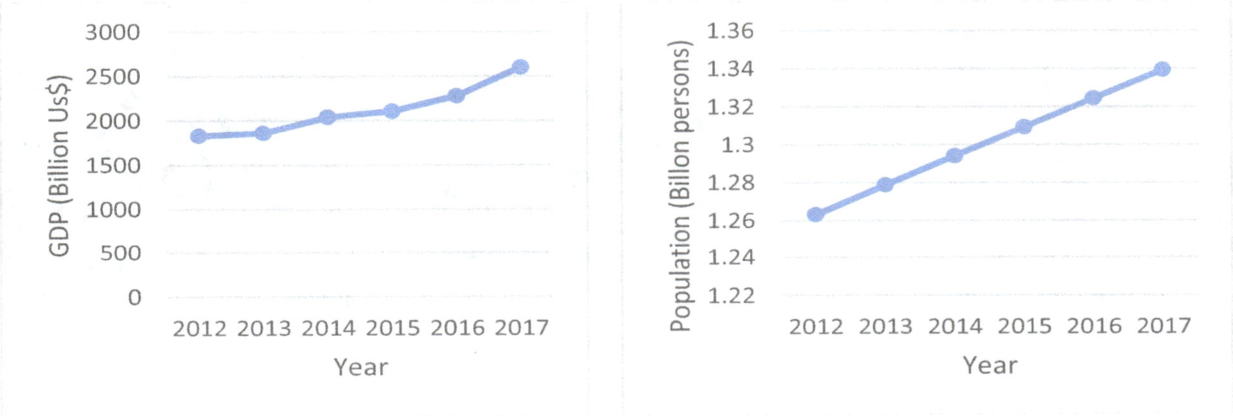

(**a**) (**b**)

Figure 8. Changes in GDP and in the total population of India during the period 2012 to 2017. (**a**) GDP current Billion US$ between 2012 and 2017. (**b**) Population total billion persons between 2012-2017.

High-resolution images from Google Earth (Figure 9) were selected to verify the results of the model. Figure 9a,b are images of steel plants in Jharkhand and West Bengal. The two open-pit minefields shown in Figure 9c,d are located in Jharkhand and Chhattisgarh. Figure 9e–h,j all show facilities related to oil and gas production, processing, and storage in Andhra Pradesh, Gujarat, Rajasthan and Assam, respectively. Figure 9i is an image of cement work named Gagal in Himachal Pradesh.

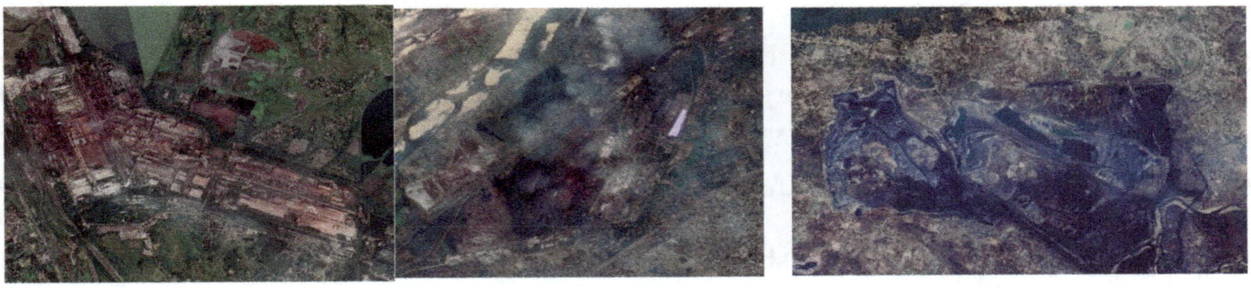

(**a**) steel plant (**b**) steel plant (**c**) open pit mine field

Figure 9. *Cont.*

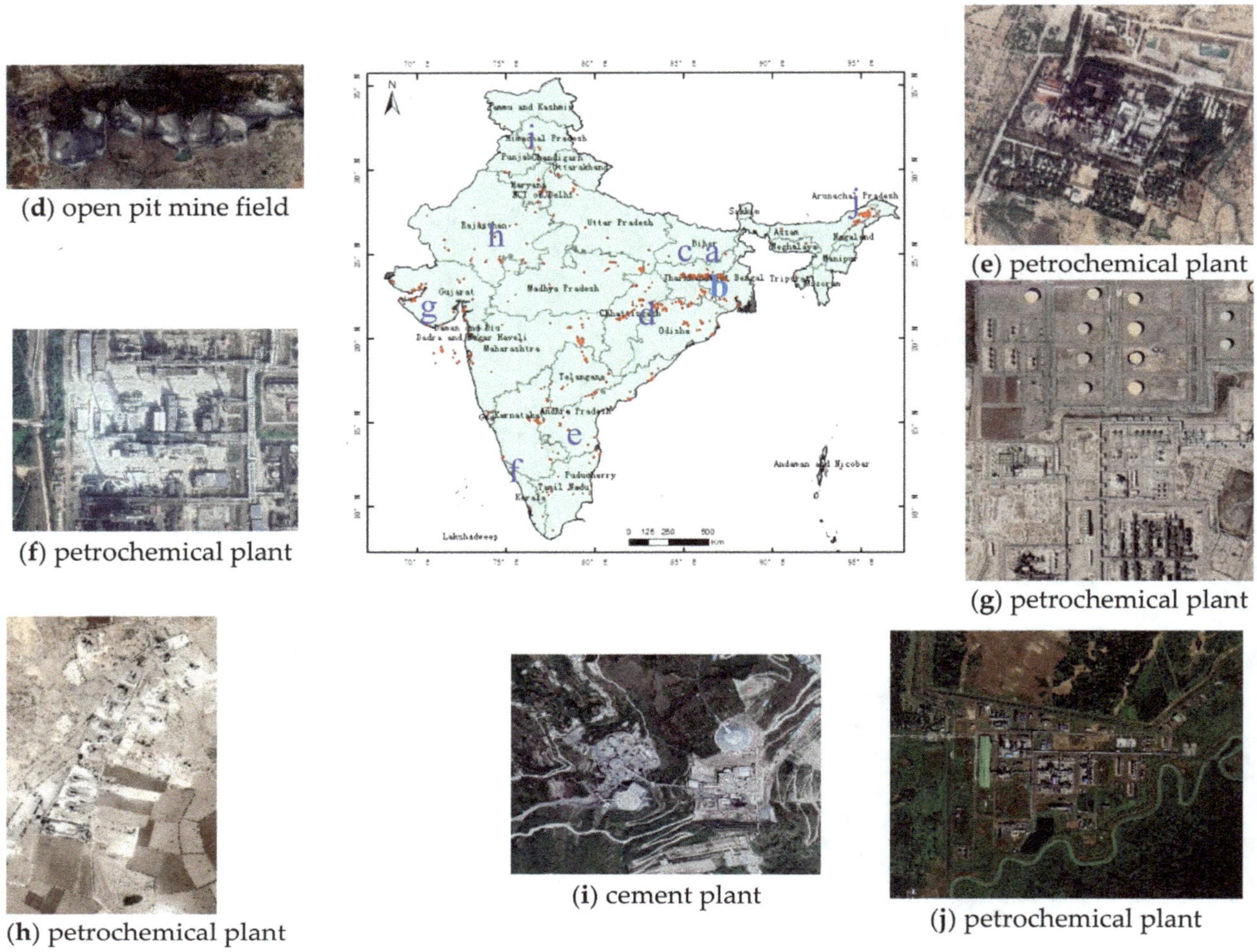

Figure 9. High-resolution imagery used to validate the model.

3.2. Characteristics of Heavy Industrial Heat Source Distribution at the State Level

The heavy industrial heat source characteristics for the 36 states and administrative regions of India were analyzed using the NWH and NFHWH values. The boundaries of the administrative divisions of these states contained land regions only, meaning that no NWH and NFHWH values for sources at sea were included.

Figure 10a shows the NWH for the states with the 20 highest values. These included Jharkhand, Chhattisgarh, and Odisha, followed by Gujarat and West Bengal. Furthermore, the NWH in Jharkhand state accounted for nearly 13.8% of the total for Indian mainland sources. The sum of the NWH in Jharkhand, Chhattisgarh, Odisha, and Gujarat accounted for 43.38% of the total; the total NWH for the top 20 states accounted for 98.37%. In addition, the NWH for Jharkhand, which is one of the richest mineral zones in the world and boasts 40% and 29% of India's mineral and coal reserves, respectively, increased continuously during the period 2012 to 2018 [34]. Chhattisgarh's heavy industry also developed due to its rich natural resources, policy incentives, and good infrastructure [35].

The NFHWH values indicate that there has been a reduction in the total amount of heavy industrial production in the administrative areas studied (Figure 10b). The largest number of fire hotspots was in Jharkhand, the same as the NWH shown in Figure 10a. However, the order of the five highest NFHWH values in 2018 was different: the order was now, Jharkhand, Odisha, Chhattisgarh, West Bengal, and then Gujarat. Moreover, the average NFHWH in Jharkhand between 2012 and 2018 accounted for nearly 29.05% of the Indian mainland total; the total NWH in Jharkhand, Odisha, and

Chhattisgarh accounted for 58.36%; and the total NWH for the top 20 states accounted for 99.78%. In addition, NFHWH values in most states increased continuously after 2012, in line with India's economic development.

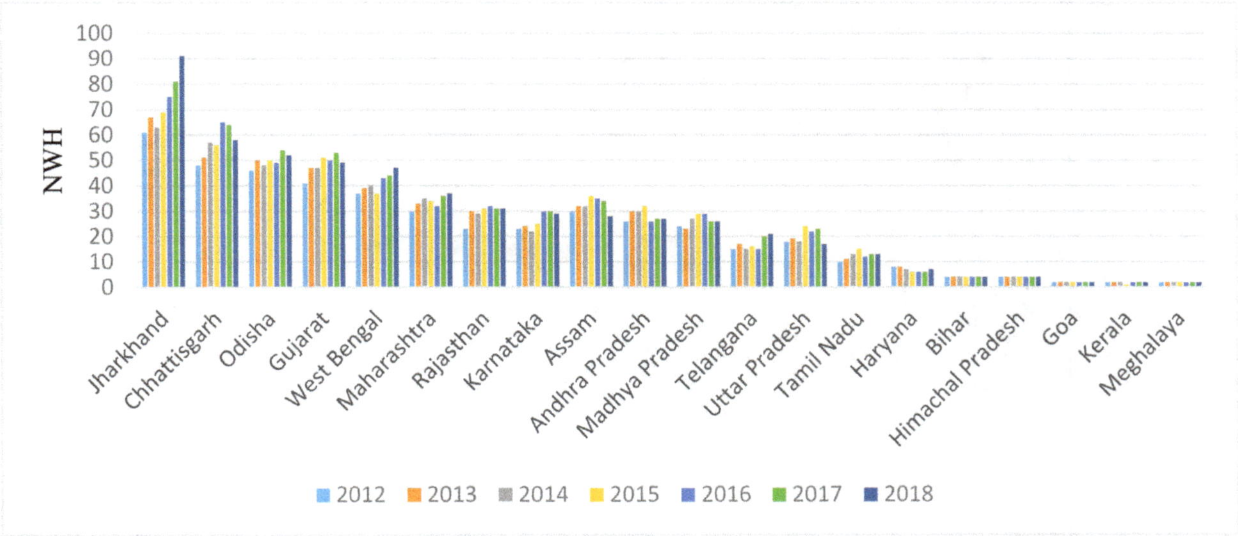

(a)

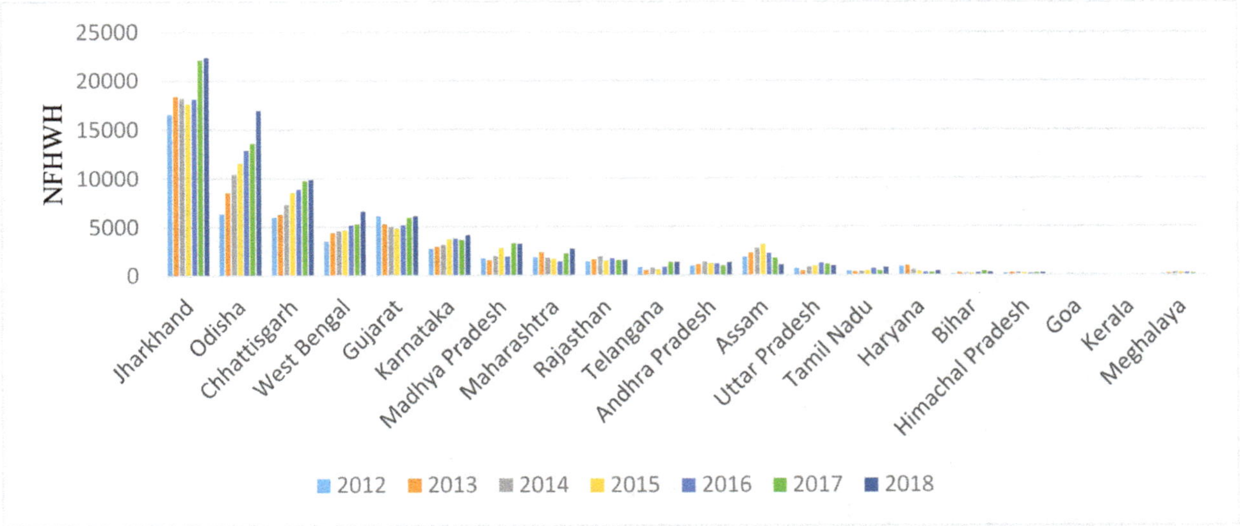

(b)

Figure 10. Changes in heavy industrial heat sources at the state level. (**a**) NWH for the top 20 states between 2012 and 2018. (**b**) NFHWH for the top 20 states between 2012 and 2018.

The distributions of *Slope_NWH* and *Slope_NFHWH* values were mapped to illustrate the changing trends for each statistical area during the seven-year period studied (Figure 11). The largest positive value of *Slope_NWH* was in Jharkhand, followed by Chhattisgarh and Gujarat, indicating that the number of heavy industry heat sources in these three states increased quickly. The smallest negative *Slope_NWH* value was found in Haryana, followed by Tripura and Andhra Pradesh. However, these negative values were all very small −0.49, −0.23, and −0.14, respectively. This means that the downward trends here were very slow. In addition, for 18 of the large states in mainland India, the values were positive and only four states had negative *Slope_NWH* values. Therefore, it can be concluded that the total number of heavy industry heat sources in India increased, as confirmed by Figure 7.

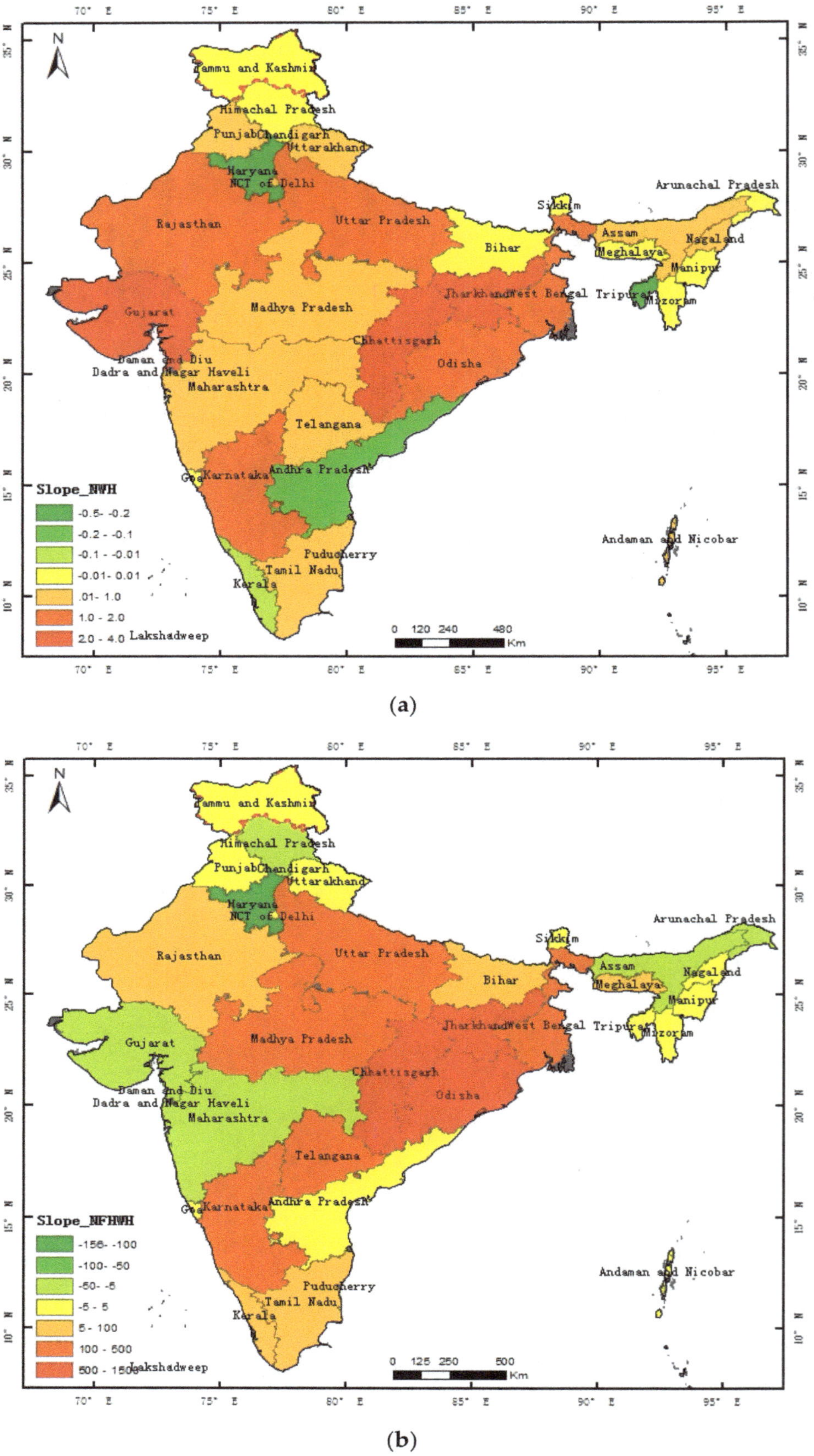

Figure 11. Changes in heavy industry heat sources at the state level from 2012 to 2018 (including Jammu and Kashmir state). (**a**) The *Slope_NWH* values for different states. (**b**) The *Slope_NFHWH* values for different states.

The *Slope_NFHWH* values are displayed in Figure 10b. These values reflect the scale of production associated with working heavy industry heat sources. The largest positive value of *Slope_NFHWH* was the value for Odisha, followed by Chhattisgarh and Jharkhand. The *Slope_NFHWH* value in Odisha was 1452.63 (the corresponding value of *Slope_NWH* was only 1.11), which was double the value in Chhattisgarh. This means that the average scale of working heavy industry heat sources in Odisha increased. The smallest negative *Slope_NWH* values were found in Haryana, followed by Gujarat and Arunachal Pradesh, showing that *Slope_NFHWH* in Gujarat was the second most negative whereas it *Slope_NWH* value was the third highest positive value. This shows that the average scale of the working heavy industry heat sources in this state was declining. It should be noted that this trend related only to the heavy industry heat sources in mainland Gujarat. In addition, there were 19 large states on the Indian mainland for which the *Slope_NFHWH* values were positive and only nine states with negative values. It can be concluded that, overall, the scale of heavy industry heat sources in all of India increased, as supported by the details shown in Figure 6.

The distribution of heavy industry heat sources was then mapped to examine the heat source characteristics for the 36 state administrative regions of India in 2018 (Figure 12). The NWH values for 2018 (Figure 12a) indicate that Jharkhand, Chhattisgarh, and Odisha have relatively large numbers of heavy enterprises, with Gujarat following. In terms of NFHWH values (Figure 12b), Chhattisgarh is followed by Jharkhand and Odisha. In addition, both NWH and NFHWH values are highest in east-central India, followed by central India; in contrast, most of the northwest and south of the country have a small number of heavy industry heat sources and fire hotspots caused by heavy enterprises.

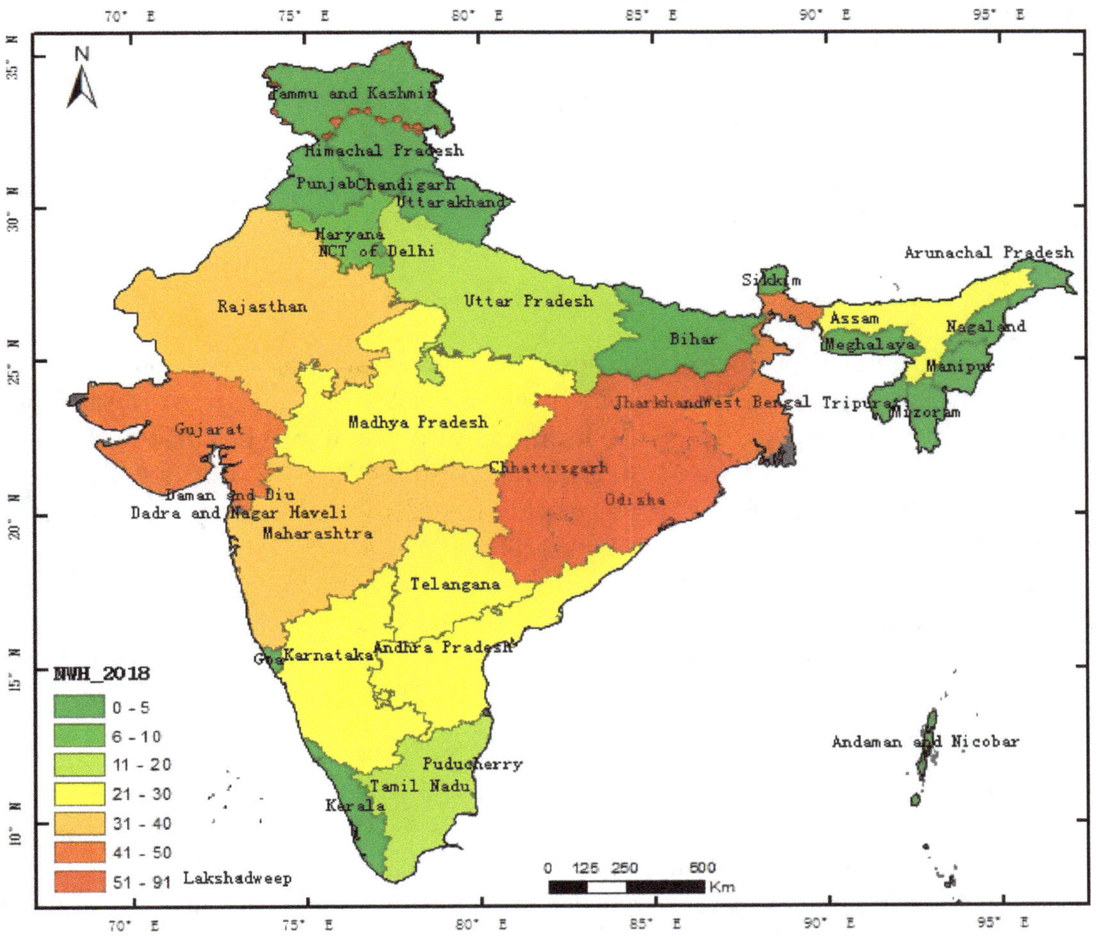

(a)

Figure 12. *Cont.*

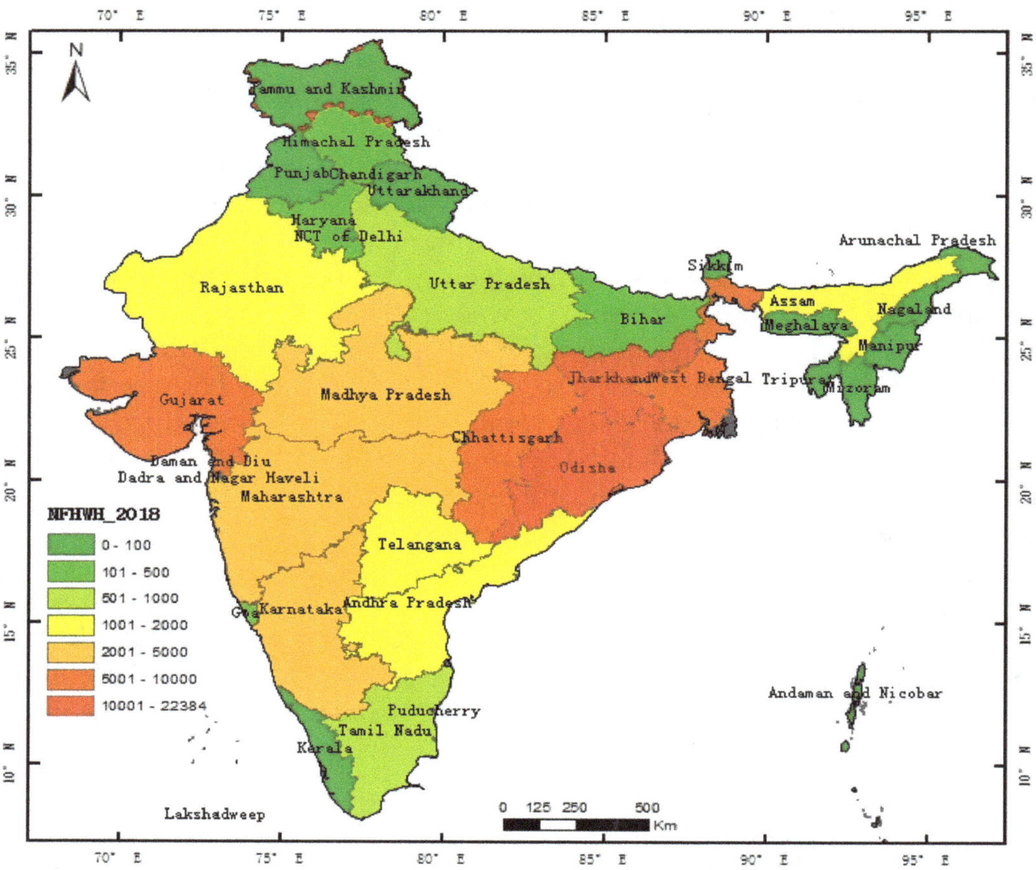

(b)

Figure 12. Distribution of heavy industry heat sources at the state level (2018) (including Jammu and Kashmir state). (**a**) The NWH values for different states (2018). (**b**) The NFHWH values for different states (2018).

4. Conclusions

India has now emerged as a global player with one of the fastest-growing major economies and is considered a newly industrialized country. Its heavy industry has grown rapidly in the past few decades. This has exacerbated pressures on the Indian environment and has also had a great impact on the world economy. The NASA's Land-SIPS VIIRS 375-m active fire product (VNP14IMG) and NPP-VIIRS night-time light data (NTL) can objectively reveal the spatiotemporal patterns of heavy industrial development in the study area. We, therefore, proposed a heavy industry heat source detection model that uses VNP14IMG and NTL. The spatial distribution and trends for heavy industry heat sources were analyzed for India at the national and state levels. The results suggest that the model is an accurate and effective means of monitoring heat sources produced by heavy industry. The accuracy of this detection model was higher than 92.7%. The following conclusions can be drawn from this study.

(1) Overall, heavy industry heat sources were found to be mainly concentrated in the north-east Assam state, ease central Jharkhand, north Chhattisgarh, and Odisha, and the coastal areas of Gujarat and Maharashtra. It is also interesting to note that a large number of heavy industrial heat sources were found concentrated around a line between Kolkata on the Eastern Indian Ocean and Mumbai on the Western Indian Ocean.

(2) The total NWH and NFHWH values for India increased throughout the period studied, especially in the case of the NFHWH. These trends were similar to those for the GDP and total population of India (Figure 7) between 2012 and 2017.

(3) The largest values of NWH and NFHWH were in Jharkhand, Chhattisgarh, and Odisha. The two largest values of *Slope_NWH* were in Jharkhand and Chhattisgarh. The smallest negative values of *Slope_NWH* and *Slope_NFHWH* were in Haryana. In addition, the *Slope_NFHWH* value for mainland Gujarat was the second most negative value, whereas it's *Slope_NWH* was the third highest positive one.

The results of this study suggest that real-time VIIRS active fire/hotspot data and NPP-VIIRS night-time light data can successfully be used for monitoring Indian heavy industrial economic development. This could be beneficial for Indian policy-makers and heavy industry regulation. Future studies should focus on distinguishing biomass fires/hotspots from other fires/hotspots, which would allow the monitoring of biomass burning related to agriculture and forest fires. Finally, we plan to add much more fire data from different satellite sensors in order to improve temporary and spatial resolutions.

Author Contributions: F.C. and J.Y. conceived and designed the experiments; Y.M. built the experimental platform, and prepared and processed the remote sensing data; C.M. designed and performed the experiments, analyzed the data and wrote the paper; J.L. and Z.N. supervised the research, and also gave comments and revised the manuscript.

Acknowledgments: The VNP14IMG was downloaded from the FIRMS website. The authors thank the editors and the three anonymous reviewers for their valuable comments that helped to improve our manuscript.

References

1. The World Bank. World Bank Open Data. Available online: https://data.worldbank.org/country/russian-federation (accessed on 9 December 2019).
2. Ma, C.; Yang, J.; Chen, F.; Ma, Y.; Liu, J.; Li, X.; Duan, J.; Guo, R. Assessing Heavy industry heat source Distribution in China Using Real-Time VIIRS Active Fire/Hotspot Data. *Sustainability* **2018**, *10*, 4419. [CrossRef]
3. Tong, D.; Zhang, Q.; Davis, S.J.; Liu, F.; Zheng, B.; Geng, G.; Xue, T.; Li, M.; Hong, C.; Lu, Z.; et al. Targeted emission reductions from global super-polluting power plant units. *Nat. Sustain.* **2018**, *1*, 59–68. [CrossRef]
4. Zhou, Y.; Zhao, F.; Wang, S.; Liu, W.; Wang, L. A Method for Monitoring Iron and Steel Factory Economic Activity Based on Satellites. *Sustainability* **2018**, *10*, 1935. [CrossRef]
5. BP (British Petroleum). BP Home-Page. Available online: https://www.bp.com/ (accessed on 9 December 2019).
6. International Energy Agency (IEA). Shaping a Secure and Sustainable Energy Future for All. Available online: https://www.iea.org/#statistics-data (accessed on 9 December 2019).
7. MEIC. Global Power Emissions Database(GPED). Available online: http://meicmodel.org/dataset-gped.html (accessed on 9 December 2019). (In Chinese).
8. Lu, Z.; Streets, D.G. Increase in NOx emissions from Indian thermal power plants during 1996–2010: Unit-based inventories and multisatellite observations. *Environ. Sci. Technol.* **2012**, *46*, 7463–7470. [CrossRef]
9. Deilami, B.R.; Ahmad, B.B.; Saffar, M.R.A.; Umar, H.Z. Review of change detection techniques from remotely sensed images. *Res. J. Appl. Sci. Eng. Technol.* **2015**, *10*, 221–229.
10. Roy, M.; Ghosh, S.; Ghosh, A. A novel approach for change detection of remotely sensed images using semi-supervised multiple classifier system. *Inf. Sci.* **2014**, *269*, 35–47. [CrossRef]
11. Liu, Y.; Hu, C.; Zhan, W.; Sun, C.; Murch, B.; Ma, L. Identifying industrial heat sources using time-series of the VIIRS Nightfire product with an object-oriented approach. *Remote Sens. Environ.* **2018**, *204*, 347–365. [CrossRef]
12. Paltridge, G.W.; Barber, J. Monitoring grassland dryness and fire potential in Australia with NOAA/AVHRR data. *Remote Sens. Environ.* **1998**, *25*, 381–394. [CrossRef]
13. Morisette, J.T.; Giglio, L.; Csiszar, I.; Justice, C.O. Validation of the MODIS active fire product over Southern Africa with ASTER data. *Int. J. Remote Sens.* **2005**, *26*, 4239–4264. [CrossRef]

14. Zhao, W. Research and evaluation of the algorithm of land surface fire detection based on FY3-VIRR data. *Fire Saf. Sci.* **2011**, *3*, 004. [CrossRef]
15. Vogeler, J.C.; Yang, Z.; Cohen, W.B. Mapping post-fire habitat characteristics through the fusion of remote sensing tools. *Remote Sens. Environ.* **2016**, *173*, 294–303. [CrossRef]
16. Schroeder, W.; Oliva, P.; Giglio, L.; Quayle, B.; Lorenz, E.; Morelli, F. Active fire detection using Landsat-8/OLI data. *Remote Sens. Environ.* **2016**, *185*, 210–220. [CrossRef]
17. Trifonov, G.M.; Zhizhin, M.N.; Melnikov, D.V.; Poyda, A.A. VIIRS Nightfire Remote Sensing Volcanoes. *Procedia Comput. Sci.* **2017**, *119*, 307–314. [CrossRef]
18. Baugh, K. Characterization of Gas Flaring in North Dakota using the Satellite Data Product, VIIRS Nightfire. In Proceedings of the AGU Fall Meeting 2015, San Francisco, CA, USA, 14–18 December 2015.
19. Sun, J.Q.; Liu, Y.X.; Dong, Y.Z. Classification of Urban Industrial Heat Sources Based on Suomi-NPP VIIRS Night-time Thermal Anomaly Products: A Case Study of the Beijing-Tianjin-Hebei Region. *Geogr. Geo-Inf. Sci.* **2018**, *34*, 13–19.
20. Schroeder, W.; Oliva, P.; Giglio, L.; Csiszar, I. The new VIIRS 375 m activefire detection data product: Algorithm description and initial assessment. *Remote Sens. Environ.* **2014**, *143*, 85–96. [CrossRef]
21. Giglio, L.; Schroeder, W.; Justice, C.O. The collection 6 MODIS active fire detection algorithm and fire products. *Remote Sens. Environ.* **2016**, *178*, 31–41. [CrossRef]
22. Dai, Z.; Hu, Y.; Zhao, G. The Suitability of Different Night-time Light Data for GDP Estimation at Different Spatial Scales and Regional Levels. *Sustainability* **2017**, *9*, 305. [CrossRef]
23. Wu, W.; Zhao, H.; Jiang, S. A Zipf's Law-Based Method for Mapping Urban Areas Using NPP-VIIRS Night-time Light Data. *Remote Sens.* **2018**, *10*, 130. [CrossRef]
24. The World Bank. The World Bank In India. Available online: http://www.worldbank.org/en/country/india (accessed on 9 December 2019).
25. Liu, F.; Zhang, Q.; Tong, D.; Zheng, B.; Li, M.; Huo, H.; He, K.B. High-resolution inventory of technologies, activities, and emissions of coal-fired power plants in China from 1990 to 2010. *Atmos. Chem. Phys.* **2015**, *15*, 13299–13317. [CrossRef]
26. VIIRS I-Band 375 m Active Fire Data. Available online: https://earthdata.nasa.gov/earth-observation-data/near-real-time/firms/viirs-i-band-active-fire-data (accessed on 18 April 2019).
27. Elvidge, C.D.; Zhizhin, M.; Baugh, K.; Hsu, F.C.; Ghosh, T. Extending Nighttime Combustion Source Detection Limits with Short Wavelength VIIRS Data. *Remote Sens.* **2019**, *11*, 395. [CrossRef]
28. VIIRS DNB Nighttime Imagery. Available online: https://maps.ngdc.noaa.gov/viewers/VIIRS_DNB_nighttime_imagery/index.html (accessed on 9 December 2019).
29. Zhu, X.; Ma, M.; Yang, H.; Ge, W. Modeling the Spatiotemporal Dynamics of Gross Domestic Product in China Using Extended Temporal Coverage Night-time Light Data. *Remote Sens.* **2017**, *9*, 626. [CrossRef]
30. Version 1 VIIRS Day/Night Band Nighttime Lights. Available online: https://eogdata.mines.edu/download_dnb_composites.html (accessed on 18 April 2019).
31. The Chinese Academy of Sciences version of the Earth Luminous Data Set (codenamed "Flint") Provides Annual Data Download Service. Available online: https://www.jianshu.com/p/5fde55a4d267?tdsourcetag=s_pcqq_aiomsg (accessed on 9 December 2019).
32. NPP_NIGHT_LIGHT. Available online: https://pan.baidu.com/s/17UqS7P66_6AMdr-a4sfUXA#list/path=%2F (accessed on 9 December 2019).
33. GADM Data. Available online: https://gadm.org/data.html (accessed on 18 April 2019).
34. India Brand Equity Foundation (IBEF). About Jharkhand: Information on Mining Industries, Economy, Agriculture & Geography. Available online: https://www.ibef.org/states/jharkhand.aspx (accessed on 9 December 2019).
35. Industrial Development & Economic Growth in Chhattisgarh. Available online: https://www.ibef.org/industry/chhattisgarh-presentation (accessed on 18 April 2019).

9

The Efficacy Analysis of Determining the Wooded and Shrubbed Area based on Archival Aerial Imagery using Texture Analysis

Przemysław Kupidura *, Katarzyna Osińska-Skotak, Katarzyna Lesisz and Anna Podkowa

Department of Photogrammetry, Remote Sensing and Spatial Information Systems, Faculty of Geodesy and Cartography, Warsaw University of Technology, 1, 00-661 Warsaw, Poland; katarzyna.osinska-skotak@pw.edu.pl (K.O.-S.); lesisz.katarzyna@gmail.com (K.L.); anna.podkowa.dokt@pw.edu.pl (A.P.)

* Correspondence: przemyslaw.kupidura@pw.edu.pl.

Abstract: Open areas, along with their non-forest vegetation, are often threatened by secondary succession, which causes deterioration of biodiversity and the habitat's conservation status. The knowledge about characteristics and dynamics of the secondary succession process is very important in the context of management and proper planning of active protection of the Natura 2000 habitats. This paper presents research on the evaluation of the possibility of using selected methods of textural analysis to determine the spatial extent of trees and shrubs based on archival aerial photographs, and consequently on the investigation of the secondary succession process. The research was carried out on imagery from six different dates, from 1971 to 2015. The images differed from each other in spectral resolution (panchromatic, in natural colors, color infrared), in original spatial resolution, as well as in radiometric quality. Two methods of textural analysis were chosen for the analysis: Gray level co-occurrence matrix (GLCM) and granulometric analysis, in a number of variants, depending on the selected parameters of these transformations. The choice of methods has been challenged by their reliability and ease of implementation in practice. The accuracy assessment was carried out using the results of visual photo interpretation of orthophotomaps from particular years as reference data. As a result of the conducted analyses, significant efficacy of the analyzed methods has been proved, with granulometric analysis as the method of generally better suitability and greater stability. The obtained results show the impact of individual image features on the classification efficiency. They also show greater stability and reliability of texture analysis based on granulometric/morphological operations.

Keywords: secondary succession monitoring; Natura 2000 threats; tree detection; archival photographs; spectro-textural classification; granulometric analysis; GLCM

---

1. Introduction

In Poland, since the 1990s, a gradual cessation of agricultural use, hence the secondary succession process, has been observed in some areas. This is also observed in valuable natural areas, including Natura 2000 habitats, and results in deterioration of biodiversity and the habitat's conservation status. This process can have different characteristics and dynamics in different areas [1]. The course of succession is influenced by, e.g., the state of abandoned land (e.g., black fallow, mowed meadow, stubble), soil fertility in nutrients, recently grown plants, the location of the ground (including proximity to the forest), slope, and insolation [2–4]. This knowledge plays an important role in proper planning of active protection of the Natura 2000 habitats.

The secondary succession process can now be effectively monitored using LiDAR (Light Detection and Ranging) data [5,6] and hyperspectral imagery (detection of succession species) [7].

However, learning about the dynamics of this phenomenon in the past (e.g., in the 1970s, 1980s, and 1990s of the 20th century) is only possible using archival materials—mainly aerial photographs, which are often the only reliable source of information about the land cover. The most common method of obtaining ranges of trees and shrubs based on archival aerial photographs is photo interpretation [8–13]. Sometimes, stereodigitalization is also used [14–18]. However, these methods are labor-intensive and time-consuming, which is why methods of automation of this process are sought. Sometimes, the classic spectral classification of the image is used [19], but in the case of archival images (often black and white), it is not always possible. Another possibility for the study of the succession process of trees and shrubs is to use the technique of dense image matching (DIM) [20]. However, according to current research in this area [20], for the DIM technique to yield good quality results, archival data must meet several conditions: (1) The scale of images should not be less than 1:13 000 for analog photos, or GSD (Ground Sample Distance) should be smaller than 25 cm for digital photos; (2) the date of obtaining the photos should provide a picture of vegetation in full development; and (3) photos should have good radiometric quality, i.e., small image blur, adequate contrast, and no mechanical damage (in the case of analogue photos).

A new approach to the study of the dynamics of the succession process based on archival aerial photographs may be the application of textural analysis of the image.

Texture is one of the most important spatial features of selected terrain coverage classes. Since it does not have an unambiguous mathematical definition, in practice, image processing uses a variety of ad hoc texture analysis methods. The gray level co-occurrence matrix (GLCM) [21–23], fractal analysis [24], discrete wavelet transformation [25], Laplace filters [26–28], Markov Random Fields [29,30], and granulometric analysis [31,32] are several such methods that should be mentioned here. Also, promising attempts have been made to use convolutional neural networks to take spatial features of images into account [33,34].

Wooded and bushy areas are a class with a distinct texture. This is mainly due to the shape of the tree crowns and also to their height. In areas with dense trees/bush coverage, typical grainy texture results from the alternate occurrence of better and worse illuminated crown fragments (due to the orientation of the surface of the tree crown relative to the light source). In areas with less dense coverage, the shadow image of individual trees or shrubs also plays an important role.

Texture is especially important for images of very high resolution. As the size of the pixel (GSD) increases, the clarity of texture decreases, and at the same time, its significance as an interpretive/classification feature does too [35,36].

The texture, although potentially significant in the classification of satellite image content, used alone might not give satisfactory results, because different LULC (Land Use/Land Cover) classes may have similar textures [37]. Therefore, the use of textural features as a complement to spectral features gives a much better result—in the spectro-textural approach. This is indicated by examples of models using different texture analysis methods in combination with spectral data [22–24,33–36,38–43].

Most examples of the spectro-textural approach are based on multispectral data, including near-infrared channels (often also mid-infrared), which are crucial for distinguishing vegetation from other terrain coverage classes, as well as for differentiating types of vegetation. In the case of archival aerial photographs, these type of data are often not available. Evidently, the repository of multispectral satellite data dates back to the 1970s, but these are medium-resolution images, unsuitable for the analysis of textures of wooded areas or for large-scale studies. On the other hand, aerial photographs usually have adequate spatial resolution, while the registered spectral band images are less favorable for texture analysis. Most often, these are images in natural colors or panchromatic (black and white) photographs.

The main objective of the presented research was to analyze the possibility of using archival aerial imagery to determine the wooded and shrubbed areas, using selected methods of texture analysis, and with limited spectral data. Two methods of texture analysis were selected for the study, the effectiveness of which has been proved by previous studies: GLCM [21,23] and granulometric analysis [31–33,36]. In the literature can be found many studies on the use of these methods in the analysis of satellite and aerial images, but there is almost no research on archival aerial photos.

2. Brief Presentation of Selected Methods of Texture Analysis

Here follows a brief presentation of two texture analysis methods used for this research: GLCM and granulometric analysis. Both methods can be used to analyze the whole image, as well as for local analysis for individual pixels. In the first case, we obtain a single value (or a set of values) describing the feature (or features) of the whole image. This may, for example, determine what the image presents (e.g., the dominant type of coverage/land use) [44,45]. The local analysis consists of analyzing a subset of the image defined by a determined neighborhood of a single pixel (e.g., all pixels in the assumed radius). Values obtained in this way are attributed to individual central pixels. This indicates the feature of texture in the environment surrounding each central pixel (defined by the range of the analyzed neighborhood). In practice, this consists of creating new images, where pixels are given values depending on the individual features obtained in the analysis.

2.1. GLCM—Grey Level Co-occurrence Matrix

The GLCM method of texture analysis was first introduced by Julesz [21]. It consists of counting the pairs of neighboring values present in the image (or in its subset) and comparing them in the (grey level co-occurrence) matrix. For the matrix thus created, selected indicators are calculated to analyze various aspects of the texture of the image (or its subset). Works of Haralick et al. [23], in which numerous indicators were presented, played an important role in developing and popularizing this method. Therefore, GLCM indicators are frequently referred to as Haralick indicators (or features). This method is highly effective [46], which has been demonstrated in many publications [47,48], although different authors suggest using different sets of Haralick features [38,39,41]. One of the main advantages of GLCM is the large number of indicators describing various aspects of the texture, enabling to analyze it from different angles. In these studies, we used a set of eight basic Haralick indicators (formulas according to the OTB CookBook [49]).

$$\mathrm{Energy} = \sum_{\mathrm{i,j}} \mathrm{g(i,j)}^2 \tag{1}$$

Energy is a description of the uniformity of texture.

$$Entropy = \sum_{i,j} g(i,j)log_2 g(i,j),\ or\ 0\ if\ g(i,j) = 0, \tag{2}$$

Entropy is a measure of randomness of pixel values distribution. It is the highest when all the frequencies g(i,j) are equal, and smaller when they are unequal. Heterogeneous images (subsets of image) will result in a higher entropy value.

$$Correlation = \sum_{i,j} \frac{(i-\mu)(j-\mu)g(i,j)}{\sigma^2} \tag{3}$$

Correlation describes how a pixel is correlated to its neighborhood. Regions of similar gray level will result in higher correlation values.

$$Inverse\ Difference\ Moment = \sum_{i,j} \frac{1}{1+(i-j)^2} g(i,j), \tag{4}$$

Inverse difference moment (IDM) measures a homogeneity of an image (its subset). Homogeneous images will result in high IDM values.

$$Inertia = \sum_{i,j} (i,j)^2 g(i,j), \tag{5}$$

Inertia (or contrast) measures a contrast of pixel values between a given pixel and its neighborhood. Images with a large amount of variation will result in high inertia.

$$Cluster\ Shade = \sum_{i,j} ((i-\mu) + (j-\mu))^3 g(i,j), \tag{6}$$

Cluster shade is a measure of the skewness of the matrix. It describes the perceptual concepts of uniformity [50].

$$Cluster\ Prominence = \sum_{i,j} ((i-\mu) + (j-\mu))^4 g(i,j), \tag{7}$$

Cluster prominence is a measure of asymmetry. The less symmetric regions will result in higher cluster prominence values.

$$Haralick's\ Correlation = \frac{\sum_{i,j} (i,j) g(i,j) - \mu_t^2}{\sigma_t^2}, \tag{8}$$

where (i,j) is the matrix cell index, $g(i,j)$ is the frequency value of the pair having index (i,j), $\mu = \sum_{i,j} i * g(i,j) = \sum_{i,j} j * g(i,j)$ (due to matrix symmetry) and means weighted pixel average, $\sigma = \sum_{i,j} (i-\mu)^2 * g(i,j) = \sum_{i,j} (j-\mu)^2 * g(i,j)$ (due to matrix symmetry) and means weighted pixel variance, and μ_t and σ_t are the mean and standard deviation of the row (or column, due to symmetry) sums.

2.2. Granulometric Analysis

The granulometric analysis in a classical form consists of sequentially executing a series of morphological openings of a binary image using structuring elements of sequentially increasing size, and then in the calculation of derivative images demonstrating the differences [31]. By specifying the size of each differential image (defined as the sum of all pixel values in the image) [51], we are able to determine the number of texture grains of specific sizes found in the source image [31]. Due to the nature of the opening operation (removal of relatively small image elements with values greater than the background value), this applies to bright texture elements. Analogically to the analysis based on the sequence of morphological openings, an analysis based on morphological closing operations can be performed. Due to the dual nature of the closing operation (in opposition to the opening operation), we obtain complementary information about the dark texture elements. This type of analysis is sometimes referred to as anti-granulometric [35]. However, in this article, both versions of the analysis (opening- and closing-based) are referred to as granulometric analysis. This type of global analysis assigns a set of features called the granulometric density function to the whole image.

The use of local granulometry is presented in [32,52]. It consists of assigning the granulometric density function to individual pixels on the basis of the analysis of their neighborhood. In practice, the effect of local granulometric analysis is usually a set of images in which pixel values represent successive values of the granulometric density function. Such images are called granulometric maps [30,50]. The usefulness of granulometric maps in the LULC classification has been proven, e.g., in [6,35,40,51]. It is worth noting that this type of granulometric analysis in many respects resembles a morphological profile [53,54]; however, it differs significantly in other respects [35,43].

Among the important advantages of granulometric analysis, a multiscale should be indicated: As a result of successive morphological operations with an increasing size of the structuring element, subsequent values (granulometric maps) present information about the presence of texture grains of different sizes.

Another important feature of this method is resistance to the so-called edge effect [55]. Most texture analysis methods are based on the analysis of the spatial frequency of the image, as a result of which the edges of the objects, even those with a low texture, are marked with values indicating high texture. This may lead to a reduced accuracy of the classification [35,36,55]. The granulometric analysis is based on a different principle: It removes whole elements of the image and analyzes their number; as a result of this, it does not show the edges of objects as areas with high texture.

3. Study Area

The study area is located in the south of Poland, in the Silesian Voivodeship, near Czestochowa city (50°45″ N; 19°17″ E) (Figure 1). This area is a part of the Olsztyn-Mirów wildlife refuge, a Natura 2000 protected area (PLH240015), established by the decision of the European Commission of 12 December 2008 No. 2009/93/EC (OJ EU L 43 of 13.02.2009) [56].

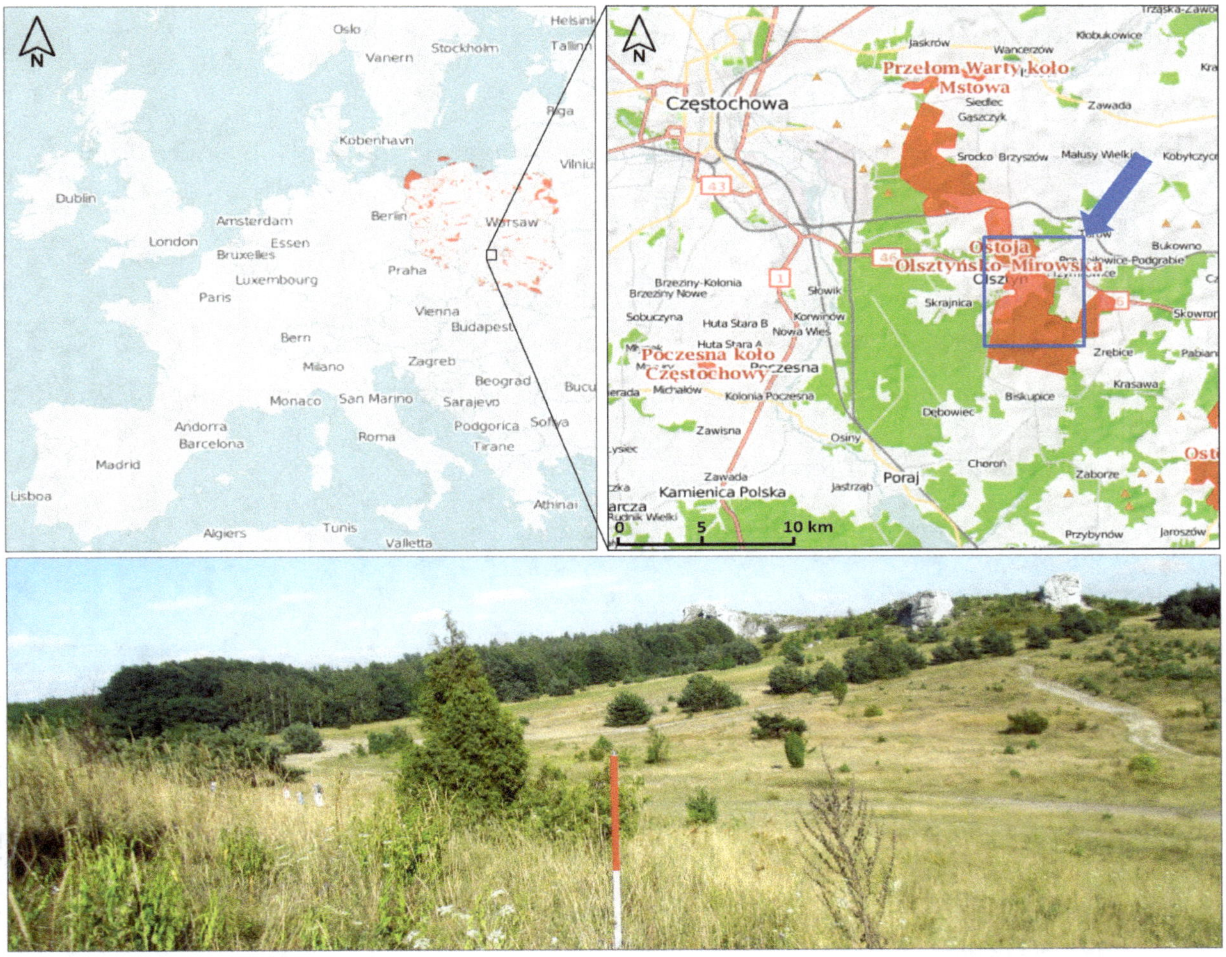

Figure 1. Location of the study area (blue rectangle) and overview. The photo shows a fragment of valuable non-forest habitats, where secondary succession is observed.

This area is characterized by large habitat diversity. Non-forest habitats present there are of particular importance. They are associated with limestone rocks, with numerous rare and endangered thermophilic species of plants and invertebrates (including the scarce large blue (*Phengaris teleius*), a species from Annex II of the Habitats Directive). A number of species reaches here the northern edge of their geographic range. In total, there are 13 types of habitats identified in the refuge with Annex I of the Habitats Directive (including 4 priority habitats) and 12 species of plants and animals from Annex II of the Habitats Directive [57].

The threat to the non-forest habitats of this area is primarily the process of secondary succession, which is the result of the abandonment of pastoralism, the lack of grazing and mowing of meadows, which has been progressing gradually since the 1990s. In order to protect valuable habitats at the Olsztyn-Mirów wildlife refuge within the framework of the LIFE11 NAT/PL/432 project "Protection of valuable non-forest habitats of the Orle Gniazda Landscape Park" (2012–2016), a number of active protection measures, such as the planting of bushes and trees and controlled sheep grazing, have been undertaken [58].

The part of the area selected for analysis (25 km^2) covers a territory which is varied in terms of terrain relief and land use, and of the nature of the secondary succession process. It allows to identify possible problems in the automation of the process of determining the succeeding ranges of trees and shrubs using textural analysis methods.

4. Materials and Methods

The main stages of the development methodology included:

1. Analysis and selection of archival aerial imagery;
2. Development of orthophotomaps;
3. Creation of simulated panchromatic (P) images;
4. Determination of the wooded and shrubbed areas using textural analysis, including:
    a. Selection of methods and parameters of texture analysis;
    b. Creation of spectro-textural data sets;
    c. Classification using support vector machine (SVM); and
    d. Additional processing.
5. Preparation of reference data and evaluation of the accuracy of determining the wooded and shrubbed areas; and
6. Analysis and comparison of results.

The methodology scheme is presented in Figure 2.

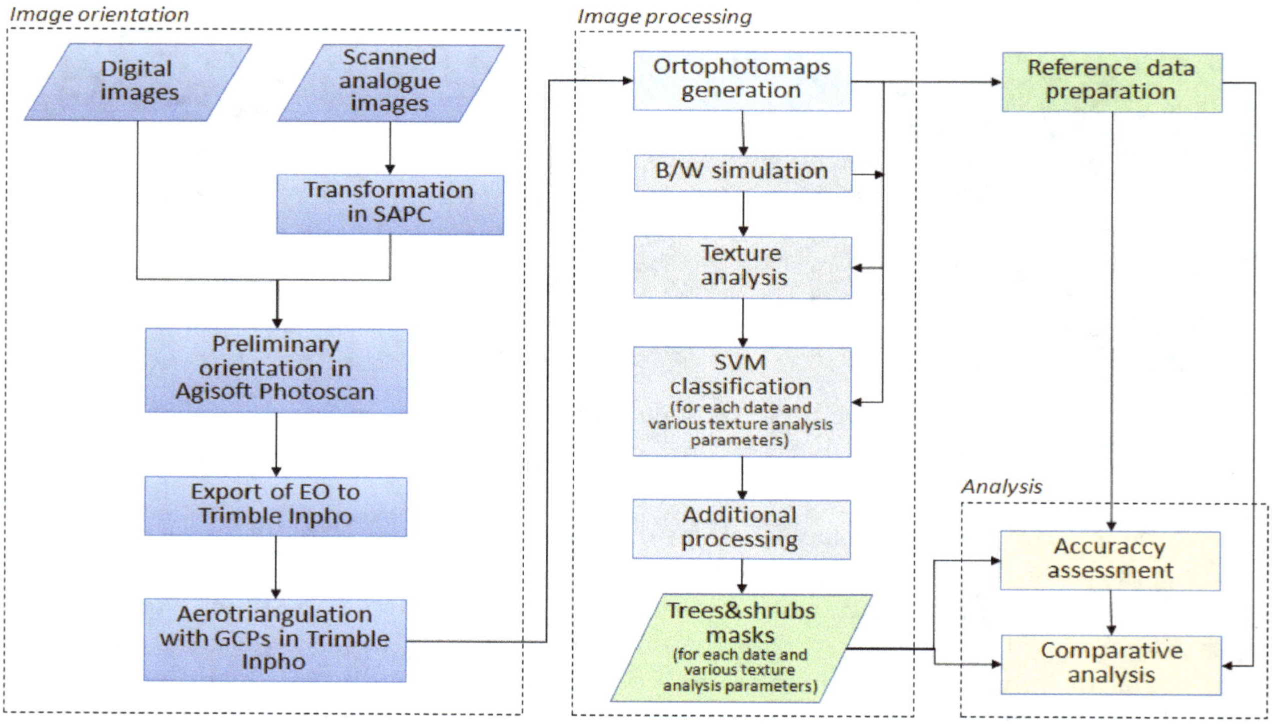

Figure 2. The methodology scheme.

4.1. Data

The aerial photographs for the study area were obtained from the State Geodetic and Cartographic Resource in Poland. They were photos from 1971, 1996, 2003, 2009, 2012, and 2015. They were both analogue and digital, black and white (panchromatic—P) and color (RGB). They differed in parameters, scale (from 1:13 000 to 1:26 000) or GSD (24 and 25 cm), and the acquisition date (different phenological periods). The basic parameters characterizing these photographs are presented in Table 1, and in Figure 3, selected parts of photos are visualized.

Table 1. The characteristics of aerial photographs used for analysis.

Date	Number of Photos	GSD or Scale	Camera	Focal Length	GPS/INS	Aerotrian-Gulation (EO)	Type
11.08.1971	12	1:18 000	RC51	210.20 mm	NO	NO	P
30.05.1996	4	1:26 000	RC20	152.97 mm	NO	NO	RGB
24.05.2003	14	1:13 000	LMK	152.30 mm	NO	NO	P
26.04.2009/ 29.04.2009	14	1:14 000	RC30	153.81 mm	YES	YES	RGB
25.03.2012	10	24 cm	UltraCamXp	100.50 mm	NO	NO	RGB
08.08.2015	10	25 cm	UltraCamXp	100.50 mm	NO	NO	RGB, CIR

Figure 3. Parts of the aerial photos used for analysis.

Based on the obtained images, orthophotomaps with GSD = 0.5 m were prepared. In the case of scanned images, first they were pre-processed in SAPC software according to the methodology developed by Salach [59]. This methodology transforms scans to a form similar to the images captured by a digital camera. Next, approximate values of exterior orientation (EO) parameters were obtained using Agisoft Photoscan software. Then, these values were used as approximate EO values in Trimble INPHO, which allowed the process of image block bundle adjustment to occur. A more detailed description of the process of developing archival materials is in [20].

4.2. *Simulation of Panchromatic Images*

Due to the importance of spectral data for spectro-textural classification, one of the aspects of this study was to compare the effectiveness of classification depending on the type of images available. However, the test pictures, apart from the type, also differ in the radiometric quality, which can undoubtedly be relevant to the accuracy of the quality of texture analysis and, as a result, to the classification performed. In order to estimate the importance of the type of photo regardless of the quality of the material itself, panchromatic images were simulated for RGB images.

Aerial panchromatic photographs images were most often made with the use of a yellow (minus-blue) filter blocking blue radiation [60,61]; so only green and red radiation was registered in panchromatic images. Therefore, simulated panchromatic images (P_s) were calculated by averaging the values of red (R) and green (G) bands:

$$P_s = \frac{R+G}{2}, \tag{9}$$

Simulated panchromatic images were generated for all RGB images—that is, for data from 1996, 2009, 2012, and 2015.

4.3. *The Determination of Wooded and Shrubbed Areas*

The determination of wooded and shrubbed areas was based on spectro-textural classification. It was made using the support vector machine (SVM) method on combined spectral data (original CIR, RGB, or P and simulated P) and texture data (Haralick indicators or granulometric maps).

4.3.1. Selection of Source Images for Texture Processing

Single grayscale images are proposed for texture processing. In the case of P images (both original and simulated), the choice was obvious—to choose these P images. In the case of multispectral images (CIR and RGB), the image of the first principal component was selected for texture processing. This image by definition represents the greatest variation of the data set. Previous research also shows that texture analysis based on the image of the first principal component gives the best distinction of the texture of selected LULC classes—in addition to the near-infrared image [62], unavailable for most of test images in this study.

4.3.2. Texture Analysis

Texture analysis was done independently using two methods: GLCM and granulometric analysis. The two proposed methods were chosen for the following reasons. GLCM is one of the most popular methods of texture analysis, according to the literature analysis still very often used in the analysis of remote sensing data [22,23,41,47]. However, granulometric analysis is a relatively well-known method of texture analysis, but its use in remote sensing is negligible. Earlier studies show that both methods are highly effective, also when compared to other texture descriptors. In addition, these methods are highly reliable and are easy to implement. In both cases, it was a local analysis performed in three variants, depending on the size of the analyzed neighborhood:

- 10-pixel radius;
- 15-pixel radius; and
- 20-pixel radius.

GLCM analysis was performed using eight Haralick features, according to the formulas given in Equations (1)–(8). As a result, eight images were obtained, which were treated as complementary texture information, basing on all four displacement vectors. The OTB Monteverdi software was used for the calculations.

The granulometric analysis based on both opening and closing operations. For each image, 6 granulometric steps (opening or closing) were made, resulting in 12 granulometric maps (6 for opening and 6 for closing), representing information about texture corresponding to the size of structuring elements (circle-shaped, from a 1-pixel to 6-pixel radius). In order to investigate which granulometric maps carry information of significance from the viewpoint of the analysis of the wooded and shrubbed areas, data sets with fewer numbers of granulometric maps were also prepared. As a result, data sets with the following numbers of granulometric maps were analyzed:

- From 1 to 3 for opening and closing (6 granulometric maps);
- From 1 to 4 for opening and closing (8 granulometric maps);
- From 1 to 5 for opening and closing (10 particle size maps); and
- From 1 to 6 for opening and closing (12 granulometric maps).

The calculations were carried out using BlueNote software [63].

4.3.3. Tested Variants

Variants (165) of data sets for classification were prepared, differing in at least one of the following features: Date of acquisition, type of image, texture analysis method, local texture analysis neighborhood size, or the number of granulometric maps used (in the case of granulometric analysis). These variants are summarized in Table 2.

Table 2. Tested variants of data sets.

Year	Image Type	GLCM			Granulometric Analysis		
		10	15	20	10	15	20
1971	P	p-1971-glcm-10	p-1971-glcm-15	p-1971-glcm-20	p-1971-x*-10	p-1971-x*-15	p-1971- *-20
1996	P	p-1996-glcm-10	p-1996-glcm-15	p-1996-glcm-20	p-1996-x*-10	p-1996-x*-15	p-1996- *-20
	RGB	c-1996-glcm-10	c-1996-glcm-15	c-1996 glcm-20	c-1996-x*-10	c-1996-x*-15	c-1996- *-20
2003	P	p-2003-glcm-10	p-2003-glcm-15	p-2003-glcm-20	p-2003-x*-10	p-2003-x*-15	p-2003- *-20
2009	P	p-2009-glcm-10	p-2009 glcm-15	p-2009-glcm-20	p-2009-x*-10	p-2009-x*-15	p-2009- *-20
	RGB	c-2009-glcm-10	c-2009-glcm-15	c-2009-glcm-20	c-2009-x*-10	c-2009-x*-15	c-2009- *-20
2012	P	p-2012-glcm-10	p-2012-glcm-15	p-2012-glcm-20	p-2012-x*-10	p-2012-x*-15	p-2012- *-20
	RGB	c-2012-glcm-10	c-2012-glcm-15	c-2012-glcm-20	c-2012-x*-10	c-2012-x*-15	c-2012- *-20
2015	P	p-2015-glcm-10	p-2015-glcm-15	p-2015-glcm-20	p-2015-x*-10	p-2015-x*-15	p-2015- *-20
	RGB	c-2015-glcm-10	c-2015-glcm-15	c-2015-glcm-20	c-2015-x*-10	c-2015-x*-15	c-2015- *-20
	CIR	cir-2015-glcm-10	cir-2015-glcm-15	cir-2015-glcm-20	cir-2015-x*-10	cir-2015-x*-15	cir-2015-x*-20

* x represents the number of subsequent granulometric maps (after opening and closing) used to create a given variant of the data set (from 3 to 6).

4.3.4. Performing the Classification

In total, a classification was made on 165 variants of data sets. Fifteen variants were tested for each of the source images: 3 variants based on GLCM analysis and 12 variants based on granulometric analysis (the larger number is due to testing 4 sub-variants depending on the number of granulometric maps used).

The classification was made using the support vector machine (SVM) using ArcGIS software. This method was chosen because of its high efficiency compared to other machine learning methods, such as random forest or maximum likelihood [64].

The training fields were developed as a result of manual digitalization based on the visual interpretation of the original spectral images. All variants developed within one date of acquisition were classified based on the same set of training data. The final effect of each classification was the binary mask of wooded and shrubbery areas, created as a result of aggregation of individual classes based on individual training fields. The basic parameters related to the implementation of the classification are presented in Table 3.

Table 3. Basic parameters of the classification of test images.

Date	Number of Variants Tested	Number of Training Fields: Wooded and Shrubbed Areas/Other Classes (Number of Pixels in Brackets)
1971	15	6/12 (1812/12180)
1996	30	4/15 (3832/4825)
2003	15	6/15 (2653/9057)
2009	30	13/32 (1379/6032)
2012	30	7/18 (8531/9520)
2015	45	11/17 (4590/5445)

4.3.5. Additional Processing

After obtaining the binary masks of wooded areas, they were additionally processed using morphological closing by reconstruction [65]. This operation was aimed at increasing the accuracy by removing any discontinuities in the mask of the wooded areas. Such discontinuities resulted mainly from the occurrence of shaded areas, i.e., those whose spectral features could significantly differ from those of other parts of wooded areas. Direct allocation of these areas to the wooded area mask could lead to an over-classification of this class in other parts of the image, and thus to a significant increase in the excess error and, as a result, to a decrease in the accuracy of the classification.

Closing by reconstruction removes (closes) relatively small discontinuities (of a size smaller than the size of the structuring element) in binary masks; at the same time, it does not change the shape of other objects, unlike other filters of a similar type (simple closure, median filter, majority, etc.).

The operation used closing by reconstruction using a 10-pixel radius structuring element. It was applied to all 165 classification variants obtained and the accuracy of newly obtained masks was independently assessed. The processing was carried out in the BlueNote software [63].

4.4. Accuracy Assessment

Evaluation of the accuracy of determining the wooded and shrubbed areas using various approaches to textural analysis was carried out, comparing their results with the reference data, which were the results of detailed visual interpretation of orthophotomaps prepared on the basis of the same archival materials. The entire analysis area was digitized for accuracy assessment purposes. Orthophotomaps were developed with GSD = 0.5 m. The smallest separation was of 2 m^2; that is, it covered about 3 × 3 pixels. Thanks to such detailed visual interpretation, it was possible—in the later stages—to indicate: (1) Which of the textural analysis variants gives the best results, and (2) what size of wooded and shrubbed areas can be automatically determined on the basis of individual archival materials.

The following accuracy indicators were used to assess the efficacy of individual approaches: Overall accuracy (OA), recall, precision, Cohen's kappa coefficient, and F1-score [66,67]. In addition, the sum of the area resulting from errors of omission and commission (EO + EC) was analyzed to indicate the variant with the smallest error area.

5. Results

The results of the analysis of the accuracy of determining the range of trees and shrubs using spectro-textural classification are presented in Table 4 and in the diagrams in Figure 4.

Table 4. Results of the accuracy assessment obtained for individual images and analysis variants (the best results are marked in green).

Variant	P					RBG (C)					CIR				
	K	OA	F1	PA	UA	K	OA	F1	PA	UA	K	OA	F1	PA	UA
								1971							
3-10	0.722	0.932	0.762	0.714	0.816	-	-	-	-	-	-	-	-	-	-
3-15	0.633	0.910	0.685	0.645	0.731	-	-	-	-	-	-	-	-	-	-
3-20	0.634	0.913	0.684	0.663	0.707	-	-	-	-	-	-	-	-	-	-
4-10	0.726	0.936	0.763	0.752	0.775	-	-	-	-	-	-	-	-	-	-
4-15	0.644	0.916	0.693	0.677	0.709	-	-	-	-	-	-	-	-	-	-
4-20	0.624	0.912	0.675	0.667	0.683	-	-	-	-	-	-	-	-	-	-
5-10	0.714	0.933	0.752	0.746	0.759	-	-	-	-	-	-	-	-	-	-
5-15	0.669	0.921	0.715	0.694	0.737	-	-	-	-	-	-	-	-	-	-
5-20	0.614	0.914	0.663	0.696	0.633	-	-	-	-	-	-	-	-	-	-
6-10	0.692	0.929	0.733	0.738	0.727	-	-	-	-	-	-	-	-	-	-
6-15	0.641	0.921	0.687	0.726	0.652	-	-	-	-	-	-	-	-	-	-
6-20	0.563	0.904	0.618	0.660	0.581	-	-	-	-	-	-	-	-	-	-
glcm-10	**0.775**	**0.945**	**0.807**	**0.754**	**0.868**	-	-	-	-	-	-	-	-	-	-
glcm-15	0.676	0.928	0.717	0.758	0.681	-	-	-	-	-	-	-	-	-	-
glcm-20	0.682	0.932	0.720	0.791	0.661	-	-	-	-	-	-	-	-	-	-
								1996							
3-10	0.724	**0.895**	0.794	0.720	0.883	**0.753**	0.904	**0.817**	**0.727**	0.931	-	-	-	-	-
3-15	**0.727**	0.891	**0.800**	0.691	0.948	0.749	0.901	0.815	0.710	0.956	-	-	-	-	-
3-20	0.621	0.850	0.720	0.630	0.840	0.690	0.877	0.772	0.671	0.907	-	-	-	-	-
4-10	0.714	0.892	0.785	**0.721**	0.863	0.744	**0.905**	0.806	0.758	0.860	-	-	-	-	-
4-15	0.671	0.863	0.761	0.633	0.955	0.735	0.894	0.806	0.693	0.963	-	-	-	-	-
4-20	0.647	0.857	0.742	0.632	0.901	0.702	0.883	0.780	0.684	0.906	-	-	-	-	-
5-10	0.703	0.884	0.780	0.691	0.894	0.733	0.893	0.804	0.695	0.952	-	-	-	-	-
5-15	0.672	0.863	0.762	0.633	**0.958**	0.744	0.898	0.812	0.702	0.963	-	-	-	-	-
5-20	0.668	0.866	0.757	0.647	0.912	0.668	0.869	0.755	0.663	0.876	-	-	-	-	-
6-10	0.695	0.878	0.776	0.669	0.923	0.723	0.889	0.797	0.685	0.951	-	-	-	-	-
6-15	0.650	0.854	0.747	0.619	0.941	0.707	0.880	0.786	0.665	0.962	-	-	-	-	-
6-20	0.611	0.846	0.713	0.622	0.835	0.642	0.860	0.735	0.648	0.850	-	-	-	-	-
glcm-10	0.527	0.796	0.661	0.533	0.869	0.597	0.820	0.714	0.561	**0.982**	-	-	-	-	-
glcm-15	0.337	0.723	0.521	0.431	0.657	0.572	0.811	0.696	0.551	0.944	-	-	-	-	-
glcm-20	0.243	0.687	0.450	0.377	0.558	0.305	0.716	0.493	0.418	0.602	-	-	-	-	-
								2003							
3-10	0.761	0.889	**0.846**	0.757	**0.957**	-	-	-	-	-	-	-	-	-	-
3-15	0.750	0.886	0.836	0.768	0.919	-	-	-	-	-	-	-	-	-	-
3-20	0.756	0.890	0.838	0.790	0.892	-	-	-	-	-	-	-	-	-	-
4-10	0.755	0.887	0.841	0.763	0.937	-	-	-	-	-	-	-	-	-	-
4-15	0.760	0.892	0.842	0.787	0.904	-	-	-	-	-	-	-	-	-	-
4-20	0.734	0.883	0.820	0.804	0.837	-	-	-	-	-	-	-	-	-	-
5-10	0.754	0.886	0.841	0.754	0.951	-	-	-	-	-	-	-	-	-	-
5-15	0.761	0.892	0.843	0.782	0.915	-	-	-	-	-	-	-	-	-	-
5-20	0.739	0.885	0.824	0.802	0.848	-	-	-	-	-	-	-	-	-	-
6-10	0.735	0.879	0.826	0.762	0.902	-	-	-	-	-	-	-	-	-	-
6-15	0.761	0.892	0.843	0.784	0.910	-	-	-	-	-	-	-	-	-	-
6-20	0.727	0.881	0.815	**0.809**	0.821	-	-	-	-	-	-	-	-	-	-
glcm-10	0.749	0.884	0.838	0.751	0.947	-	-	-	-	-	-	-	-	-	-
glcm-15	0.748	0.885	0.835	0.767	0.917	-	-	-	-	-	-	-	-	-	-
glcm-20	0.700	0.862	0.805	0.730	0.898	-	-	-	-	-	-	-	-	-	-

Table 4. *Cont.*

Variant	P					RBG (C)					CIR				
	K	OA	F1	PA	UA	K	OA	F1	PA	UA	K	OA	F1	PA	UA
								2009							
3-10	0.756	0.879	0.869	0.837	0.903	0.763	0.881	0.875	0.826	**0.929**	-	-	-	-	-
3-15	0.735	0.868	0.859	0.820	0.901	0.802	0.902	**0.892**	0.876	0.908	-	-	-	-	-
3-20	0.744	0.872	0.865	0.816	0.919	0.781	0.891	0.883	0.843	0.928	-	-	-	-	-
4-10	0.773	0.887	0.877	0.848	0.908	0.770	0.886	0.873	0.867	0.879	-	-	-	-	-
4-15	0.757	0.879	0.870	0.833	0.909	0.786	0.894	0.883	0.864	0.904	-	-	-	-	-
4-20	0.739	0.869	0.862	0.815	0.914	0.777	0.889	0.880	0.850	0.912	-	-	-	-	-
5-10	0.764	0.883	0.872	0.848	0.897	0.780	0.890	0.881	0.853	0.911	-	-	-	-	-
5-15	0.757	0.879	0.870	0.838	0.904	0.773	0.887	0.878	0.849	0.909	-	-	-	-	-
5-20	0.728	0.864	0.854	0.822	0.887	0.756	0.879	0.869	0.836	0.905	-	-	-	-	-
6-10	0.767	0.884	0.875	0.842	0.910	0.782	0.892	0.881	0.862	0.902	-	-	-	-	-
6-15	0.752	0.877	0.866	0.841	0.892	0.772	0.887	0.876	0.858	0.894	-	-	-	-	-
6-20	0.747	0.874	0.863	0.838	0.889	0.762	0.882	0.871	0.847	0.896	-	-	-	-	-
glcm-10	**0.791**	**0.896**	**0.888**	0.852	**0.927**	**0.816**	**0.909**	**0.899**	**0.890**	0.908	-	-	-	-	-
glcm-15	0.776	0.889	0.878	**0.857**	0.901	0.796	0.899	0.888	0.875	0.901	-	-	-	-	-
glcm-20	0.768	0.885	0.873	0.856	0.892	0.793	0.897	0.886	0.875	0.898	-	-	-	-	-
								2012							
3-10	0.578	0.785	0.784	0.692	0.903	0.569	0.779	0.784	0.676	0.934	-	-	-	-	-
3-15	0.672	0.836	0.825	0.763	0.897	0.688	0.844	0.832	0.779	0.892	-	-	-	-	-
3-20	0.653	0.828	0.808	0.776	0.843	**0.698**	**0.849**	**0.837**	0.782	0.902	-	-	-	-	-
4-10	0.556	0.774	0.774	0.680	0.897	0.550	0.768	0.777	0.663	0.939	-	-	-	-	-
4-15	0.650	0.825	0.812	0.756	0.876	0.585	0.790	0.785	0.702	0.891	-	-	-	-	-
4-20	0.581	0.794	0.762	0.758	0.766	0.656	0.830	0.809	0.785	0.835	-	-	-	-	-
5-10	0.537	0.762	0.770	0.660	0.923	0.581	0.786	0.789	0.684	0.933	-	-	-	-	-
5-15	0.661	0.831	0.817	0.764	0.878	0.602	0.799	0.793	0.713	0.893	-	-	-	-	-
5-20	0.636	0.819	0.800	0.765	0.838	0.658	0.832	0.808	0.793	0.824	-	-	-	-	-
6-10	0.598	0.795	0.795	0.698	0.923	0.608	0.801	0.798	0.710	0.910	-	-	-	-	-
6-15	0.644	0.824	0.803	0.774	0.835	0.481	0.732	0.745	0.630	0.912	-	-	-	-	-
6-20	0.684	0.844	0.824	0.801	0.848	0.684	0.845	0.820	**0.821**	0.818	-	-	-	-	-
glcm-10	**0.701**	**0.849**	**0.843**	0.762	**0.944**	0.613	0.802	0.805	0.698	**0.950**	-	-	-	-	-
glcm-15	0.675	0.836	0.831	0.745	0.939	0.643	0.819	0.816	0.726	0.932	-	-	-	-	-
glcm-20	0.659	0.830	0.816	0.764	0.876	0.597	0.802	0.772	0.765	0.780	-	-	-	-	-
								2015							
3-10	0.414	0.711	0.763	0.661	0.902	0.549	0.777	0.813	0.716	0.941	0.704	0.853	**0.870**	0.800	0.955
3-15	0.372	0.691	0.758	0.372	0.936	0.452	0.730	0.783	0.452	0.943	0.698	0.850	**0.870**	0.698	0.970
3-20	0.321	0.667	0.748	0.614	0.957	0.467	0.738	0.790	0.672	0.958	0.691	0.847	0.868	0.782	0.974
4-10	0.422	0.715	0.764	**0.667**	0.894	0.498	0.752	0.794	0.695	0.925	0.678	0.840	0.860	0.786	0.950
4-15	0.408	0.708	0.763	0.656	0.913	0.526	0.766	0.805	0.705	0.938	0.648	0.826	0.852	0.758	0.973
4-20	0.431	0.719	**0.774**	0.662	0.931	0.483	0.745	0.795	0.679	0.959	0.656	0.830	0.856	0.761	**0.977**
5-10	0.388	0.698	0.749	0.655	0.873	0.469	0.737	0.780	0.687	0.902	0.668	0.835	0.857	0.775	0.959
5-15	0.433	0.721	0.773	0.666	0.921	0.518	0.762	0.801	0.704	0.930	0.647	0.825	0.851	0.761	0.965
5-20	0.434	0.721	**0.776**	0.662	0.937	0.494	0.751	0.799	0.683	0.963	0.623	0.813	0.844	0.743	0.976
6-10	0.416	0.713	0.772	0.654	0.942	0.491	0.749	0.797	0.683	0.958	0.669	0.836	0.860	0.768	**0.977**
6-15	0.402	0.706	0.767	0.648	0.940	0.489	0.748	0.796	0.684	0.953	0.634	0.819	0.847	0.752	0.970
6-20	0.377	0.694	0.763	0.635	0.956	0.485	0.746	0.797	0.679	0.963	0.642	0.823	0.848	0.759	0.962
glcm-10	0.398	0.704	0.770	0.643	0.961	**0.593**	**0.798**	**0.827**	**0.741**	0.935	0.691	0.847	0.867	0.785	0.967
glcm-15	0.246	0.631	0.730	0.587	0.967	0.425	0.717	0.778	0.654	0.962	**0.709**	**0.856**	**0.874**	**0.794**	0.974
glcm-20	0.264	0.640	0.736	0.592	**0.970**	0.397	0.704	0.770	0.643	0.959	0.591	0.798	0.833	0.727	0.976

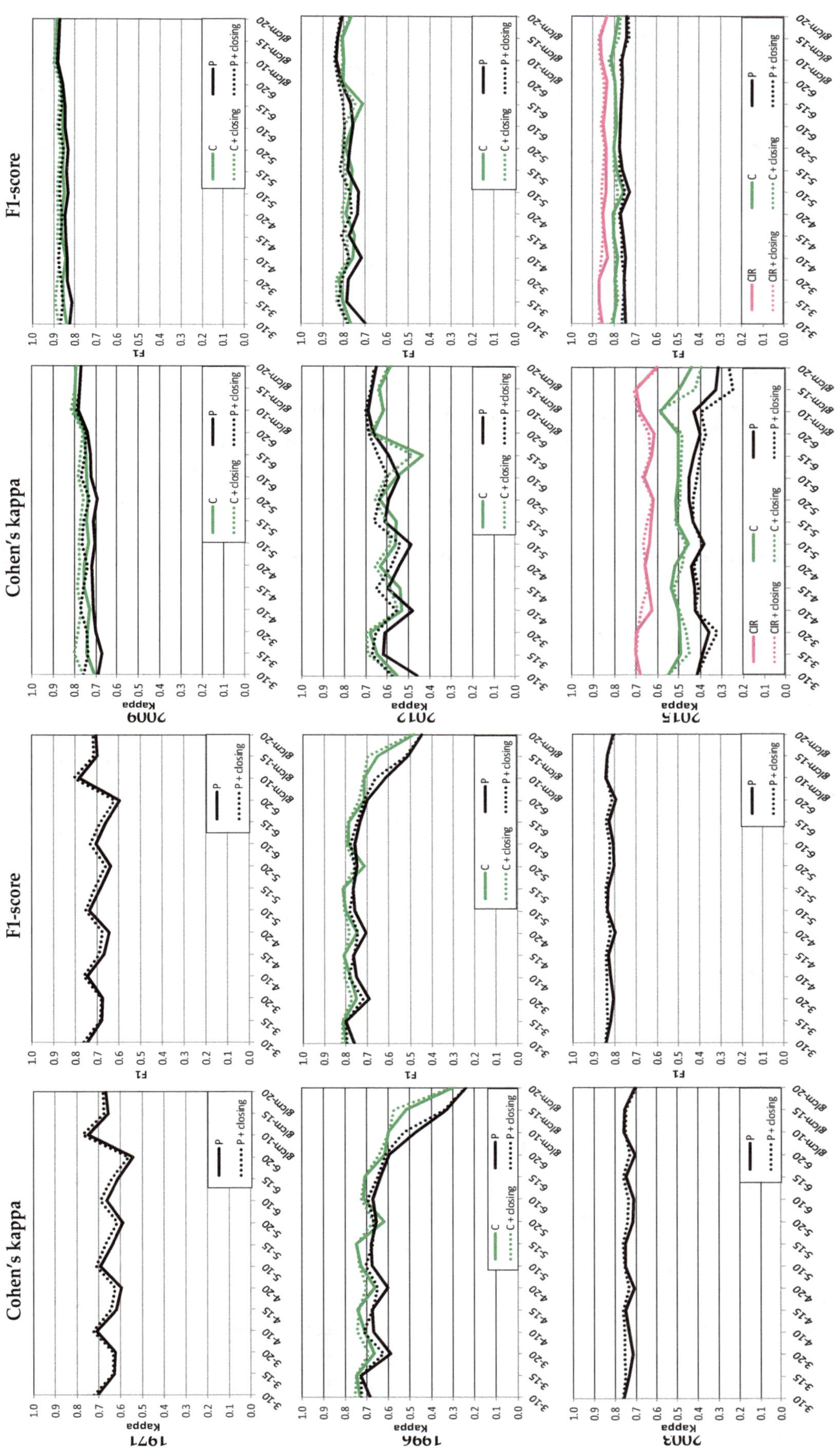

Figure 4. Comparison of accuracy results for different variants of textural analysis (Cohen's kappa coefficient of agreement and F1-score).

Generally, it was found that the best results for most archival materials were obtained using a local texture analysis radius of 10 or 15 pixels. A larger radius of the area of analysis (20 pixels) gave better results only in 2012 (early spring). Visualizations of the results of the texture analysis operation at various analysis parameters (before closing) for a fragment of the research area are presented in the figures included in Supplementary 1 (Figures S1–S14).

In the case of black and white aerial photographs (P) from 1971, the highest accuracy was obtained using the GLCM method and a radius of texture analysis of 10 pixels (Table 4, Figure 4). Slightly lower values of individual accuracy parameters were obtained for granulometric analysis p-1971-4-10, p-1971-3-10, and p-1971-5-10.

For 1996, the best results were obtained based on granulometric analysis: c-1996-3-15, c-1996-3-10, p-1996-3-15, and p-1996-3-10. The GLCM method for these photographs is characterized by the lowest values of all analyzed accuracy parameters (Table 4, Figure 4).

In the case of 2003, relatively small differences were observed in the evaluation of the accuracy of individual variants of granulometric analysis and GLCM (Kappa coefficient from 0.70 to 0.76) (Table 4). The variants with the highest accuracy were: p-2003-3-10, p-2003-4-15, p-2003-5-15, and p-2003-6-15.

In turn, as far as 2009 is concerned, the best results were obtained using the GLCM method, and this applies to the analysis of both the simulated image P and the original RGB images (Table 4, Figure 4). The highest accuracy was obtained for the following variants: c-2009-glcm-10, c-2009-3-15, and p-glcm-2009-10.

The GLCM method proved to be the most effective in relation to the imagery from 2012—the variant p-2012-glcm-10 was characterized by the highest accuracy (OA, kappa, F1, UA) (Table 4, Figure 4). In the case of granulometric analysis, the best results were obtained by analyzing RGB images in the variant c-2012-3-20.

Imagery from 2015 enabled to carry out the widest list of experiments—using the simulated P picture, as well as RGB and CIR images. Definitely the highest accuracy was obtained on the basis of CIR image analysis (Table 4, Figure 4). The highest values of individual accuracy indicators were characterized by the following variants: cir-2015-glcm-15, cir-2015-3-10, cir-2015-3-15, and c-2015-3-20.

Comparing the results obtained on the basis of RGB images and of the P images simulated on their basis (Figure 4) in most cases, higher accuracy was obtained for RGB images.

6. Discussion

The tested variants differed in many features: Type of spectral data (P, RGB, CIR), type of textural data (method and selected processing parameters), but also in the quality of the source spectral data related to the type of images (analog, digital), original image scale, original spatial resolution, and general radiometric quality of photos. However, the analysis of the results allows one to see some patterns.

Firstly, the comparison between variants differing only with the type of spectral data used in the classification (P/RGB/CIR; images from 1996, 2009, 2012, 2015) unambiguously indicate a strong relationship between spectral resolution or the type of spectral bands used, and the accuracy of classification, also using textural analysis. Variants using RGB images were more accurate compared to P in almost all cases (Table 4, Figure 4). The only case in which this relationship is not completely unambiguous is the classification of the photo from 2012; however, this is an early spring photo, and at the time of exposure, the deciduous plants were free of (visible) leaves (see Figure 3), which caused the spectral differences between them and the other classes of land cover to be much smaller than in the case of photos from other periods (with visible leaves). This affected both the texture of wooded and bushy areas, as well as the areas of meadows that dominate in this area.

In addition, the case of the imagery from 2015 confirmed the thesis about the importance of the near-infrared range (Figure 5, Figure 6). The accuracy of classification of selected variants of this imagery was generally low (measured by the Cohen's kappa coefficient, the F1-score achieves quite high values); however, variants using CIR images were significantly better than the other variants:

The kappa coefficient values were at 0.1–0.2 higher than for RGB variants and 0.2–0.3 higher than for P variants (Table 4). Also, PA and UA are the highest in the case of CIR image analysis (Figure 5). The difference in accuracy (measured by the Cohen's kappa coefficient and the F1-score) between the P and RGB variants is lower than between RGB and CIR, and this is the case for each analyzed variant of textual analysis (Figure 5). It shows the important role of near-infrared range in the analysis of vegetation. This spectral band, as more sensitive to the diversity of plant cover, allows better detection of trees and shrubs. In the case of P and RGB photographs, there is an overestimation of the ranges of trees and shrubs for herbaceous vegetation (Figure 6), with the largest overestimation for the P imagery.

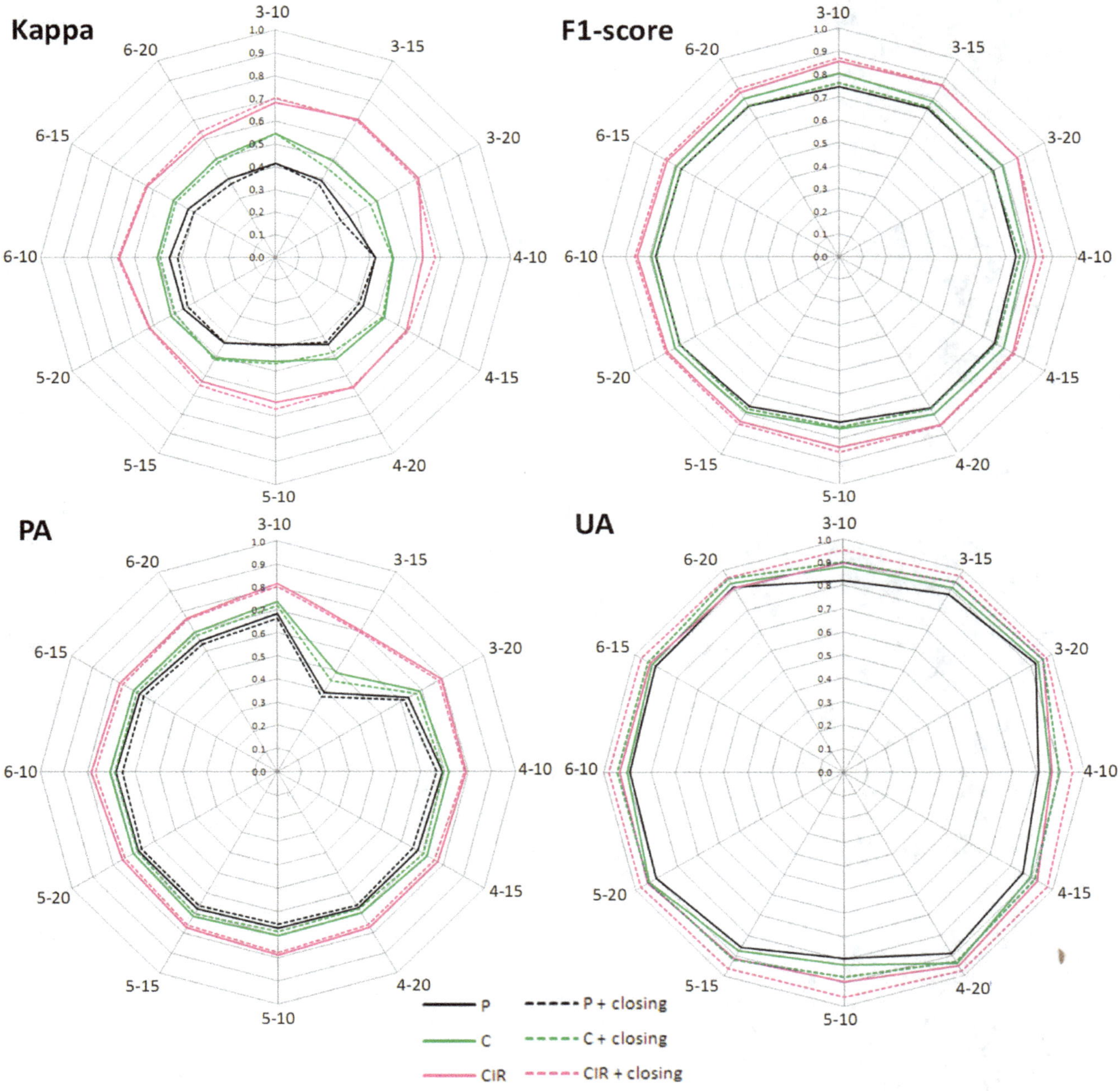

Figure 5. Comparison of accuracy indicators (Cohen's kappa, F1-score, Producer's Accuracy—PA, User's Accuracy—UA) for the classification based on individual granulometric analysis variants obtained for 2015.

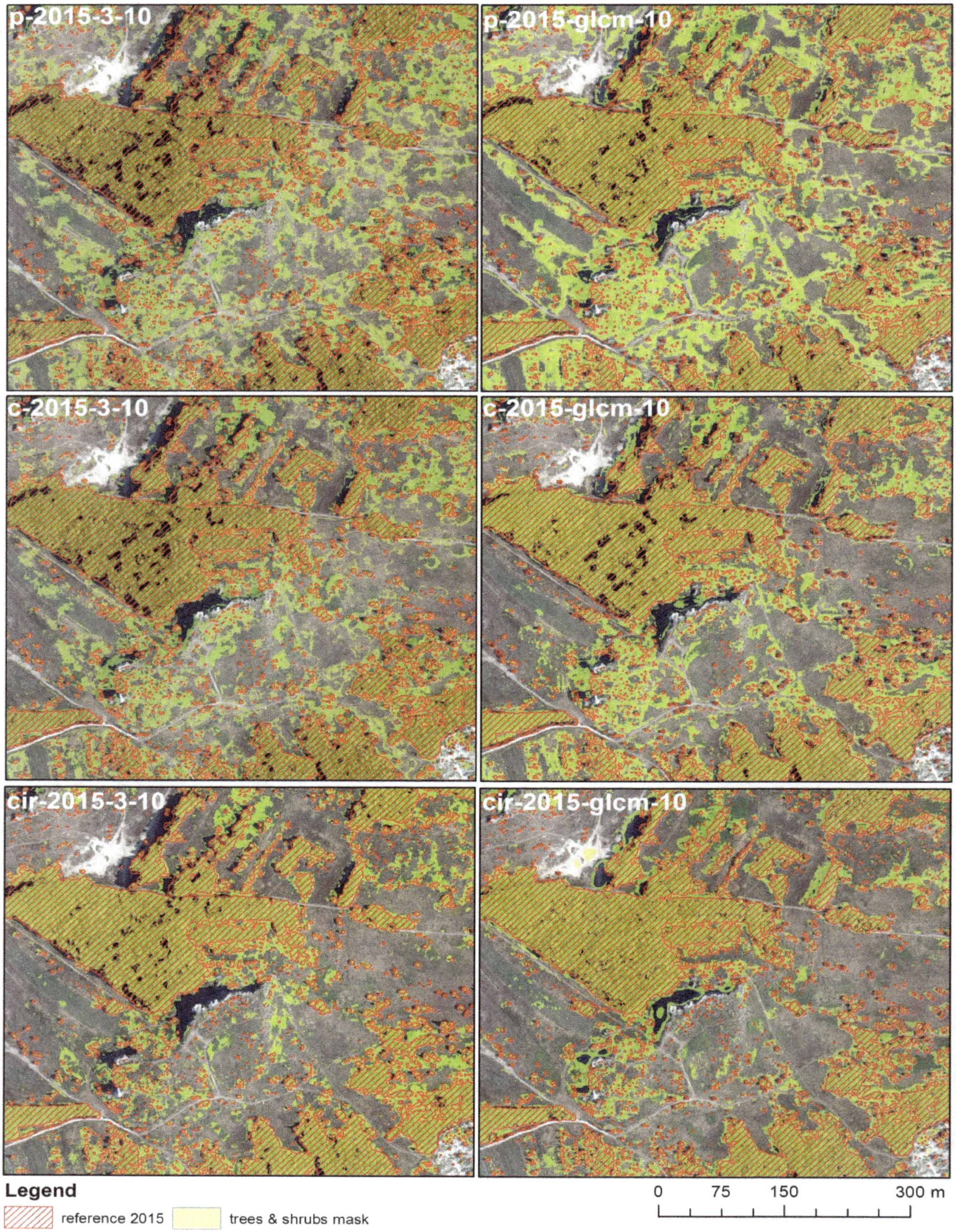

Figure 6. Comparison of tree/shrub masks obtained on the basis of images from 2015: P, RGB, and CIR in two variants of texture analysis—granulometry and GLCM.

The above conclusions correspond with the results obtained in previous studies, showing the importance of appropriate spectral resolution in the classification using both spectral data and texture analysis results [68,69], and the importance of image quality for the results of texture analysis itself [70].

Another basically unambiguous conclusion is the usefulness of the proposed additional processing that is closing by reconstruction, applied to individual masks of wooded areas obtained as a result of classification. Closing is an extensive operation, which means that in its result, the mask can only increase (or remain unchanged). At the same time, the use of the operation by reconstruction does not lead to a change in the shape of the original mask, but only to filling of relatively small holes. According to a detailed accuracy analysis, the use of the closing by reconstruction operation led to a decrease in UA, but at the same time, to a more significant increase in PA and, as a result, to an increase in the overall accuracy of classification. This is presented in Figure 7. Unfortunately, in the case of small trees and shrubs growing in a large dispersion, this may lead to overestimation of the area occupied by them, which is also seen in the analysis of texture for 2012 presented for two variants (granulometry and GLCM) in Figure 8.

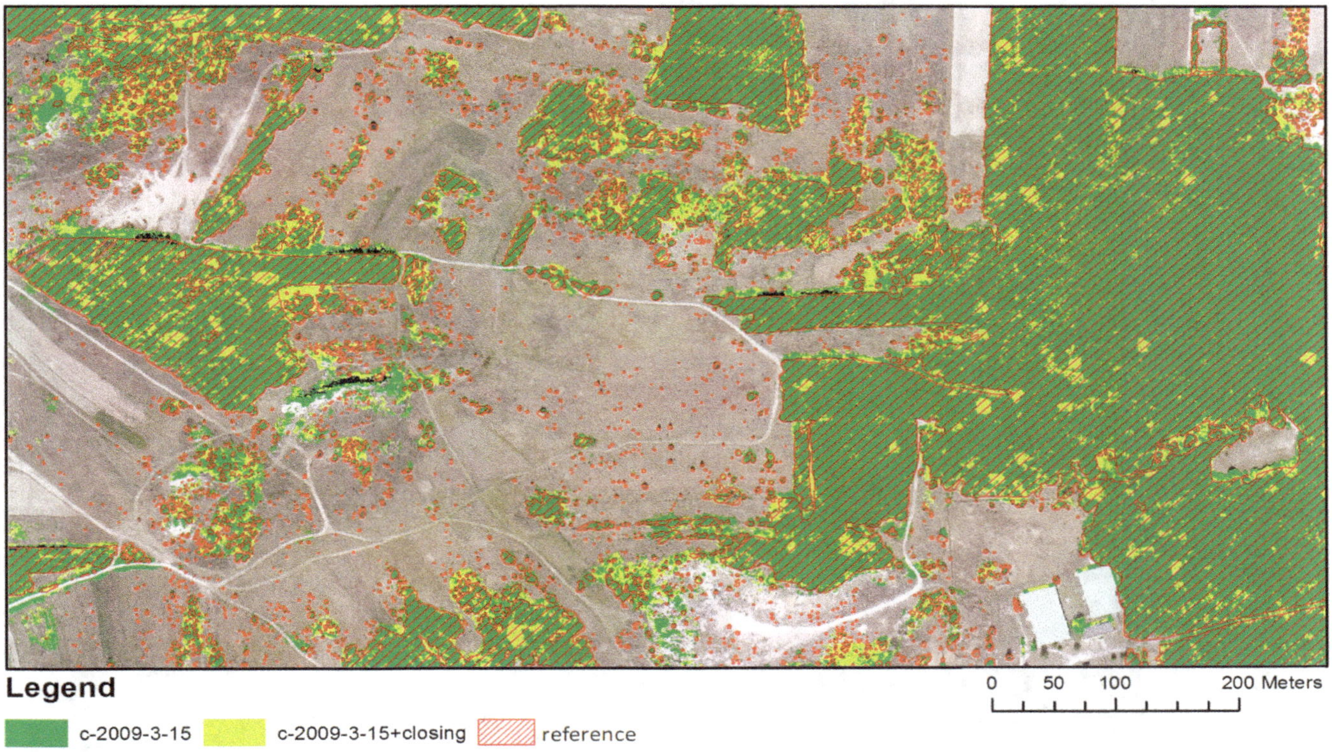

Figure 7. Comparison of tree and shrub masks for 2009, variant c-2009-3-15 before closing and after closing, with a reference mask imposed.

The comparison of the effectiveness of two methods: GLCM and granulometric analysis do not give a clear indication of the better one. In two cases (1971 and 2009), better results were obtained by using the GLCM method, although the differences between the two methods were relatively small (the difference between kappa values between the best variants was around 0.05 for 1971 and 0.01 for 2009, in favor of GLCM). In one case (2015), the accuracy obtained for the best variants measured by the kappa coefficient and other general indicators were basically identical; the difference in favor of one of the GLCM variants was 0.005, which can be considered as negligible. In the remaining three cases, greater accuracy was observed for variants based on granulometric analysis. In two cases (2003 and 2012), the differences in accuracy were very small (differences between the kappa coefficient values for the best variants amounted to approximately 0.01, in favor of granulometric analysis). However, in the case of 1996, these were very large differences (between 0.2 and 0.5) in favor of granulometric analysis, which is mainly due to the very low accuracy of GLCM based variants. This is a case worth a separate commentary.

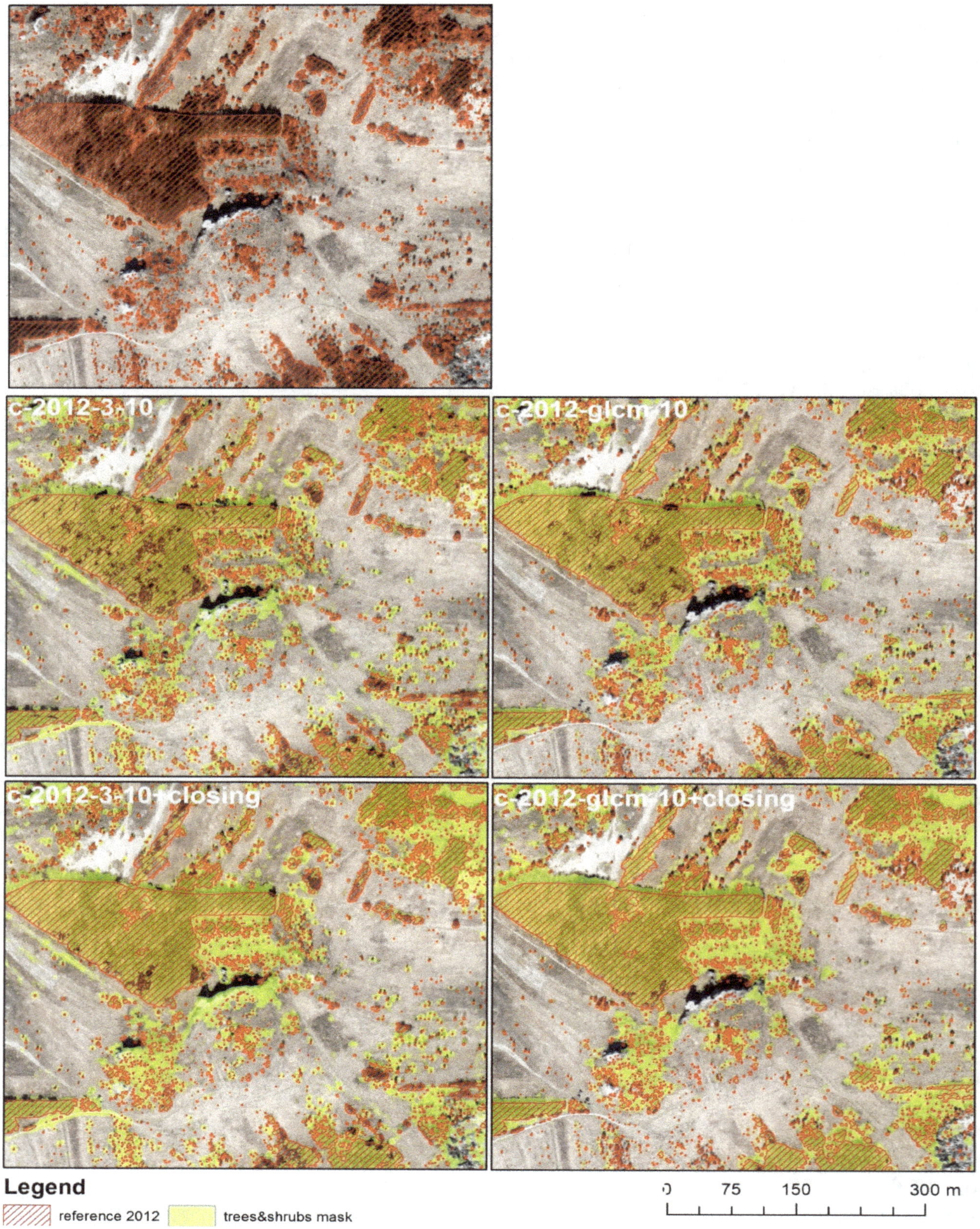

Figure 8. Comparison of tree and shrub masks for 2012, variants c-2012-3-10 and c-2012-glcm-10 before closing and after closing, with a reference mask imposed.

Analyzing Figure 4 in the context of the dates of imagery acquisition, it can be seen that the best results for photos from the period of dense foliage (May–September, with the exception of data from 1971 with lower radiometric quality) gave the use of granulometric analysis using six granulometric maps and a radius of 10–15 pixels (variants marked as x-xxxx-3-10 and x-xxxx-3-15). However, in the case of photos from the spring period (March–April), with a less dense foliage, the highest accuracy was obtained by using the GLCM method with a radius of 10–15 pixels.

The masks obtained on the basis of the GLCM analysis or granulometric analysis are presented in Figure 9. The low accuracy of variants based on the GLCM analysis results mainly from very low user accuracy. This is due to the high texture of agricultural areas (resulting from agrotechnical activities, including mowing meadows) and the areas on which the trees were cleared. At the same time, granulometric analysis results in a much smaller error of commission (Figure 9, Table 4). Probably, this is due to its multi-scale: Areas characterized by high but different textures are marked differently on different granulometric maps. This increases the mutual distinction of these areas. In addition, the accuracy of the GLCM-based variants for 1996 clearly decreases as the area of the local texture analysis increases. For an area with a radius of 10 pixels, the kappa value is 0.597, for a radius of 15 pixels is 0.572, while for an area with a radius of 20 pixels is only 0.305. As can be seen in Table 4, the decrease in the overall accuracy is mainly caused by a decrease in PA, and thus an increase in the error of omission. Analysis of the images of individual GLCM indices shows that part of the tree-covered areas obtains identical values as unripen areas. The area of these types of surfaces increases with the radius of the analysis area. This can be seen in Figure 9: The boundary of the wooded area detected on the basis of the selected GLCM indices "moves back" inside the area, along with the increase of the analysis area. In the case of granulometric analysis, this effect is not so pronounced, although it also occurs. Here too, the importance of multiscale granulometric analysis can be seen. Different values in selected granulometric maps indicate the "activation" in response to a texture with grains of different sizes. In GLCM indicator images, areas with a larger grain texture simply resemble areas with no clear texture. Hence the observed decrease in accuracy for the classification based on the GLCM analysis.

Interestingly, in the case of other images from different dates, this effect is not so important. It seems that this can be explained by the relatively low (in some areas) clearness of the texture in the 1996 picture compared to, for example, the picture from 2009 (compare Figure 3), which in turn may result from the small scale of the original image from 1996, 1:26 000.

Based on the above, it can be concluded that granulometric analysis is more stable and, in general, slightly better as a texture descriptor for image classification. There is no extensive literature on this subject, but existing publications also show greater efficacy of granulometric analysis in this area [36,71].

The analysis of the influence of the size of a texture analysis area on the accuracy of the classification indicates a general (though not unequivocal) tendency for the classification quality to decrease along with the increase of the size of the analysis environment (figures for all variants may be found in the Supplementary Materials). In most cases, the worst results were obtained for the analysis performed for a 20-pixel environment. The only exception is the classification of images from 2012; these, however, differ significantly from the others when it comes to the image of wooded and bushy areas (early spring photo and lack of leaves on deciduous trees and shrubs). However, when analyzing areas covered with coniferous trees, a similar tendency can be noticed as described above—a decrease in accuracy when the area of analysis increases. One can also observe a certain consistency in terms of individual variants within one source image—the relationships between variants differing only in the size of area of a texture analysis are similar for different texture analysis methods and for a different number of granulometric maps used (in the case of the granulometric analysis).

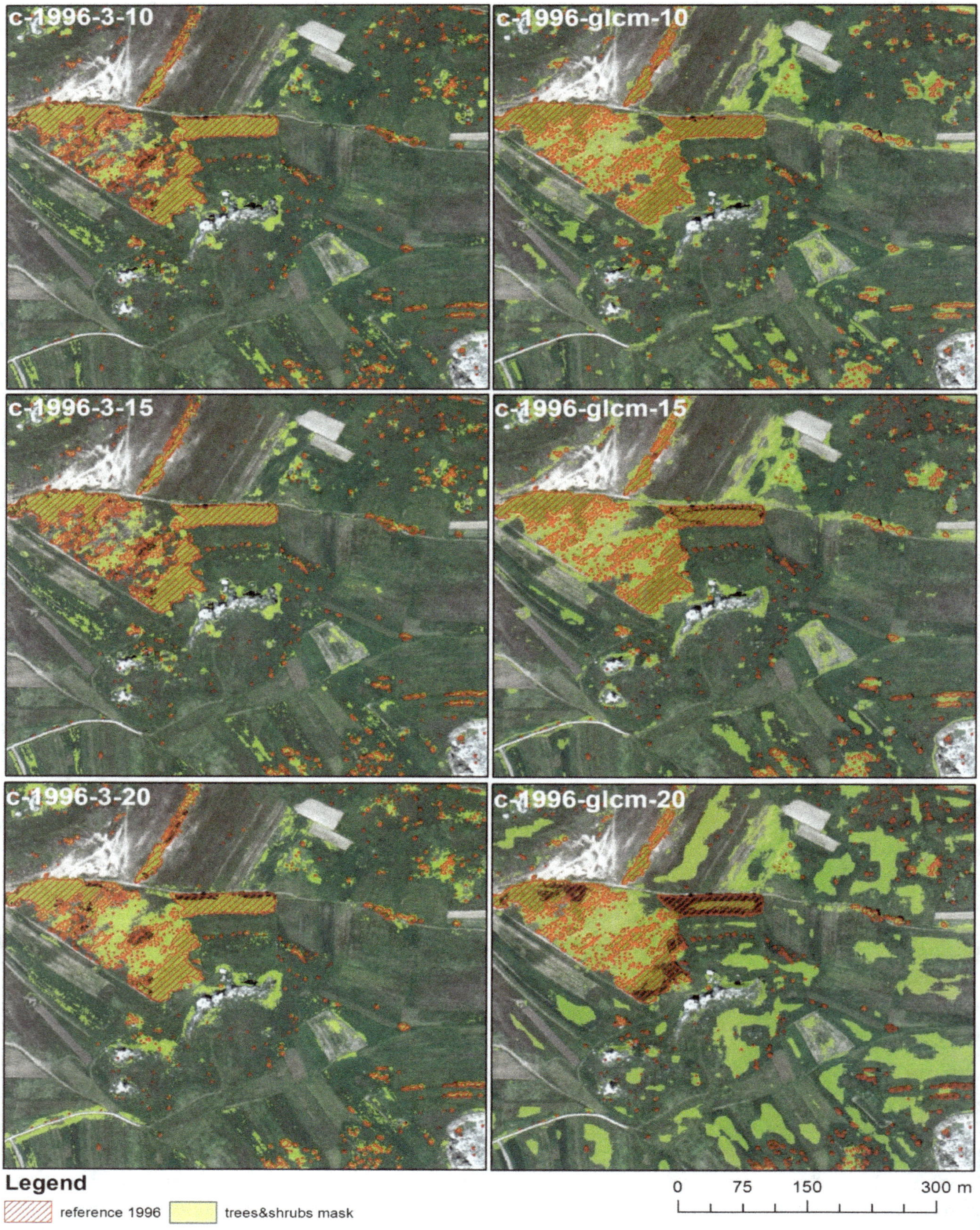

Figure 9. Comparison of tree and shrub masks for 1996, variants c-1996-3-10, c-1996-3-15, c-1996-3-20, c-1996-glcm-10, c-1996- glcm-15, and c-1996-glcm-20, with a reference mask imposed.

A comparison of the results of classification variants based on a granulometric analysis (differing in the number of granulometric maps used) allows to draw unambiguous conclusions. For this type of texture and with this pixel size (0.5 m), the use of granulometric maps based on a structuring element with a radius of more than 3 pixels (a size of approximately 3.5 m) does not improve the classification efficacy, and in most cases on the contrary: The use of granulometric maps showing the presence of textures with larger grains (larger than 3.5 m) can lead to a decrease in efficacy. This is due to the fact that this type of information is no longer related to the occurrence of wooded or shrubbed areas.

Analyzing the course of the border of wooded and bushy areas for particular dates and variants of texture analysis, the impact of the date of image acquisition is clearly visible. The problem is the shadow, which in the pictures acquired in early spring (2012) is the longest and increased significantly the detected surface of wooded areas (Figure 10). On the other hand, the shadow increases the texture of the image of deciduous trees—leafless at this time of the year. Due to this effect, they were correctly classified. In other dates, this effect is weaker—the borders were best reproduced in the photographs from August (2015) and May (1996 and 2003), i.e., the period when the sun is highest above the horizon, and the shadows are short.

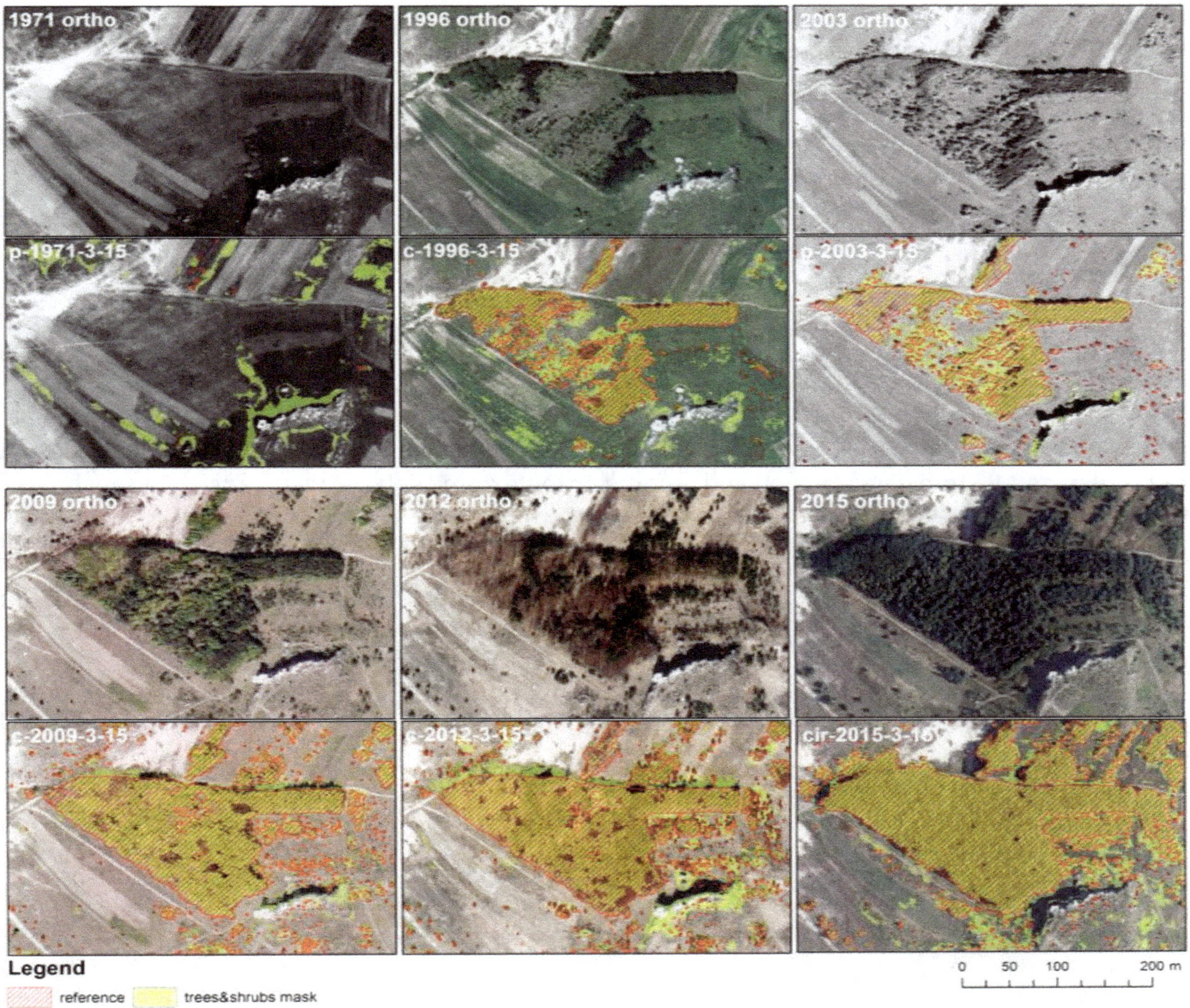

Figure 10. The effect of the date of image acquisition on the accuracy of determining the boundaries of wooded and shrubbed areas.

The analysis of the results also shows that regardless of the archival data and the texture analysis method used, individual trees and shrubs are difficult to detect—only larger trees and shrubs are possible to detect with these methods.

The importance of impact of the scale/GSD of aerial photographs on the effectiveness of determining the range of trees and shrubs has not been observed. The most important were the impact of the time period and radiometric quality of the source materials. Some of the errors in the case of photos from 1971, including film damage (scratches were seen in the images), resulted from the poor quality of these sources.

Summing up, in general, the highest accuracy in determining the range of trees and shrubs for photos recorded in the summer was obtained using granulometric analysis with six granulometric maps and a analysis radius of 10–15 pixels. For photos from the spring period, the GLCM method gave better results, also with a 10–15 pixel radius of analysis.

When assessing the tested methods from the point of view of their possible application, it should be noted that they are easy to implement and automate the classification process (also in open-source software). Importantly, compared to the dense image matching (DIM) technique (one of the most effective methods of analyzing aerial photographs in the context of research), they enable the determination of the range of trees and shrubs based on images obtained during the period when the trees and shrubs are not fully covered by leafs (early spring). DIM, although a method characterized by high accuracy in determining object boundaries and the ability to analyze in three dimensions, does not allow the determination of the range of trees and shrubs in the leafless period [20]. During this time, it is only possible to analyze the range of coniferous trees and shrubs, which may be of interest, e.g., in the context of conducting research on the impact of climate change on the range of occurrence of individual species of trees and shrubs [72].

7. Conclusions

The aim of the research was to assess the applicability of texture analysis for the analysis of the dynamics of succession of trees and shrubs. The research was carried out using aerial imagery (characterized by various technical parameters) acquired in six different years (1971, 1996, 2003, 2009, 2012, and 2015) in various phenological periods.

A total number of 330 classification variants, which differ in source data (date of image, spectral resolution) and texture data (method and parameters of texture analysis) were tested. The analysis of the obtained results showed great effectiveness in determining the wooded and shrubbed areas using texture analysis.

In the majority of cases tested, the results obtained indicate a similar efficacy of both methods of texture analysis (in the analyzed range), except for one case (1996), in which the efficacy of granulometric analysis was significantly better than the efficacy of GLCM analysis, due to the high texture in areas of arable land and meadows and to the relatively small scale of the photos. The advantage of the granulometric analysis resulted mainly from multi-scality, thanks to which the wooded/shrubbed areas were distinguished better from agricultural areas with a relatively high texture. Therefore, it appears from all the analyzed examples that granulometric analysis is much more stable in terms of the analyzed application. This confirms the results of previous studies which compared, among others, these two above-mentioned methods [36,71].

The accuracy of determining the wooded and shrubbed areas is also influenced by the date of image acquisition—in the case of early spring images, the observed long shadows, typical for this period, caused a significant extension of the area of tree/shrub masks. However, the role of the shadows was also positive, because thanks to them, the texture of the areas with deciduous trees, which were leafless at this time of year, was strengthened. It is therefore possible to designate tree and shrubbed areas based on images from leafless periods using texture analysis, which is much more difficult when using dense image matching methods [20].

In addition, in summer time, wooded and shrubbed areas may be overestimated due to the presence of herbaceous plants, which can also be characterized by high textures.

The operation of closing by reconstruction proved to be an effective method of filling gaps in wooded areas. However, in the case of large scattering of trees and shrubs, it leads to overestimation of the range of such areas. Therefore, it is necessary to consider the legitimacy of its use in areas where the succession of trees and shrubs can be thus characterized.

Summing up the conducted analyses, it can be concluded that texture analysis is an effective method for determining the range of wooded and shrubbed areas, in particular those with considerable density of trees and shrubs. The granulometric analysis showed generally greater suitability for this purpose than GLCM. In the case of detection of individual trees and shrubs, the effectiveness of such analyses is smaller—large trees and shrubs are generally correctly detected, while small-sized objects are not. This means that these methods can be an effective tool for monitoring only the later stages of succession of trees and shrubs.

Supplementary Materials:
Figure S1: Comparison of tree and shrub masks for 1971 obtained using P image, granulometric analysis variants, with a reference mask imposed; Figure S2: Comparison of tree and shrub masks for 1996 obtained using P image, granulometric analysis variants, with a reference mask imposed; Figure S3: Comparison of tree and shrub masks for 1996 obtained using RGB image, granulometric analysis variants, with a reference mask imposed; Figure S4: Comparison of tree and shrub masks for 2003 obtained using P image, granulometric analysis variants, with a reference mask imposed; Figure S5: Comparison of tree and shrub masks for 2009 obtained using P image, granulometric analysis variants, with a reference mask imposed; Figure S6: Comparison of tree and shrub masks for 2009 obtained using RGB image, granulometric analysis variants, with a reference mask imposed; Figure S7: Comparison of tree and shrub masks for 2012 obtained using P image, granulometric analysis variants, with a reference mask imposed; Figure S8: Comparison of tree and shrub masks for 2012 obtained using RGB image, granulometric analysis variants, with a reference mask imposed; Figure S9: Comparison of tree and shrub masks for 2015 obtained using P image, granulometric analysis variants, with a reference mask imposed; Figure S10: Comparison of tree and shrub masks for 2015 obtained using RGB image, granulometric analysis variants, with a reference mask imposed; Figure S11: Comparison of tree and shrub masks for 2015 obtained using CIR image, granulometric analysis variants, with a reference mask imposed; Figure S12: Comparison of tree and shrub masks for 1971, 1996, and 2003, GLCM variants, with a reference mask imposed; Figure S13: Comparison of tree and shrub masks for 2009 and 2012, GLCM variants, with a reference mask imposed; Figure S14: Comparison of tree and shrub masks for 2015, GLCM variants, with a reference mask imposed.

Author Contributions: The aim of the study and the methodology of research: P.K. and K.O.-S.; texture analysis and classification: P.K. and K.L.; reference data preparation: A.P. and K.O.-S.; accuracy assessment: K.O.-S.; results analysis, preparation of tables and charts, synthesis of the study data: K.O.-S. and P.K.; writing—original draft preparation: P.K. and K.O.-S.; writing—review and editing: K.O.-S. and P.K.; visualization preparation: K.O.-S.; WP5 leading and supervision: K.O.-S.

Acknowledgments: We would like to express our sincere gratitude to Bożena Michna and Agnieszka Jakubowska for administrative support during HabitARS project.

References

1. Falińska, K. *Ekologia Roślin*, 3rd ed.; Wydawnictwo Naukowe PWN: Warszawa, Poland, 2004; pp. 1–520.
2. Benjamin, K.; Domon, G.; Bouchard, A. Vegetation composition and succession of abandoned farmland: Effects of ecological, historical and spatial factors. *Landsc. Ecol.* **2005**, *20*, 627–647. [CrossRef]
3. Pueyo, Y.; Beguería, S. Modelling the rate of secondary succession after farmland abandonment in a Mediterranean mountain area. *Landsc. Urban. Plan.* **2007**, *8*, 245–254. [CrossRef]
4. Weiner, J. *Życie I Ewolucja Biosfery*; Wydawnictwo Naukowe PWN: Warszawa, Poland, 2003; pp. 1–610.
5. Falkowski, M.J.; Evans, J.S.; Martinuzzi, S.; Gessler, P.E.; Hudak, A.T. Characterizing forest succession with LIDAR data: An evaluation for the inland Northwest, USA. *Remote Sens. Environ.* **2009**, *113*, 946–956. [CrossRef]
6. Kolecka, N.; Kozak, J.; Kaim, D.; Dobosz, D.; Ginzler, C.; Psomas, A. Mapping secondary forest succession on abandoned agricultural land with LiDAR point clouds and terrestrial photography. *Remote Sens.* **2015**, *7*, 8300–8322. [CrossRef]

7. Radecka, A.; Michalska-Hejduk, D.; Osińska-Skotak, K.; Kania, A.; Górski, K.; Ostrowski, W. Mapping secondary succession species in agricultural landscape with the use of hyperspectral and ALS data. *J. Appl. Remote Sens.* **2019**, *13*, 034502. [CrossRef]
8. Próchnicki, P. The implementation of GIS and remote sensing to analysis of shrub succession in the Narew National Park. *Rocz. Geomatyki* **2006**, *I*, 127–134.
9. Maryniak, D.; Drzewiecki, W. Land cover changes in Błędowska Desert area between 1926 and 2005. *Arch. Fotogram. Kartografii i Teledetekcji* **2010**, *21*, 245–256.
10. Rahmonov, O.; Oleś, W. Vegetation succession over an area of a medieval ecological disaster. The case of the Błędów Desert, Poland. *Erkunde* **2010**, *64*, 241–255. [CrossRef]
11. Bryś, H.; Gołuch, P. Pustynia Błędowska dawniej i dziś—Interpretacja wieloczasowych zdjęć lotniczych i obrazów satelitarnych. *Acta Scientiarum Polonorum Geodesia et Descriptio Terrarum* **2011**, *10*, 5–16.
12. Oikonomakis, N.; Ganatsas, P. Land cover changes and forest succession trends in a site of Natura 2000 network (Elatia forest), in northern Greece. *For. Ecol. Manag.* **2012**, *285*, 153–163. [CrossRef]
13. Kolecka, N.; Dobosz, M.; Ostafin, K. Forest Cover Change and Secondary Forest Succession Since 1977 in Budzów Commune, the Polish Carpathians. *Prace Geograficzne* **2016**, *146*, 51–65.
14. Holopainen, M.; Jauhiainen, S. Detection of peatland vegetation types using digitized aerial photographs. *Can. J. Remote Sens.* **1999**, *25*, 475–485. [CrossRef]
15. Miller, M.E. Use of historic aerial photography to study vegetation change in the Negrito Creek watershed, southwestern New Mexico. *Southwest Nat.* **1999**, *44*, 121–131.
16. Pitt, D.; Runesson, U.; Bell, F.W. Application of large- and medium-scale aerial photographs to forest vegetation management: A case study. *For. Chron.* **2000**, *76*, 903–913. [CrossRef]
17. Ligocki, M. Zastosowanie zdjęć lotniczych do badania sukcesji wtórnej na polanach śródleśnych. *Teledetekcja Środowiska* **2001**, *32*, 143–151.
18. Jauhiainen, S.; Holopainen, M.; Rasinmäki, A. Monitoring peatland vegetation by means of digitized aerial photographs. *Scand. J. For. Res.* **2007**, *22*, 168–177. [CrossRef]
19. Szostak, M.; Wężyk, P.; Hawryło, P.; Puchała, M. Monitoring the secondary forest succession and land cover/use changes of the błędów desert (Poland) using geospatial analyses. *Quaest. Geogr.* **2016**, *35*, 5–13. [CrossRef]
20. Osińska-Skotak, K.; Jełowicki, Ł.; Bakuła, K.; Michalska-Hejduk, D.; Wylazłowska, J.; Kopeć, D. Analysis of using dense image matching techniques to study the process of secondary succession in Non-forest Natura 2000 habitats. *Remote Sens.* **2019**, *11*, 893. [CrossRef]
21. Julesz, B. Visual pattern discrimination. *IRE Trans. Inf. Theory* **1962**, *8*, 84–92. [CrossRef]
22. Darling, E.M.; Joseph, R.D. Pattern recognition from satellites altitudes. *IEEE Trans. Syst. Man Cybern.* **1968**, *4*, 30–47. [CrossRef]
23. Haralick, R.M.; Shanmugam, K.; Dinstein, I. Textural Features for Image Classification. *IEEE Trans. Syst. Man Cybern.* **1973**, *4*, 610–621. [CrossRef]
24. Lam, N.S.N. Description and measurement of Landsat TM using fractals. *Photogramm. Eng. Remote Sens.* **1990**, *56*, 187–195.
25. Mallat, S.G. A Theory for Multiresolution Signal Decomposition: The Wavelet Representation. *IEEE Trans. Pattern Anal. Mach. Intell.* **1989**, *11*, 674–693. [CrossRef]
26. Marr, D. *Vision*; Freeman and Company: New York, NY, USA, 1982; Chapter 2; pp. 54–78.
27. Horn, B. *Robot Vision*; MIT Press: Cambridge, MA, USA, 1986.
28. Haralick, R.M.; Shapiro, L. *Computer and Robot Vision*; Addison-Wesley Publishing Company: Berkshire, UK, 1992; Volume 1, pp. 346–351.
29. Spitzer, F. *Random Fields and Interacting Particle Systems*; M.A.A. Summer Seminar Notes; Mathematical Association of America: Washington, DC, USA, 1971.
30. Preston, C.J. *Gibbs States on Countable Sets*; Cambridge University Press: London, UK, 1974.
31. Haas, A.; Matheron, G.; Serra, J. Morphologie Mathématique et granulométries en place. *Ann. Mines* **1967**, *12*, 768–782.
32. Dougherty, E.R.; Pelz, J.B.; Sand, F.; Lent, A. Morphological image segmentation by local granulometric size distributions. *J. Electron. Imaging* **1992**, *1*, 46–60.
33. Cheng, G.; Li, Z.; Han, J.; Yao, X.; Guo, L. Exploring Hierarchical Convolutional Features for Hyperspectral Image Classification. *IEEE Trans. Geosci. Remote Sens.* **2018**, *56*, 6712–6722. [CrossRef]

34. Zhou, P.; Han, J.; Cheng, G.; Zhang, B. Learning compact and discriminative stacked autoencoder for hyperspectral image classification. *IEEE Trans. Geosci. Remote Sens.* **2019**, *57*, 1–11. [CrossRef]
35. Kupidura, P. *Wykorzystanie granulometrii obrazowej w klasyfikacji treści zdjęć satelitarnych*; Prace Naukowe Politechniki Warszawskiej; Warsaw University of Technology Publishing House: Warsaw, Poland, 2015.
36. Kupidura, P. The Comparison of different methods of texture analysis for their efficacy for land use classification in satellite imagery. *Remote Sens.* **2019**, *11*, 1233. [CrossRef]
37. Kupidura, P.; Popławski, W.; Sitko, P. Comparison of efficiency of extraction of built-up areas in aerial images using fractal analysis and morphological granulometry. *Teledetekcja Środowiska* **2015**, *52*, 29–37.
38. Weszka, J.S.; Dyer, C.R.; Rosenfeld, A. A Comparative Study of Texture measures for Terrain Classification. *IEEE Trans. Syst. Man Cybern.* **1976**, *6*, 269–285. [CrossRef]
39. Conners, R.W.; Harlow, C.A. A Theoretical Comaprison of Texture Algorithms. *IEEE Trans. Pattern Anal. Mach. Intell.* **1980**, *2*, 204–222. [CrossRef] [PubMed]
40. Mering, C.; Chopin, F. Granulometric maps from high resolution satellite images. *Image Anal. Stereol.* **2002**, *21*, 19–24. [CrossRef]
41. Bekkari, A.; Idbraim, S.; Elhassouny, A.; Mammass, D.; El Yassa, M.; Ducrot, D. SVM and Haralick Features for Classification of High Resolution Satellite Images from Urban Areas. In *Image and Signal Processing*; Elmoataz, A., Mammass, D., Lezoray, O., Nouboud, F., Aboutajdine, D., Eds.; ICISP 2012. Lecture Notes in Computer Science; Springer: Berlin, Heidelberg, 2012; Volume 7340, pp. 17–26.
42. Wawrzaszek, A.; Krupiński, M.; Aleksandrowicz, S.; Drzewiecki, W. Fractal and multifractal characteristics of very high resolution satellite images. In Proceedings of the 2013 IEEE International Geoscience and Remote Sensing Symposium—IGARSS, Melbourne, Australia, 21–26 July 2013; pp. 1501–1504.
43. Kupidura, P.; Skulimowska, M. Morphological profile and granulometric maps in extraction of buildings in VHR satellite images. *Arch. Photogramm. Cartogr. Remote Sens.* **2015**, *27*, 83–96.
44. Aleksandrowicz, S.; Wawrzaszek, A.; Drzewiecki, W.; Krupiński, M. Change detection using global and local multifractal description. *IEEE Geosci. Remote Sens. Lett.* **2016**, *13*, 1183–1187. [CrossRef]
45. Drzewiecki, W.; Wawrzaszek, A.; Krupiński, M.; Aleksandrowicz, S.; Bernat, K. Applicability of multifractal features as global characteristics of WorldView—2 panchromatic satellite images. *Eur. J. Remote Sens.* **2016**, *49*, 809–834. [CrossRef]
46. Humeau-Heurtier, A. Texture feature extraction methods: A survey. *IEEE Access* **2019**, *7*, 8975–9000. [CrossRef]
47. Baraldi, A.; Parmiggiani, F. An investigation of the textural characteristics associated with gray level coocurrence matrix statistical parameters. *IEEE Trans. Geosci. Remote Sens.* **1995**, *33*, 293–304. [CrossRef]
48. Pathak, V.; Dikshit, O. A new approach for finding appropriate combination of texture parameters for classification. *Geocarto Int.* **2010**, *25*, 295–313. [CrossRef]
49. OTB CookBook. Available online: https://www.orfeo-toolbox.org/CookBook/recipes/featextract.html (accessed on 19 July 2019).
50. Unser, M. Sum and difference histograms for texture classification. *IEEE Trans. Pattern Anal. Mach. Intell.* **1986**, *8*, 118–125. [CrossRef]
51. Kupidura, P.; Koza, P.; Marciniak, J. *Morfologia Matematyczna w teledetekcji*; Wydawnictwo Naukowe PWN: Warsaw, Poland, 2010.
52. Vincent, L. Opening Trees and Local Granulometries. In *Mathematical Morphology and its Applications to Image and Signal Processing*; Springer: Boston, MA, USA, 1996; pp. 273–280.
53. Mura, D.A.; Benediktsson, J.A.; Waske, B.; Bruzzone, L. Morphological attribute profiles for the analysis of very high resolution images. *IEEE Trans. Geosci. Remote Sens.* **2010**, *48*, 3747–3762. [CrossRef]
54. Mura, D.A.; Benediktsson, J.A.; Bruzzone, L. Self-dual Attribute Profiles for the Analysis of Remote Sensing Images. In *Mathematical Morphology and Its Applications to Image and Signal Processing*; Soille, P., Pesaresi, M., Ouzounis, G.K., Eds.; ISSM 2011. Lecture Notes in Computer Science; Springer: Heidelberg/Berlin, Germany, 2011; Volume 6671, pp. 320–330.
55. Ruiz, L.A.; Fdez-Sarria, A.; Recio, J.A. Texture feature extraction for classification of remote sensing data using wavelet decomposition: A comparative study. *ISPRS Archives* **2004**, *35*, 1109–1114.
56. RDOŚ Katowice (Regional Directorate for Environmental Protection in Katowice). Ostoja Olsztyńsko-Mirowska. Available online: http://katowice.rdos.gov.pl/files/artykuly/25790/ostoja_olsztynsko_mirowska.pdf (accessed on 5 October 2019).

57. Upper Silesia Nature Heritage Center. Available online: http://przyroda.katowice.pl/pl/ochrona-przyrody/natura-2000/ostoje-siedliskowe/300-ostoja-olsztysko-mirowska (accessed on 5 October 2019).
58. Regional Directorate for Environmental Protection in Katowice, LFE11 NAT/PL/432 Protection of valuable natural non-forest habitats typical of the Orle Gniazda Landscape Park. Available online: http://lifezpkws.pl (accessed on 5 October 2019).
59. Salach, A. SAPC—Application for adapting scanned analogue photographs to use them in structure from motion technology. *Int Arch. Photogramm. Remote Sens Spat. Inf. Sci.* **2017**, *XLII-1/W1*, 197–204. [CrossRef]
60. Blakeman, R.H. The Identification of Crop Disease and Stress by Aerial Photography. In *Applications of Remote Sensing in Agriculture*; Steven, M.D., Clark, J.A., Eds.; Elsevier: Amsterdam, The Netherlands, 1990; pp. 229–254.
61. Schulte, W.O. The use of panchromatic, infrared, and color aerial photography in the study of plant distribution. *Photogramm. Eng.* **1951**, *XVII*, 688–714.
62. Staniak, K. Badanie Wpływu Rodzaju Obrazu Źródłowego Na Efektywność Analizy Granulometrycznej. Master's Thesis, Warsaw University of Technology, Warsaw, Poland, 2016.
63. BlueNote Software. Available online: https://sourceforge.net/projects/bluenote (accessed on 19 July 2019).
64. Niemyski, S. Comparison of Chosen Decision Rules in Classification of Multispectral Satellite Images. Master's Thesis, Warsaw University of Technology, Warsaw, Poland, 2018.
65. Nieniewski, M. *Segmentacja Obrazów Cyfrowych. Metody Segmentacji Wododziałowej*; Akademicka Oficyna Wydawnicza EXIT: Warszawa, Poland, 2005; pp. 1–184.
66. Congalton, R.G.; Green, K. *Assessing the Accuracy of Remotely Sensed Data: Principles and Practices*; CRC Press, Taylor & Francis Group: Boca Raton, FL, USA, 2008.
67. Powers, D.M.W. Evaluation: From precision, recall and F-measure to ROC, informedness, markedness & correlation. *J. Mach. Learn. Technol.* **2011**, *2*, 37–63.
68. Li, Z.; Hayward, R.; Zhang, J.; Jin, H.; Walker, R. Evaluation of spectral and texture features for object-based vegetation species classification using support vector machines. *ISPRS Archives* **2010**, *38*, 122–126.
69. Mirzapour, F.; Ghassemian, H. Improving hyperspectral image classification by combining spectral, texture and shape features. *Int. J. Remote Sens.* **2015**, *36*, 1070–1096.
70. Staniak, K.; Kupidura, P. Analysis of the impact of the source image type on the efficacy of texture analysis. *Teledetekcja Środowiska* **2017**, *57*, 1–16.
71. Kupidura, P.; Uwarowa, I. The comparison of GLCM and granulometry for distinction of different classes of urban area. In Proceedings of the 2017 Joint Urban Remote Sensing Event (JURSE), Dubai, UAE, 6–8 March 2017; pp. 1–4. [CrossRef]
72. Farjon, A. Picea abies. The IUCN Red List of Threatened Species 2017: E.T42318A71233492. 2017. Available online: http://dx.doi.org/10.2305/IUCN.UK.2017-2.RLTS.T42318A71233492.en (accessed on 15 August 2019).

10

Oil Flow Analysis in the Maritime Silk Road Region using AIS data

Yijia Xiao [1,2], Yanming Chen [1,2,*], Xiaoqiang Liu [1,2], Zhaojin Yan [1,2], Liang Cheng [1,2] and Manchun Li [1,2]

1 Jiangsu Provincial Key Laboratory of Geographic Information Science and Technology, Key Laboratory for Land Satellite Remote Sensing Applications of Ministry of Natural Resources, School of Geography and Ocean Science, Nanjing University, Nanjing 210023, China; MG1827078@smail.nju.edu.cn (Y.X.); MG1727062@smail.nju.edu.cn (X.L.); DG1727032@smail.nju.edu.cn (Z.Y.); lcheng@nju.edu.cn (L.C.); limanchun@nju.edu.cn (M.L.)

2 Collaborative Innovation Center of South China Sea Studies, Nanjing 210023, China

* Correspondence: chenyanming@nju.edu.cn

Abstract: Monitoring maritime oil flow is important for the security and stability of energy transportation, especially since the "21st Century Maritime Silk Road" (MSR) concept was proposed. The U.S. Energy Information Administration (EIA) provides public annual oil flow data of maritime oil chokepoints, which do not reflect subtle changes. Therefore, we used the automatic identification system (AIS) data from 2014 to 2016 and applied the proposed technical framework to four chokepoints (the straits of Malacca, Hormuz, Bab el-Mandeb, and the Cape of Good Hope) within the MSR region. The deviations and the statistical values of the annual oil flow from the results estimated by the AIS data and the EIA data, as well as the general direction of the oil flow, demonstrate the reliability of the proposed framework. Further, the monthly and seasonal cycles of the oil flows through the four chokepoints differ significantly in terms of the value and trend but generally show an upward trend. Besides, the first trough of the oil flow through the straits of Hormuz and Malacca corresponds with the military activities of the U.S. in 2014, while the second is owing to the outbreak of the Middle East Respiratory Syndrome in 2015.

Keywords: automatic identification system data; 21st Century Maritime Silk Road region; oil flow analysis; maritime oil chokepoint; Middle East Respiratory Syndrome

1. Introduction

"The Belt and Road Initiative", comprising the "Silk Road Economic Belt" and "21st Century Maritime Silk Road" (MSR), was proposed by China to promote economic trade and deepen the connection between China and its associated countries [1–4]. The MSR region is not only the lifeline of China's energy transportation [5] but also a key region in terms of maritime oil transportation [6]. The maritime oil chokepoint (MOC) is defined as the narrow channels along the widely used global sea routes by the U.S. Energy Information Administration (EIA) [7,8]. Within the MSR and its surrounding regions, the straits of Malacca, Hormuz, Bab el-Mandeb, and the Cape of Good Hope are four MOCs. These four MOCs control the oil import of the largest oil importer, China [6,9], and the oil export of the Persian Gulf, as well as other major oil importers and exporters. Therefore, the oil flow through these four chokepoints can show the real energy transportation within the MSR and its surrounding regions directly. Furthermore, transportation systems are often exposed to risks ranging from natural disasters to hazardous events caused by man [4,10,11]. Therefore, the monitoring of oil flow through those MOCs can show anomalies and help countries to respond promptly. Presently, the data on the available average daily oil flows through MOCs from 2011 to 2016, released by the EIA, are calculated from the annual oil flow [7,8]. However, the information on the EIA data is too limited to meet the requirements for analysis on a smaller timescale. Therefore, obtaining and analyzing the oil flow in these four MOCs on multiple timescales is significant for ensuring China's maritime oil transportation security and strategic energy reserves.

To determine and analyze the oil flow, it is necessary to first assess the number of oil tankers passing through the MOCs at a certain time and generate statistics accordingly. In existing research, statistical methods for the ship traffic volume are divided into two categories: statistics based on imaging systems and statistics based on laser sensors. Statistics based on imaging systems employ images covering MOCs from radar, closed-circuit television (CCTV), infrared, and other imaging systems. Although the radar imaging system [12,13] presents the advantages of a large observation range and is not affected by the weather, not all areas can be observed by the system. In addition, the radar image can only provide spatial information about the ship. As supplements to the radar imaging system, the CCTV imaging system [14,15] and infrared imaging system [16,17] can obtain intuitive images, but they are extremely limited when it comes to monitoring wide ranges. For the statistics based on laser sensors [18,19], a laser beam is emitted to the target by a laser-ranging sensor, and it is received by the photoelectric element after being reflected by the target. Therefore, the position of the target is calculated based on the time elapsed from the emission to the reception of the laser beam and the speed of laser propagation, which is key to detecting ships through the MOCs. This method presents the advantages of round-the-clock operation, low cost, and high accuracy. However, it only obtains the spatial information about the ship and not the load information. Although the aforementioned methods can provide statistics regarding ship traffic volume, the imaging and laser sensors are only deployed near the port and can only obtain statistical information near the port. Furthermore, they present a common problem regarding the lack of load information in the calculation of the oil flow.

In 2000, the International Maritime Organization (IMO) adopted a regulation stating that internationally voyaging cargo ships of 300 gross tonnage or more, non-internationally voyaging cargo ships of 500 gross tonnage or more, and all passenger ships regardless of size are required to be equipped with an automatic identification system (AIS) [20–22]. Being a new type of spatio-temporal data, AIS data provide the potential for the oil flow analysis of the MOCs across four timescales (daily, monthly, seasonal, and annual). The AIS data can be received via shore-based facilities or satellite-AIS [20,23], which enables the worldwide monitoring of ship activities. These data, which include information such as the deadweight tonnage, IMO number (ship number issued by IMO), and the Maritime Mobile Service Identity (MMSI) number (ship number issued by MMSI), can be used to acquire unique ship identification and oil flow statistics. Previous studies mainly focused on ship spatial feature mining [6,24,25], ship anomaly detection [26–28], ship collisions [29–31], ship main route extraction [5,32–34], geospatial pattern analysis [35,36], catching assessment [37], and environmental pollution assessment [38–41]. However, studies on the oil flow statistics of MOCs are rare.

Considering the aforementioned problem, a maritime oil flow analysis technical framework is proposed herein. This framework promotes the statistics of oil flow to the scale of a single ship, and the temporal resolution and spatial resolution are improved when compared with the statistical data. Furthermore, the framework proposed in this study can calculate the transport volume of oil, which addresses the disadvantage that the previous methods have in only being able to calculate the deadweight of the oil tankers. We apply the framework to the straits of Malacca, Hormuz, Bab el-Mandeb, and the Cape of Good Hope by using the AIS data from 2014 to 2016. Therefore, we calculated and analyzed the oil flow through the MOCs, thus making available statistical data for ensuring the security and stability of oil transportation within the MSR and its surrounding region. Furthermore, with the support of real-time data, the oil flow through these choke points can be monitored for countries within the MSR and its surrounding regions to master and respond to special situations.

2. Study Area and Data

2.1. Study Area

The study area covers the MSR and its surrounding region, including the straits of Malacca, Hormuz, Bab el-Mandeb, and the Cape of Good Hope (see Figure 1). These four MOCs are our study targets due to their important strategic positions.

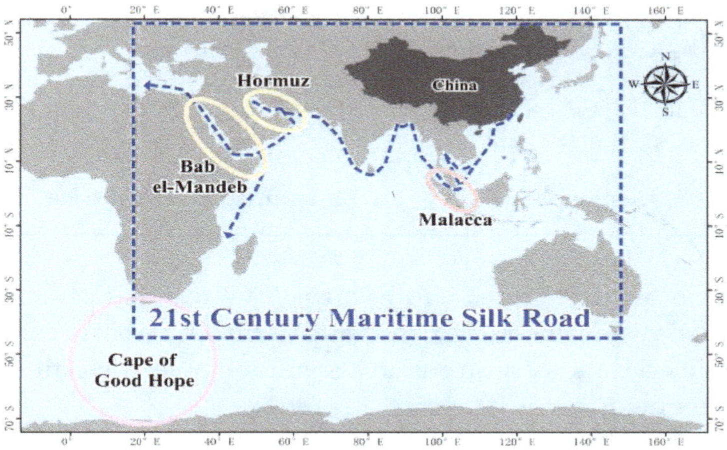

Figure 1. The study area.

The Strait of Malacca, which links the Indian Ocean and the South China Sea, is an important channel for West Asia to transport oil to East Asia. China, Japan, and South Korea, and some other countries regard it as the "lifeline" of energy transportation. Flow through the Strait of Malacca rose to 16 million barrels per day (b/d) in 2016, whereby the Strait of Malacca retained its position as the second busiest MOC [8]. The Strait of Hormuz, which links the Persian Gulf and the Arabian Sea, controls the oil export from the Persian Gulf. It is the busiest MOC worldwide, through which millions of barrels of oil travel globally every day [8]. The Bab el-Mandeb Strait, which links the Red Sea and the Arabian Sea, is an important waterway for maritime traffic and trade between Europe, Asia, and Africa. More than 20,000 ships pass through it annually, making it one of the most important and busiest straits. Suez tankers traveling from the Persian Gulf to Europe generally choose to travel through that strait. The Cape of Good Hope, which links the Atlantic Ocean and the Indian Ocean, was the best choice for ships to travel between Asia and Europe before the Suez Canal was navigated. It still works for supertankers that are unable to travel through the Suez Canal today.

2.2. Study Data

Limited by our existing research data, this study used the AIS data from 1 January 2014 to 31 December 2016. This study stored 1096 database files for an interval of one day. The number of the original AIS data is more than 7.1 billion, which accounts for 1594G. The data contain information for 30 attributes (presented in Table 1), which can be divided into three categories: static information, dynamic information, and voyage-related information [23]. Static information relates to the fixed physical characteristics of the ship itself. This information is recorded manually; thus, it is prone to missing data and errors. Dynamic information is the information that changes over time on a voyage. The navigation status information in the dynamic information is manually inputted, which has the values of "underway using the engine", "at anchor", "moored", "underway sailing", etc. The rest is generated automatically via the sensor connected to the AIS; therefore, it is of high reliability. The longitude and latitude information are in full precision (1/10,000 degree) [42]. Voyage-related information refers to the information that must be manually inputted before each voyage. This includes the details of the "estimated arrival time", "destination", and "draft." Such information is generally reported to countries along the way via ship-shore data exchange.

Table 1. Field of automatic identification system (AIS) data.

Categories	Attribute		
Static information	1. MMSI	2. IMO	3. Vessel_name
	4. Callsign	5. Vessel_type	6. Vessl_type_code
	7. Vessel_type_cargo	8. Vessel_class	9. Length
	10. Width	11. Flag	12. Flag_code
	13. Vessel_type_main	14.Vessel_type_sub	
Dynamic information	1. Longitude	2. Latitude	3. Speed overground
	4. Course overground	5. Rate of turning	6. Heading
	7. Nav_status	8. Nav_status_code	9. Source
	10.Ts_pos_utc	11. Ts_static_utc	12.DT_pos_utc
	13.DT_static_utc		
Voyage-related information	1. Destination	2. Estimated arrival time	3. Draft

The load information was obtained from https://www.myshiptracking.com/, https://www.vesselfinder.com/, https://www.marinetraffic.com/, http://ship.chinaports.com/, and http://marinelike.com/en/vessels/. The obtained data were stored in a separate table, including three attributes: MMSI, Vessel_Type, and DeadWeight.

3. Study Method

In this study, a maritime oil flow analysis technical framework, which consists of data preprocessing, extraction of the ship point pairs, and discrimination of the oil tanker load condition, is proposed (see Figure 2). To develop this framework, this study used the AIS spatio-temporal massive data. The framework is applied to the straits of Malacca, Hormuz, Bab el-Mandeb, and the Cape of Good Hope.

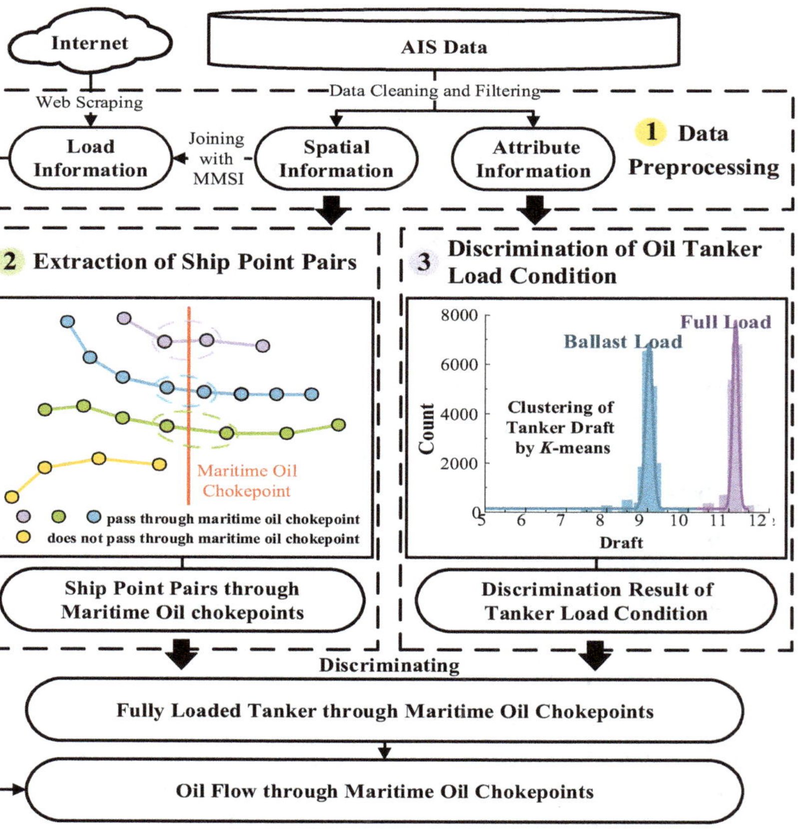

Figure 2. Main technical framework.

3.1. Data Preprocessing

The original AIS data inevitably contain errors and missing information due to the manual input of the data. Therefore, data cleaning is necessary to ensure the quality of the AIS data. The data that need to be cleaned are collectively called "dirty data." Dirty data consist of three types of data: incomplete data, erroneous data, and duplicate data [43–45]. The standards for data cleaning are presented in Table 2.

Table 2. Data cleaning standards.

Data Type	Cleaning Standard
Incomplete data	Data whose IMO, MMSI, Draft, or TS_pos_utc are empty
Erroneous data	Data whose MMSI is not 9 digits long
Duplicate data	Data with the same property values for all fields

The AIS data contain non-tanker data and data outside the MOCs, which are not required for this study. Therefore, after data cleaning, data filtering is required to eliminate any redundant information. In this study, the criterion of data filtering is that the value of "Vessel_type_sub" is "crude oil tanker" or "oil products tanker" and that the AIS data is within four MOCs. The result of attribute filtering is used in Section 3.3. Because the oil flow statistics require load information, the result of attribute and spatial filtering is connected to the load information through the public key of the MMSI number, which is then used in Section 3.2. The process of data preprocessing is shown in Figure 3.

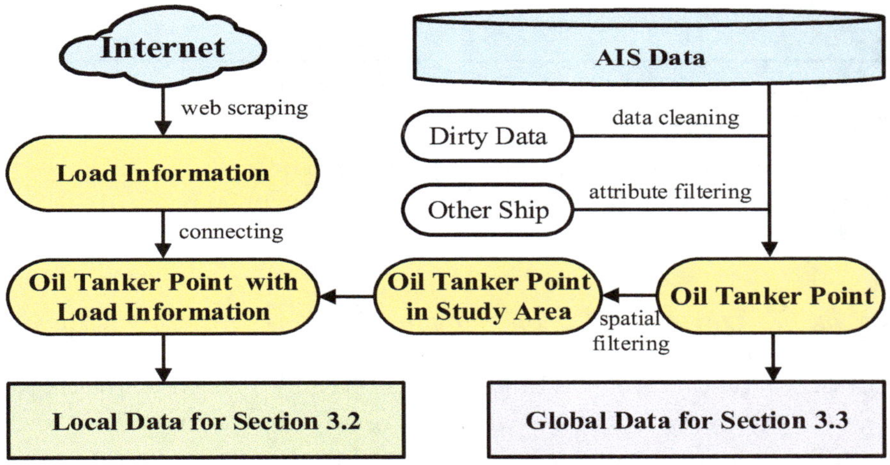

Figure 3. Data preprocessing.

3.2. Extraction of Ship Point Pairs Based on Ship Trajectory

It is necessary to generate the ship trajectory to determine whether the ship passes through a MOC or not. The ship trajectory is formed by connecting the set of ship points with the same MMSI number in chronological order (see Figure 4a). Its mathematical expression is $S_{MMSI} = \{V_1, V_2, \ldots, V_k, \ldots, V_{m-1}, V_m\}$, where MMSI in S_{MMSI} can identify the ship trajectory uniquely and V_k is the shipping point reported at time t_k.

The generated ship trajectory does not distinguish voyages, and it still has redundant information. Therefore, it is necessary to extract the ship point pair around the MOC from the trajectory. If adjacent trajectory points V_k and V_{k+1} are on both sides of the chokepoint, then they are extracted as the ship point pair $\{V_k, V_{k+1}\}$; otherwise, they would be dropped (refer to Figure 4b). As each ship point pair corresponds to the ship passing the MOC in a voyage, extracting the ship point pair can identify the voyage. This can simultaneously eliminate redundant information.

The adjacent points in some ship point pairs have significant time differences, which cannot be considered to be the same voyage. Therefore, we set the threshold as 24 h to remove these anomalies [5].

If the time difference between the adjacent points is more than 24 h, the ship point pair will be deleted; otherwise, it will be retained (see Figure 4c). The retained ship point pairs will be used for the subsequent oil flow statistics (see Figure 4d).

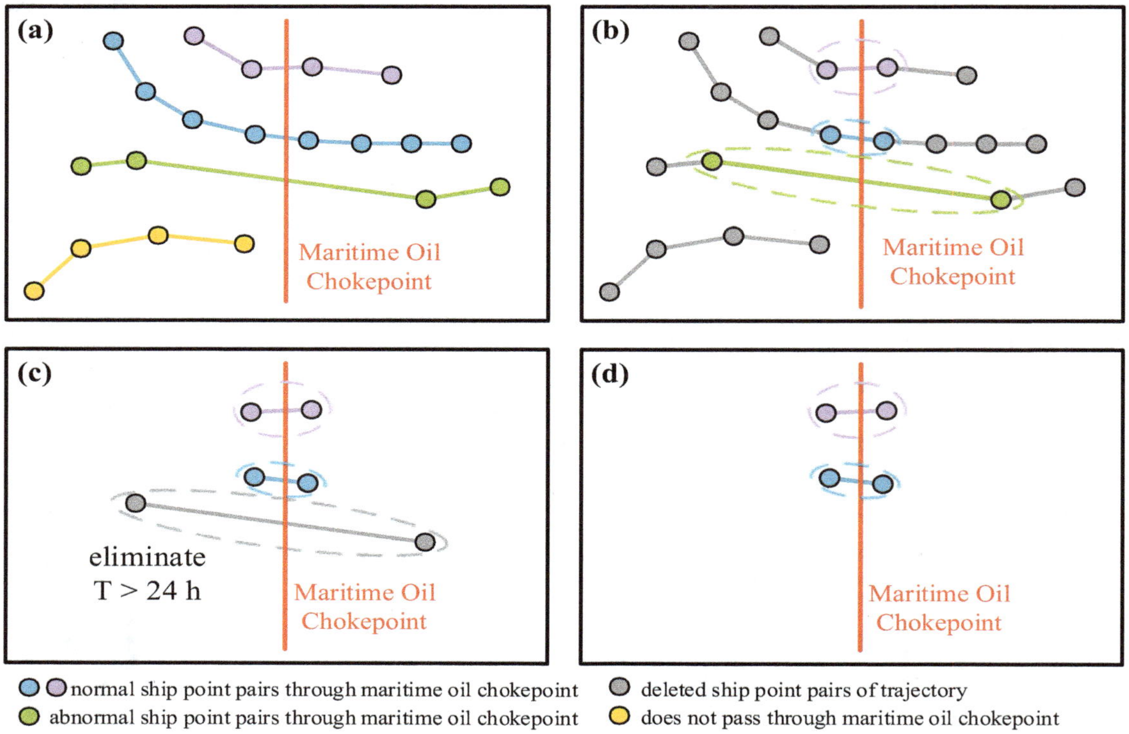

Figure 4. The extraction of ship point pairs. (**a**) Original ship trajectories; (**b**) The process of extracting the ship point pairs; (**c**) The process of eliminating abnormal ship point pairs; (**d**) The extracting result of ship point pairs.

3.3. Discrimination of the Oil Tanker Load Condition by the K-Means Clustering Method

The load condition of the oil tanker while passing through the MOC needs to be determined to develop the statistics for the oil flow. Park et al. mentioned that for the tanker and bulk carriers, the two most common operating conditions are the full load and ballast conditions [46]. Therefore, we only considered the full load and the ballast load of the tanker in the calculations, regardless of the other conditions.

For every oil tanker, we counted the record number under the different drafts and obtained the statistical results. The frequency histogram would be drawn by taking the draft as the abscissa and the record number as the ordinate (see Figure 5b). The frequency distribution map of the oil tanker draft presents a bimodal distribution structure, and the two peaks correspond to the ballast load and full load, respectively.

This fact enables us to discriminate the load condition of every oil tanker using the k-means clustering method [47,48]. The k-means clustering method is conducted by taking the draft as the clustering distance and the record number as the weight. The category with a small value of clustering results is considered to be the load condition of the "ballast load", while the other category is considered to be the "full load" (see Figure 5c). There is a large difference between the draught under the full load and that under the ballast load. By using this method, the draught with a small deviation will not lead to a misjudgment of the load condition, which ensures that the result will not be affected by the unreliable draught.

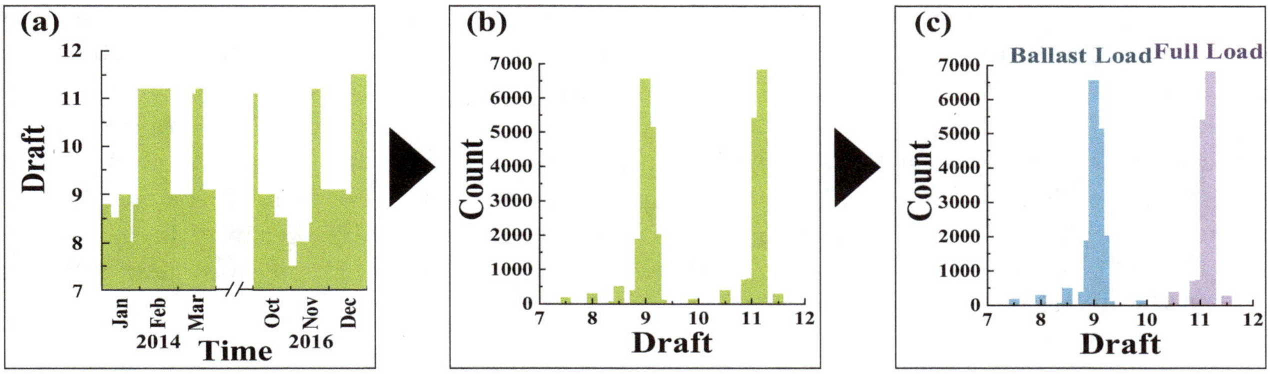

Figure 5. Discrimination of the oil tanker load condition. (**a**) The variation of the oil tanker draft over time; (**b**) The frequency distribution map of the oil tanker draft; (**c**) The clustering result of the oil tanker draft.

4. Results and Discussion

Combined with the discriminant result of the oil tanker load condition, the oil tanker point pair passing through the MOC can be identified as a ballast load or full load. For the condition of the full load, we used a deadweight to indicate the capacity of the tankers in a voyage. By contrast, for the condition of the empty load, zero was used instead. Therefore, the oil flow through the chokepoint can be obtained and analyzed across four timescales: the day, month, season, and year.

4.1. Annual Variation in Oil Flow

We compared the annual oil flow through the four MOCs estimated from the AIS data with the annual oil flow released by the EIA (see Figure 6 and Table 3). The comparison shows that the two sets of data are highly consistent. This indicates that the framework proposed in this study is reliable.

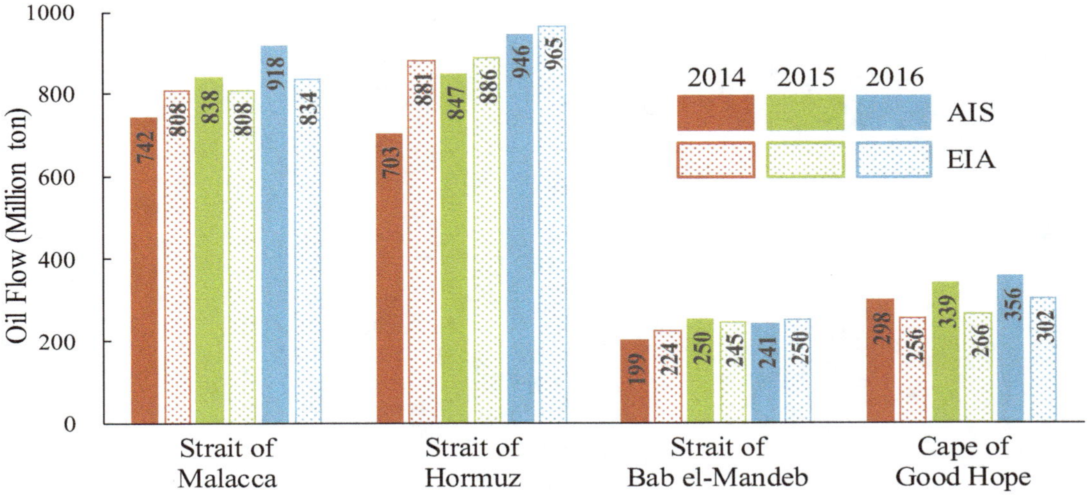

Figure 6. Annual oil flow through maritime oil chokepoints (MOCs) compared with the Energy Information Administration (EIA) data.

Table 3. Ratio (%) of the annual oil flow estimated in this study compared to the EIA data of the MOCs.

Year	Strait of Malacca	Strait of Hormuz	Strait of Bab el-Mandeb	Cape of Good Hope
2014	91.80	79.82	88.82	116.66
2015	103.68	95.53	102.09	127.29
2016	109.98	98.02	96.17	117.67

The estimated annual oil flow values through the straits of Malacca, Hormuz, and Bab el-Mandeb do not differ much from the EIA data. Except for the value in the Strait of Hormuz in 2014, which is approximately 80% of the EIA data, the rest are approximately between 90% and 110% of the EIA data (See Table 3). In the Cape of Good Hope, because it is a different statistical region, the estimated annual oil flow value is at least 116% of the EIA data, which is significantly higher than the EIA data. We statistically analyzed all of the oil tankers passing between the Cape of Good Hope and Antarctica in this study, while the EIA data only considered the oil tankers within a certain range around the Cape of Good Hope. Therefore, the annual oil flow in the Cape of Good Hope estimated in this study has a higher value.

The annual oil flow data estimated in this study and the EIA data show an upward trend from 2014 to 2016. In the Strait of Malacca, Strait of Hormuz, and the Cape of Good Hope, the annual oil flow data estimated and the EIA data both show an upward trend year by year. In the Strait of Bab el-Mandeb, the annual oil flow data estimated shows an overall increase from 2014 to 2016 with a similar growth trend to the EIA data. They both increased significantly from 2014 to 2015, and they stabilized from 2015 to 2016, with only a slight increase or decrease.

The scatter plot is shown in Figure 7 by taking the EIA data as the abscissa and the oil flow estimated from the AIS data as the ordinate. As shown in Figure 7, the slope of the fit line is close to 1. The mean of the estimated oil flow is 560 million tons, while that of the EIA data is 536 million tons, showing a small difference of 4%. R^2 has a large value of 0.9517, and the root mean square error (RMSE) is 68.6889 million tons, which is very small relative to the annual oil flow. Furthermore, we calculated the correlation coefficient between the estimated annual oil flow and the EIA data, which is 0.9756. All of the statistical results show that the oil flow estimated from the AIS data and EIA data are similar and that they have a strong correlation.

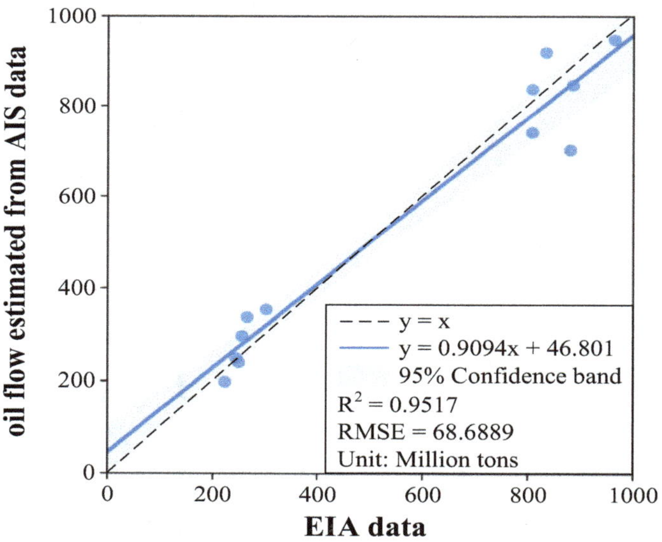

Figure 7. A scatter plot and linear fitting of the estimated oil flow and the EIA data.

4.2. Two-Way Annual Average Oil Flow

With the aim of checking the direction of crossing through the MOC and the correlation with the load, the two-way annual average oil flow of the four MOCs from 2014 to 2016 is calculated (see Figure 8). This study revealed that the oil flow is different for different directions and the general direction of the oil flow calculated in this study is consistent with the actual direction of oil flow. This shows that the framework proposed in this study is reliable.

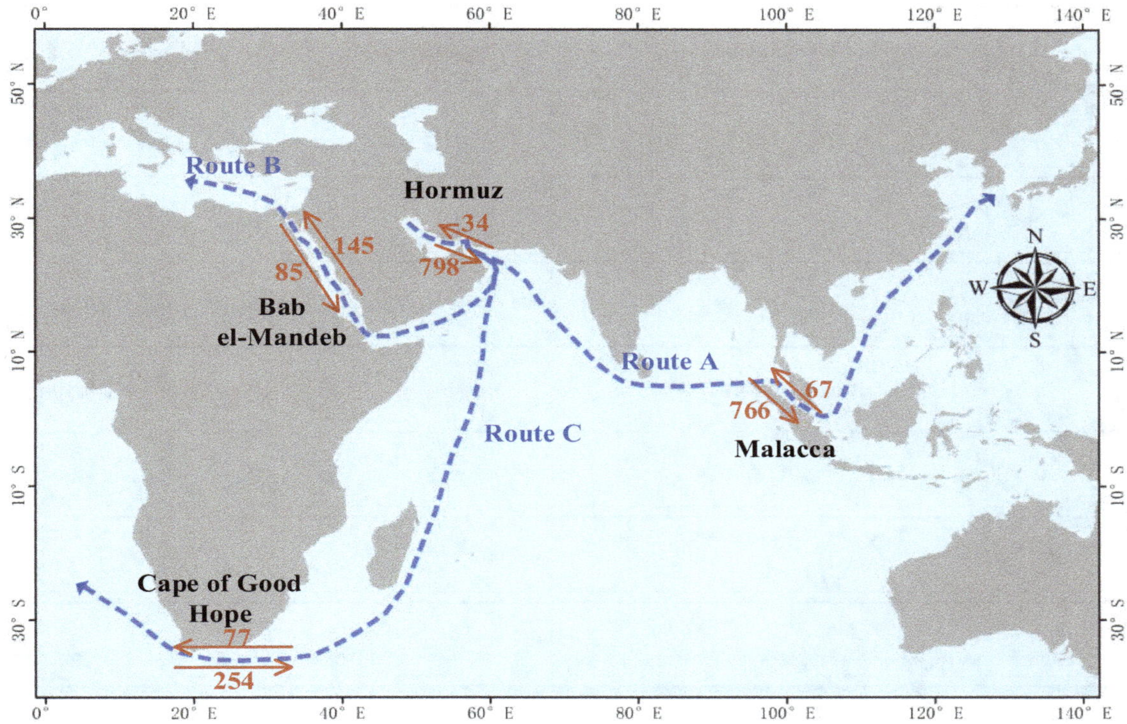

Figure 8. Two-way annual average oil flows of the four MOCs (units: 1 million tons).

As shown in Figure 8, in the Strait of Hormuz, the average annual amount of oil exported from the Persian Gulf is 798 million tons, while the average annual amount of oil transported into the Persian Gulf is only 34 million tons. This is reasonable, because the strait of Hormuz controls oil export from the Persian Gulf, which is the world's largest oil exporter. There are three main routes for transporting oil from the Persian Gulf (see Route A, Route B, and Route C in Figure 8). In the Strait of Bab el-Mandeb and Malacca, the main direction of the oil flow calculated in this study is consistent with the routes. Meanwhile, in the Cape of Good Hope, this is not consistent with the routes.

The reasons for inconsistency in the Cape of Good Hope are as follows: ① Route C is the least busy of the three main routes; hence, only 77 million tons of oil are transported from the Indian Ocean to the Pacific Ocean in the Cape of Good Hope. ② There are many big oil exporters in western Africa, such as Angola and Nigeria, which rank ninth and tenth in the world. These countries will export oil to other countries such as China through the Cape of Good Hope. Therefore, 254 million tons of oil are transported from the Pacific Ocean to the Indian Ocean in the Cape of good Hope. As a result, the general direction of oil flow from the Pacific Ocean to the Indian Ocean in the Cape of Good Hope is reasonable.

4.3. Daily, Monthly, and Seasonal Variation in Oil Flow

The daily oil flows through the four MOCs estimated from the AIS data are decomposed into the cycle, trend, and residual flows based on the monthly and seasonal cycles by using STL (Seasonal and Trend decomposition using Loess) (see Figure 9). STL is a decomposition method based on locally weighted regression. By setting different parameters, STL can be used to decompose the data according to the different cycles [49]. As illustrated in Figure 9, the red time series diagram represents the daily oil flow, the green time series decomposition diagram represents the monthly oil flow, and the blue time series decomposition diagram represents the seasonal oil flow. They have a smaller timescale than the EIA data; therefore, they can provide more information for analysis.

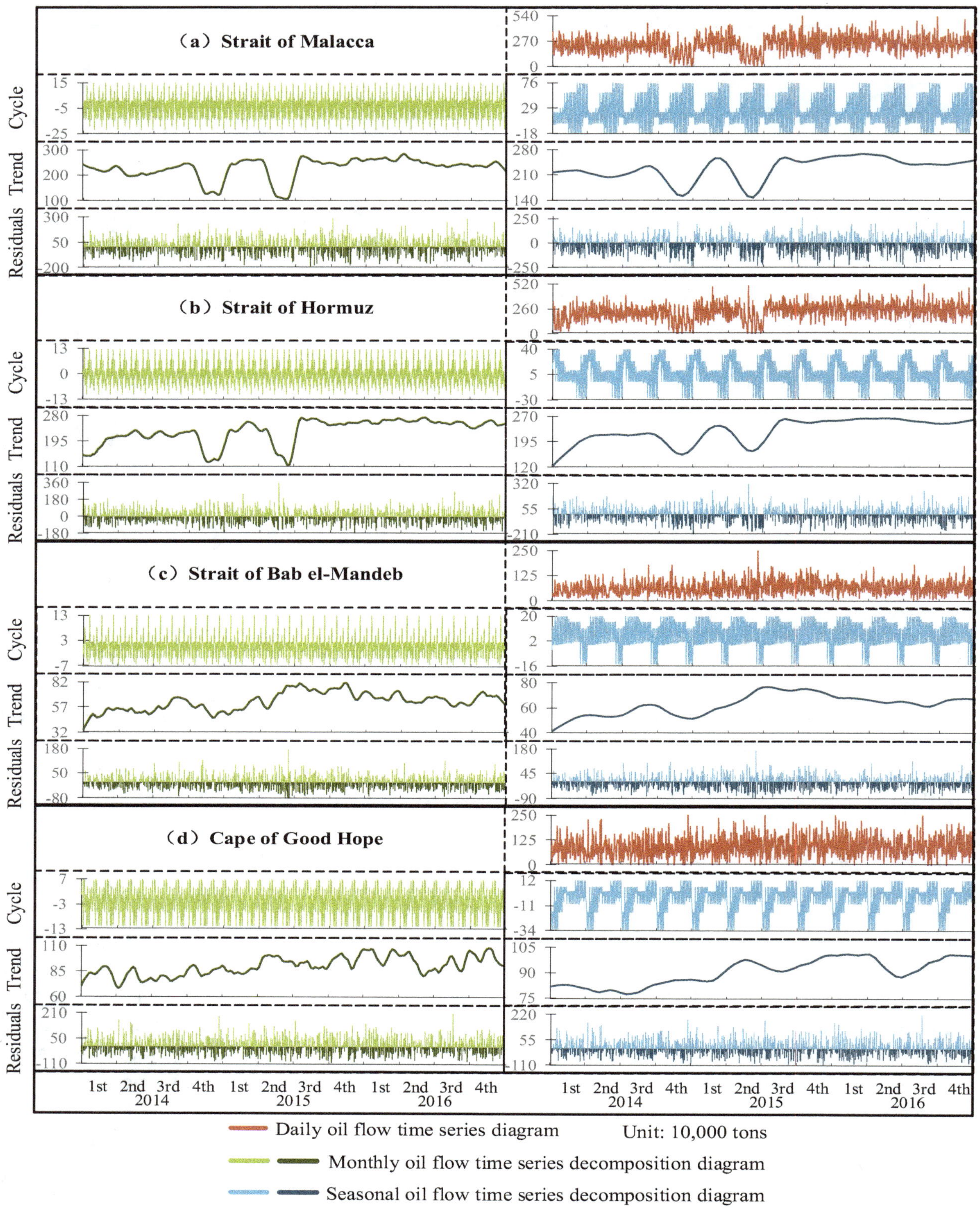

Figure 9. Daily oil flow time series diagram and the monthly and seasonal oil flow time series decomposition diagram in the four MOCs.

The cycles of these four chokepoints are significantly different in terms of the value and variation trend (see the cycle in Figure 9). The maximum values for the cycles of the four chokepoints in the seasonal cycle range from 12 to 76, the minimum values range from −34 to −18, and the variation range is from 36 to 94. All these statistical values show the large numerical differences between the cycles of the four chokepoints. The seasonal cycle of the oil flow through the Strait of Malacca shows an overall upward trend, with violent fluctuations in one cycle. The Strait of Hormuz decreases at first and then stabilizes and continues to decrease later in one cycle. The Strait of Bab el-Mandeb fluctuates at first and then decreases suddenly in one cycle. Finally, the Cape of Good Hope increases at first and then stabilizes and continues to increase later in one cycle. The variation trends of the seasonal cycles of the four chokepoints have different forms. The monthly cycles show a pattern similar to that of the seasonal cycles; that is, significant differences exist between the cycles of the four chokepoints.

Although the cycles of the four MOCs are different, their trends are consistent with each other's. Two obvious troughs exist in the Strait of Malacca and Strait of Hormuz, which will be explained in Section 4.3. Ignoring the troughs, the oil flows through the four chokepoints and there is an increase in the fluctuations at first. Then, they reach their peaks in mid-late 2015 without a further increase or with a slight decrease. In 2014, the oil flow through the Strait of Malacca decreases at first and then it increases. In 2016, the flow through the straits of Malacca and Bab el-Mandeb shows a slight downward trend. In 2016, the oil flows through the Strait of Hormuz and the Cape of Good Hope, but these flows do not show an upward or downward trend. In contrast, the flow through the Strait of Hormuz is more stable, while the flow through the Cape of Good Hope fluctuates more widely.

4.4. Events Corresponding to the Troughs in the Oil Flow

As demonstrated in Figure 9, two obvious troughs exist in the Strait of Malacca and the Strait of Hormuz, and we were able to find some events that correspond to these troughs. We believe that the first trough in the Strait of Hormuz in the fourth quarter of 2014 has a connection with the military activities of the U.S. During this period, the U.S. successively sent two aircraft carriers on 18 October 2014, and laser artillery warships on 10 December 2014, to the Persian Gulf. They were sent to combat the extremist armed forces during the civil wars in Iraq and Syria (see Figure 10a,b). These series of military activities are highly related to the first trough in the Strait of Hormuz in space and time. The second trough in the Strait of Hormuz in the second quarter of 2015 is associated with the Middle East Respiratory Syndrome (MERS) outbreak in 2015 [50] (see Figure 10c). The MERS originated in the Middle East, broke out on 18 May 2015, and ended approximately on 14 July 2015. Since it was the origin of the outbreak, the Middle East was naturally affected by it; thus, this caused the formation of the trough during the outbreak. The troughs in the Strait of Malacca relate to the troughs in the Strait of Hormuz, which can be explained as follows.

The oil flows through the straits of Malacca and Hormuz have two obvious troughs, while those in the Bab el-Mandeb Strait and the Cape of Good Hope have none. Furthermore, the troughs in the straits of Malacca and Hormuz are highly consistent in terms of the duration and the extent. This indicates that the events corresponding to the troughs had a considerable impact on Route A but that they had little impact on Routes B and C (see Figure 10c). We have developed the following hypotheses in terms of why the two troughs mainly affected Route A. (1) The oil passing through the Strait of Hormuz is transported more to Asian markets via Route A than to Britain and the U.S. via Routes B and C. As estimated by the EIA, 76% of the crude oil and condensate that moved through the Strait of Hormuz went to the Asian markets in 2018 [7]. (2) The aircraft carriers and warships sent to the Persian Gulf by the U.S. influenced the oil tankers of the other countries to a certain extent. Therefore, Route A was affected greatly, while the remaining two routes, i.e., Routes B and C, to Britain and the U.S., were less affected. (3) During the MERS epidemic in 2015, apart from the Middle East, only countries in Southeast Asia—including the Republic of Korea and China—found MERS cases. More information is listed in Table 4. All of these countries would more or less reduce their oil import. Therefore, the volume of oil transportation to Southeast Asian countries via Route A would be reduced, while Routes B and C would not be affected by the MERS epidemic. Thus, Route A is highly sensitive to the changes in the oil flow through the Strait of Hormuz. As a result, the two troughs of oil flow in the Strait of Malacca are highly consistent with those in the Strait of Hormuz in terms of time and duration.

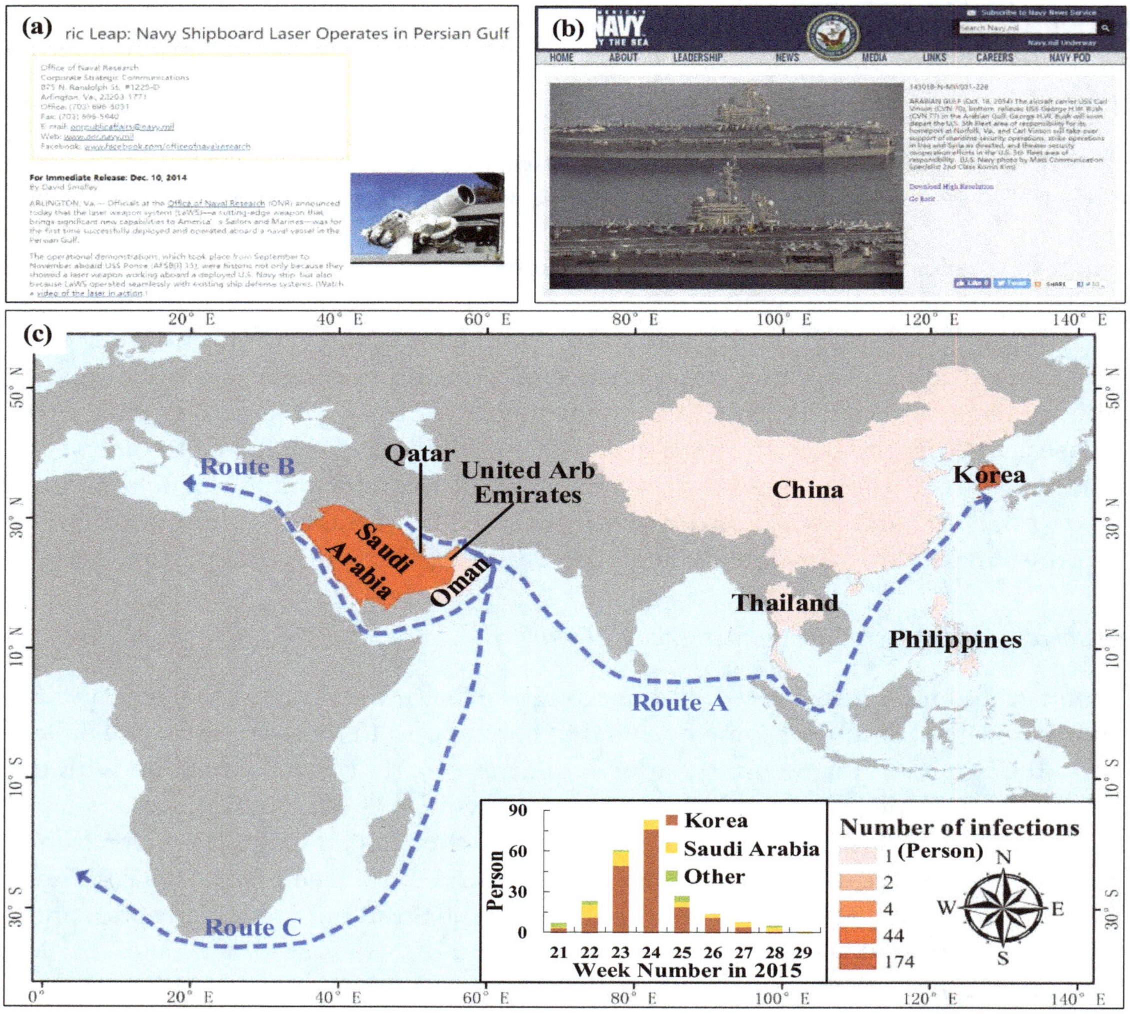

Figure 10. Events corresponding to the abnormal troughs. (**a**) Military activities in the Persian Gulf on 18 October 2014 (https://www.onr.navy.mil/en/Media-Center/Press-Releases/2014/LaWS-shipboard-laser-uss-ponce); (**b**) Military activities in the Persian Gulf on 10 December 2014 (https://www.navy.mil/view_image.asp?id=186243); (**c**) The situation of countries involved in the MERS outbreak in 2015 (https://www.who.int/csr/don/archive/disease/coronavirus_infections/en/).

Table 4. Number of cases in various countries during the MERS epidemic in 2015 (https://www.who.int/csr/don/archive/disease/coronavirus_infections/en/).

Date	Country	Number	Date	Country	Number
18 May 2015	United Arb Emirates	1	20 May 2015	Korea	2
21 May 2015	Qatar	1	21 May 2015	Korea	1
22 May 2015	Qatar	2	24 May 2015	Saudi Arabia	2
26–29 May 2015	Korea	8	26–30 May 2015	Saudi Arabia	9
29 May 2015	China	1	29 May 2015	Oman	1
30 May 2015	Korea	1	31 May 2015	Korea	2
1–3 June 2015	Korea	15	1–4 June 2015	Saudi Arabia	5
3 June 2015	United Arb Emirates	1	4 June 2015	Korea	6
5 June 2015	Korea	5	5–8 June 2015	Saudi Arabia	8
6 June 2015	Korea	9	7 June 2015	Korea	14
8–12 June 2015	Korea	62	9–12 June 2015	Saudi Arabia	3
13–16 June 2015	Korea	28	13–17 June 2015	Saudi Arabia	5
15 June 2015	United Arb Emirates	1	17–19 June 2015	Korea	1
18 June 2015	Thailand	1	19–30 June 2015	Saudi Arabia	6
20–23 June 2015	Korea	9	21 June 2015	United Arb Emirates	2
24–26 June 2015	Korea	4	27–30 June 2015	Korea	1
1–3 July 2015	Korea	2	1–14 July 2015	Saudi Arabia	6
4–7 June 2015	Korea	2	6 July 2015	Philippines	1

5. Conclusions

Oil flow assessment through the MOCs can help increase the security and stability of energy transportation and also enable the discovery of anomalies. To generate statistics on oil flow through MOCs, a maritime oil flow analysis technical framework is proposed. Using the AIS data from 1 January 2014 to 31 December 2016, the framework was applied to four MOCs within the MSR and its surrounding regions. The oil flows through the four chokepoints across four timescales (daily, monthly, seasonal, and annual) were determined and analyzed. The following conclusions were drawn from this study:

1. The annual oil flow estimated from the AIS data is similar to the EIA data in terms of the value and variation trends, except for the Cape of Good Hope. The statistics, such as R^2 and RMSE, show that the two sets of data are similar and have a strong correlation. The general directions of the oil flow in the four MOCs are consistent with the actual directions. Therefore, the framework proposed in this study is reliable.
2. Compared with the EIA data, the daily, monthly, and seasonal oil flow estimated in this study has smaller timescales, which can provide more information. The cycles of the four MOCs differed greatly, but their trends are consistent with each other's. Furthermore, there were two obvious troughs in the oil flow through the straits of Malacca and Hormuz.
3. The two troughs in the straits of Malacca and Hormuz were consistent in terms of the duration and extent. We believe that the first trough in the Strait of Hormuz is related to the military activities of the U.S. within the Persian Gulf. The second trough is associated with the worldwide MERS outbreak that occurred in 2015. The troughs in the Strait of Malacca are related to the troughs in the Strait of Hormuz.

By analyzing the above data, we discovered certain patterns as well as anomalies. These discoveries can provide support for national energy transportation strategy formulation, and the prevention and control of abnormalities. However, immense potential exists that can still be mined to provide significant value. Furthermore, with the support of real-time data, the real-time oil flow through the MOCs can be determined. This would be valuable regarding the detection of anomalies in time to aid national emergency responses, among other uses. The framework is suitable for ships carrying only one type of cargo in all MOCs in the world. However, it still has shortcomings, such as not using new technology. In our future research, we will introduce a new time series method to mine deeper information. We will also consider additional variables such as the nationality of the ship and the ship size to expand the breadth of the analysis.

Author Contributions: Conceptualization, Yijia Xiao, Yanming Chen, and Manchun Li; Data curation, Yijia Xiao and Zhaojin Yan; Formal analysis, Xiaoqiang Liu, Zhaojin Yan and; Funding acquisition, Yanming Chen, Liang Cheng, and Manchun Li; Investigation, Yanming Chen, Liang Cheng, and Manchun Li; Methodology, Yijia Xiao, Xiaoqiang Liu, and Zhaojin Yan; Project administration, Yanming Chen and Manchun Li; Resources, Liang Cheng; Supervision, Yanming Chen and Manchun Li; Visualization, Yijia Xiao and Zhaojin Yan; Writing—original draft, Yijia Xiao, Xiaoqiang Liu, and Zhaojin Yan; Writing—review & editing, Yanming Chen, Liang Cheng and Manchun Li. All authors have read and agreed to the published version of the manuscript.

References

1. Jia, H. Scientific collaborations shine on Belt and Road. *Natl. Sci. Rev.* **2017**, *4*, 652–657. [CrossRef]
2. Huang, Y. Understanding China's Belt & Road initiative: Motivation, framework and assessment. *China Econ. Rev.* **2016**, *40*, 314–321.
3. Lai, L.; Guo, K. The performance of one belt and one road exchange rate: Based on improved singular spectrum analysis. *Phys. A Stat. Mech. Appl.* **2017**, *483*, 299–308. [CrossRef]
4. Wan, C.; Yang, Z.; Zhang, D.; Yan, X.; Fan, S. Resilience in transportation systems: A systematic review and future directions. *Transp. Rev.* **2018**, *38*, 479–498. [CrossRef]

5. Cheng, L.; Yan, Z.J.; Xiao, Y.J.; Chen, Y.M.; Zhang, F.L.; Li, M.C. Using big data to track marine oil transportation along the 21st-century Maritime Silk Road. *Sci. China Technol. Sci.* **2019**, *62*, 677–686. [CrossRef]
6. Yan, Z.; Xiao, Y.; Cheng, L.; Chen, S.; Zhou, X.; Ruan, X.; Li, M.; He, R.; Ran, B. Analysis of global marine oil trade based on automatic identification system (AIS) data. *J. Transp. Geogr.* **2020**, *83*, 102637. [CrossRef]
7. EIA. The Strait of Hormuz Is the World's Most Important Oil Transit Chokepoint. Available online: https://www.eia.gov/todayinenergy/detail.php?id=39932 (accessed on 24 November 2019).
8. EIA. World Oil Transit Chokepoints. Available online: https://www.eia.gov/beta/international/regions-topics.php?RegionTopicID=WOTC (accessed on 24 November 2019).
9. Zhang, Z.X. China's energy security, the Malacca dilemma and responses. *Energy Policy* **2011**, *39*, 7612–7615. [CrossRef]
10. Faturechi, R.; Miller-Hooks, E. Measuring the performance of transportation infrastructure systems in disasters: A comprehensive review. *J. Infrastruct. Syst.* **2015**, *21*, 04014025. [CrossRef]
11. Omer, M.; Mostashari, A.; Nilchiani, R.; Mansouri, M. A framework for assessing resiliency of maritime transportation systems. *Marit. Policy Manag.* **2012**, *39*, 685–703. [CrossRef]
12. Leng, X.; Ji, K.; Yang, K.; Zou, H. A Bilateral CFAR Algorithm for Ship Detection in SAR Images. *IEEE Geosci. Remote Sens. Lett.* **2015**, *12*, 1536–1540. [CrossRef]
13. Wang, Y.; Liu, H. A hierarchical ship detection scheme for high-resolution SAR images. *IEEE Trans. Geosci. Remote Sens.* **2012**, *50*, 4173–4184. [CrossRef]
14. Xiao, L.; Xu, M.; Hu, Z. Real-Time Inland CCTV Ship Tracking. *Math. Probl. Eng.* **2018**, *2018*, 1205210. [CrossRef]
15. Lee, J.M.; Lee, K.H.; Nam, B.; Wu, Y. Study on Image-Based Ship Detection for AR Navigation. In Proceedings of the 2016 6th International Conference on IT Convergence and Security (ICITCS), Prague, Czech Republic, 26 September 2016; IEEE: Piscataway, NJ, USA, 2016; pp. 1–4.
16. Wang, H.; Zou, Z.; Shi, Z.; Li, B. Detecting ship targets in spaceborne infrared image based on modeling radiation anomalies. *Infrared Phys. Technol.* **2017**, *85*, 141–146. [CrossRef]
17. Jiang, B.; Ma, X.; Lu, Y.; Li, Y.; Feng, L.; Shi, Z. Ship detection in spaceborne infrared images based on Convolutional Neural Networks and synthetic targets. *Infrared Phys. Technol.* **2019**, *97*, 229–234. [CrossRef]
18. Misović, D.S.; Milić, S.D.; Durović, Ž.M. Vessel Detection Algorithm Used in a Laser Monitoring System of the Lock Gate Zone. *IEEE Trans. Intell. Transp. Syst.* **2016**, *17*, 430–440. [CrossRef]
19. Kim, Y.-K.; Kim, Y.; Jung, Y.S.; Jang, I.G.; Kim, K.-S.; Kim, S.; Kwak, B.M. Developing accurate long-distance 6-DOF motion detection with one-dimensional laser sensors: Three-beam detection system. *IEEE Trans. Ind. Electron.* **2012**, *60*, 3386–3395.
20. Kroodsma, D.A.; Mayorga, J.; Hochberg, T.; Miller, N.A.; Boerder, K.; Ferretti, F.; Wilson, A.; Bergman, B.; White, T.D.; Block, B.A.; et al. Tracking the global footprint of fisheries. *Science* **2018**, *359*, 904–908. [CrossRef]
21. Goerlandt, F.; Montewka, J.; Zhang, W.; Kujala, P. An analysis of ship escort and convoy operations in ice conditions. *Saf. Sci.* **2017**, *95*, 198–209. [CrossRef]
22. Boerder, K.; Miller, N.A.; Worm, B. Global hot spots of transshipment of fish catch at sea. *Sci. Adv.* **2018**, *4*, eaat7159. [CrossRef]
23. Cervera, M.A.; Ginesi, A.; Eckstein, K. Satellite-based vessel Automatic Identification System: A feasibility and performance analysis. *Int. J. Satell. Commun. Netw.* **2011**, *29*, 117–142. [CrossRef]
24. Etienne, L.; Devogele, T.; Buchin, M.; McArdle, G. Trajectory Box Plot: A new pattern to summarize movements. *Int. J. Geogr. Inf. Sci.* **2016**, *30*, 835–853. [CrossRef]
25. Wen, Y.; Zhang, Y.; Huang, L.; Zhou, C.; Xiao, C.; Zhang, F.; Peng, X.; Zhan, W.; Sui, Z. Semantic modelling of ship behavior in harbor based on ontology and dynamic bayesian network. *ISPRS Int. J. Geo-Inf.* **2019**, *8*, 107. [CrossRef]
26. Pallotta, G.; Vespe, M.; Bryan, K. Vessel pattern knowledge discovery from AIS data: A framework for anomaly detection and route prediction. *Entropy* **2013**, *15*, 2218–2245. [CrossRef]
27. Mascaro, S.; Nicholson, A.; Korb, K. Anomaly detection in vessel tracks using Bayesian networks. *Int. J. Approx. Reason.* **2014**, *55*, 84–98. [CrossRef]
28. Rong, H.; Teixeira, A.P.; Guedes Soares, C. Data mining approach to shipping route characterization and anomaly detection based on AIS data. *Ocean Eng.* **2020**, *198*, 106936. [CrossRef]

29. Zhang, W.; Goerlandt, F.; Montewka, J.; Kujala, P. A method for detecting possible near miss ship collisions from AIS data. *Ocean Eng.* **2015**, *107*, 60–69. [CrossRef]
30. Silveira, P.A.M.; Teixeira, A.P.; Soares, C.G. Use of AIS data to characterise marine traffic patterns and ship collision risk off the coast of Portugal. *J. Navig.* **2013**, *66*, 879–898. [CrossRef]
31. Zhang, W.; Goerlandt, F.; Kujala, P.; Wang, Y. An advanced method for detecting possible near miss ship collisions from AIS data. *Ocean Eng.* **2016**, *124*, 141–156. [CrossRef]
32. Vettor, R.; Soares, C.G. Detection and analysis of the main routes of voluntary observing ships in the North Atlantic. *J. Navig.* **2015**, *68*, 397–410. [CrossRef]
33. Zhen, R.; Jin, Y.; Hu, Q.; Shao, Z.; Nikitakos, N. Maritime anomaly detection within coastal waters based on vessel trajectory clustering and Naïve Bayes Classifier. *J. Navig.* **2017**, *70*, 648–670. [CrossRef]
34. Chen, R.; Chen, M.; Li, W.; Wang, J.; Yao, X. Mobility modes awareness from trajectories based on clustering and a convolutional neural network. *ISPRS Int. J. Geo-Inf.* **2019**, *8*, 208. [CrossRef]
35. Zhang, X.; Chen, Y.; Li, M. Research on geospatial association of the urban agglomeration around the South China Sea based on marine traffic flow. *Sustainability* **2018**, *10*, 3346. [CrossRef]
36. Gourmelon, F.; Le Guyader, D.; Fontenelle, G. A dynamic GIS as an efficient tool for integrated coastal zone management. *ISPRS Int. J. Geo-Inf.* **2014**, *3*, 391–407. [CrossRef]
37. Natale, F.; Gibin, M.; Alessandrini, A.; Vespe, M.; Paulrud, A. Mapping fishing effort through AIS data. *PLoS ONE* **2015**, *10*, e0130746. [CrossRef] [PubMed]
38. Coello, J.; Williams, I.; Hudson, D.A.; Kemp, S. An AIS-based approach to calculate atmospheric emissions from the UK fishing fleet. *Atmos. Environ.* **2015**, *114*, 1–7. [CrossRef]
39. Merico, E.; Dinoi, A.; Contini, D. Development of an integrated modelling-measurement system for near-real-time estimates of harbour activity impact to atmospheric pollution in coastal cities. *Transp. Res. Part D Transp. Environ.* **2019**, *73*, 108–119. [CrossRef]
40. Jia, H.; Adland, R.; Prakash, V.; Smith, T. Energy efficiency with the application of Virtual Arrival policy. *Transp. Res. Part D Transp. Environ.* **2017**, *54*, 50–60. [CrossRef]
41. Wang, C.; Hao, L.; Ma, D.; Ding, Y.; Lv, L.; Zhang, M.; Wang, H.; Tan, J.; Wang, X.; Ge, Y. Analysis of ship emission characteristics under real-world conditions in China. *Ocean Eng.* **2019**, *194*, 106615. [CrossRef]
42. Tu, E.; Zhang, G.; Rachmawati, L.; Rajabally, E.; Huang, G. Bin Exploiting AIS Data for Intelligent Maritime Navigation: A Comprehensive Survey from Data to Methodology. *IEEE Trans. Intell. Transp. Syst.* **2018**, *19*, 1559–1582. [CrossRef]
43. Wu, X.; Zhu, X.; Wu, G.-Q.; Ding, W. Data mining with big data. *IEEE Trans. Knowl. Data Eng.* **2013**, *26*, 97–107.
44. Yin, Y.; Gong, G.; Han, L. Theory and techniques of data mining in CGF behavior modeling. *Sci. China Inf. Sci.* **2011**, *54*, 717–731. [CrossRef]
45. Zhu, J.; Chen, J.; Hu, W.; Zhang, B. Big learning with Bayesian methods. *Natl. Sci. Rev.* **2017**, *4*, 627–651. [CrossRef]
46. Park, D.M.; Kim, Y.; Seo, M.G.; Lee, J. Study on added resistance of a tanker in head waves at different drafts. *Ocean Eng.* **2016**, *111*, 569–581. [CrossRef]
47. MacQueen, J. Some Methods for Classification and Analysis of Multivariate Observations. In Proceedings of the Fifth Berkeley Symposium on Mathematical Statistics and Probability, Berkeley, CA, USA, 21 June–18 July 1965 and 27 December 1965–7 January 1966; University of California Press: Berkeley, CA, USA, 1967; Volume 1, pp. 281–297.
48. Hartigan, J.A.; Wong, M.A. Algorithm AS 136: A k-means clustering algorithm. *J. R. Stat. Soc. Ser. C Appl. Stat.* **1979**, *28*, 100–108. [CrossRef]
49. Cleveland, R.B.; Cleveland, W.S.; McRae, J.E.; Terpenning, I. STL: A seasonal-trend decomposition. *J. Off. Stat.* **1990**, *6*, 3–73.
50. Donnelly, C.A.; Malik, M.R.; Elkholy, A.; Cauchemez, S.; Van Kerkhove, M.D. Worldwide Reduction in MERS Cases and Deaths since 2016. *Emerg. Infect. Dis.* **2019**, *25*, 1758. [CrossRef]

11

Multi-Scale Validation of MODIS LAI Products based on Crop Growth Period

Ting Wang [1], Yonghua Qu [2,3,4], Ziqing Xia [1], Yiping Peng [1] and Zhenhua Liu [1,*]

1 College of Natural Resources and Environment, South China Agricultural University, Guangzhou 510642, China; wt@stu.scau.edu.cn (T.W.); xzq@stu.scau.edu.cn (Z.X.); pyppyp@stu.scau.edu.cn (Y.P.)

2 State Key Laboratory of Remote Sensing Science, Jointly Sponsored by Beijing Normal University and the Institute of Remote Sensing Applications of CAS, Beijing 100875, China; qyh@bnu.edu.cn

3 Beijing Key Laboratory for Remote Sensing of Environment and Digital Cities, Beijing Normal University, Beijing 100875, China

4 School of Geography, Beijing Normal University, Beijing 100875, China

* Correspondence: zhenhua@scau.edu.cn.

Abstract: Leaf area index (LAI) is one of the most important canopy structure parameters utilized in process-based models of climate, hydrology, and biogeochemistry. In order to determine the reliability and applicability of satellite LAI products, it is critical to validate satellite LAI products. Due to surface heterogeneity and scale effects, it is difficult to validate the accuracy of LAI products. In order to improve the spatio-temporal accuracy of satellite LAI products, we propose a new multi-scale LAI product validation method based on a crop growth cycle. In this method, we used the PROSAIL model to derive Advanced Spaceborne Thermal Emission and Reflection Radiometer (ASTER) LAI data and Gaofen-1 (GF-1) for the study area. The Empirical Bayes Kriging (EBK) interpolation method was used to perform a spatial multi-scale transformation of Moderate Resolution Imaging Spectroradiometer (MODIS) LAI products, GF-1 LAI data, and ASTER LAI data. Finally, MODIS LAI satellite products were compared with field measured LAI data, GF-1 LAI data, and ASTER LAI data during the growing season of crop field. This study was conducted in the agricultural oasis area of the middle reaches of the Heihe River Basin in northwestern China and the Conghua District of Guangzhou in Guangdong Province. The results suggest that the validation accuracy of the multi-scale MODIS LAI products validated by ASTER LAI data were higher than those of the GF-1 LAI data and the reference field measured LAI data, showing a R^2 of 0.758 and relative mean square error (RRMSE) of 28.73% for 15 m ASTER LAI and a R^2 of 0.703 and RRMSE of 30.80% for 500 m ASTER LAI, which imply that the 15 m MODIS LAI product generated by the EBK method was more accurate than the 500 m and 8 m products. This study provides a new validation method for satellite remotely sensed products.

Keywords: multi-scale LAI product validation; PROSAIL model; EBK; crop growth period

1. Introduction

Leaf area index (LAI) plays an important role in the biophysical processes of vegetation canopies and exchange processes of matter and energy between the Earth and atmosphere [1–4]. With advances in remote sensing technology, a number of satellite LAI products have been generated to monitor regional and global vegetation. Among them, the Moderate Resolution Imaging Spectroradiometer (MODIS) LAI product has been widely used. However, the availability of satellite products for scientific research and practical applications presents uncertainties due to various effects such as model algorithm, observation conditions, and sensor specifications [5]. Thus, it is important to validate the accuracy of LAI satellite products for assuring their effective utility in various disciplines [6].

Methods for the validation of LAI products are divided into three types: product intercomparison, direct validation, and indirect validation [7,8]. Product intercomparisons are conducted to evaluate the relative accuracy of satellite LAI products by comparing the spatio-temporal consistency among different satellite LAI products with similar time phase, same projection coordinate system, and spatial resolution [9]. Garrigues et al. [7] investigated the performances and spatio-temporal consistencies of four major global LAI products at 1/11.2° spatial sampling and a monthly time step: ECOCLIMAP climatology, GLOBCARBON (GLOBal Biophysical Products Terrestial CARBON Studies), CYCLOPES (Carbon Cycle and Change in Land Observational Products from an Ensemble of Satellites), and MODIS Collection 4. Yang et al. [9] compared and analyzed the spatio-temporal consistency of GEOV1 (GEOLAND2 Version 1), GLASS (Global Land Surface Satellite), and MODIS LAI products in southwest China's mountainous area. A product intercomparison can easily be achieved over a large number of sites representing the global distribution of land surface types and over a complete vegetation cycle. However, the product intercomparison method does not utilize concurrent field measurement data to evaluate the spatio-temporal consistencies of different products [7,8,10]. Thus, this approach cannot guarantee the absolute accuracy of satellite products [11].

The direct verification method considers ground measurements as the reference against which to evaluate the precision of satellite products. Sea et al. [12] compared MODIS LAI collections 4.8 and 5.0 with ground-based measurements taken along a 900 km north–south transect through the savanna in the Northern Territory of Australia. Fang et al. [8] verified MODIS LAI and CYCLOPES LAI using global field measured data and suggested that the direct comparison method is promising when a sufficient number of ground points are used and the field is homogeneous over a large area. However, it is difficult to satisfy this surface homogeneity requirement when validating estimates from remote sensing images with larger pixel size because vegetation cover is rarely uniform over large areas [13–15].

The indirect validation method is a more reasonable approach for validating low-resolution satellite products because it can avoid the spatial scale-mismatch problem [16]. The method builds a "bridge" between ground measured data and low-resolution remote sensing products using relatively high resolution satellite or aerial remote sensing images. Yang et al. [17] verified the accuracy of MODIS LAI products in the Qinghai Lake Basin by using Landsat 8 OLI data combined with field measurement data. They resampled the 30 m Landsat 8 LAI products into 1 km pixels to compare them to the MOD15A2 product, pixel by pixel. However, few studies [18,19] have been able to downscale the lower spatial resolution satellite LAI products into the spatial resolution of the bridging LAI results and then conduct a comparison, as it is difficult to generate LAI products with lower resolution without a sufficient number of homogeneous samples.

These existing validation efforts suggest that the present methods cannot determine the applicable scale of the LAI satellite product only based on the accuracy of the validation of the satellite product with the original resolution. As one of the most important canopy structure parameters in process-based models of crop biogeochemistry, LAI reflects crop growth characteristics. In order to quickly and efficiently monitor crop growth, it is very important to validate MODIS LAI products on crop growth period. The existing validation methods also rarely consider the crop growth cycle, which reduces the reliability of the verification accuracy. The objective of this study was to build a new validation method based on the Empirical Bayes Kriging (EBK) interpolation and the crop growth cycle to verify the accuracy of multi-scale MODIS LAI products. Taking an oasis agricultural area in the middle reaches of the Heihe River Basin and the Conghua District of Guangzhou in Guangdong Province as examples, the method used the PROSAIL model and Advanced Spaceborne Thermal Emission and Reflection Radiometer (ASTER) images and the Gaofen-1 (GF-1) images to produce ASTER LAI and GF-1 LAI reference maps. Then, the EBK interpolation method was driven to perform spatial upscaling transformations on ASTER LAI from 15 m spatial resolution to 500 m, GF-1 LAI from 8 m to 500 m and spatial downscaling transformation on the MODIS LAI images from 500 m spatial resolution to 15 m and 8 m. Finally, the multitemporal and spatial scale LAI product validation was completed on the crop during its crop growth cycle.

2. Materials and Methods

2.1. Study Area

This study dealt with both the agricultural oasis area of the middle reaches of the Heihe River Basin (Figure 1b,c) and the Conghua District of Guangzhou in Guangdong Province (Figure 1d,e). The agricultural oasis area of the middle reaches of the Heihe River Basin is located in northwestern China with the latitude and longitude ranges of 38° to 42° N and 98° to 101°30′ E. It has a temperate continental climate, with short and hot summers, long and cold winters, an annual precipitation of only 54.9~436.2 mm, and annual evaporation of 1700 mm. Due to pressure on water resources in this fragile ecological environment, it is critical to monitor surface vegetation cover and vegetation growth of this agricultural oasis district in the Heihe River Basin. Satellite LAI products offer a means to efficiently monitor vegetation cover over larger regions, but to do this reliably, it is necessary to validate their accuracy. Additionally, Conghua District of Guangzhou City, China (11,317′–11,404′ E, 2322′–2356′ N) is located in the transition zone from the Pearl River Delta to the northern mountain area of Guangzhou. It has a humid subtropical monsoon climate characterized by warm winters, hot summers, little frost and snow, and sufficient rain and sunshine. The annual average temperature is 21.6 °C, and the annual precipitation is 2176.3 mm. Rice is one of the major crops in the study area and a year can be planted in two seasons.

2.2. Data and Pre-Processing

A total of 121 samples were collected from corn during the 2012 growing season on 24 May, 15 June, 23 June, 10 July, 3 August, 12 August, 18 August, 27 August, and 20 September. The measured samples including LAI, leaf angle distribution (LAD), chlorophyll a and b content (Cab) spectral data were provided by Heihe Plan Science Data Center, National Natural Science Foundation of China [20–22]. In the Conghua District of Guangzhou City, a total of 100 samples (green points in Figure 1e) were collected on 3 August, 17 September, 26 September, 15 October, and 24 October in 2015a. In order to meet the validation requirements for the MODIS LAI product at 500 m spatial resolution, field sampling plots of 0.5 km by 0.5 km were designed and within each plot, the LAI were collected by LAI-2000 Plant Canopy Analyzers (LI-COR, Inc., Lincoln, NE, USA) at five locations. Additionally, the values at the five locations were averaged for each sampling plot. One of the five points was located at the center of the plot, and the other four points were located at equal distances along the diagonal. Other parameters used in the PROSAIL model were derived from the Leaf Optical Properties Experiment 93 (LOPEX93) database including leaf structure parameter, pigment content, water content, and content of other components. Figure 1 shows the distribution of the sample points collected on 10 July 2012.

To validate the performance of our approach, we obtained images from the MOD15A2H LAI products, GF-1, and ASTER. Additionally, to avoid the combined effect of soil background value and vegetation, the MODIS LAI, GF-1, and ASTER images were selected for the same time phase with the measured LAI data We used the MODIS Reprojection Tool to transform the MODIS LAI data from Sinusoidal to UTM-WGS84 projection and applied the scaling factor of 0.1 [23] to obtain the standard MODIS LAI. The ASTER data and GF-1 data were geometrically corrected with a total of 24 ground control points measured using global positioning system (GPS) and applying a least squares transformation. The root mean square error (RMSE) between the estimated and measured ground control points was less than 0.5 pixel. Atmospheric corrections of ASTER data and GF-1 data were conducted using the FLAASH model. Geometric corrections for the MODIS products were performed with the MODIS Tools software.

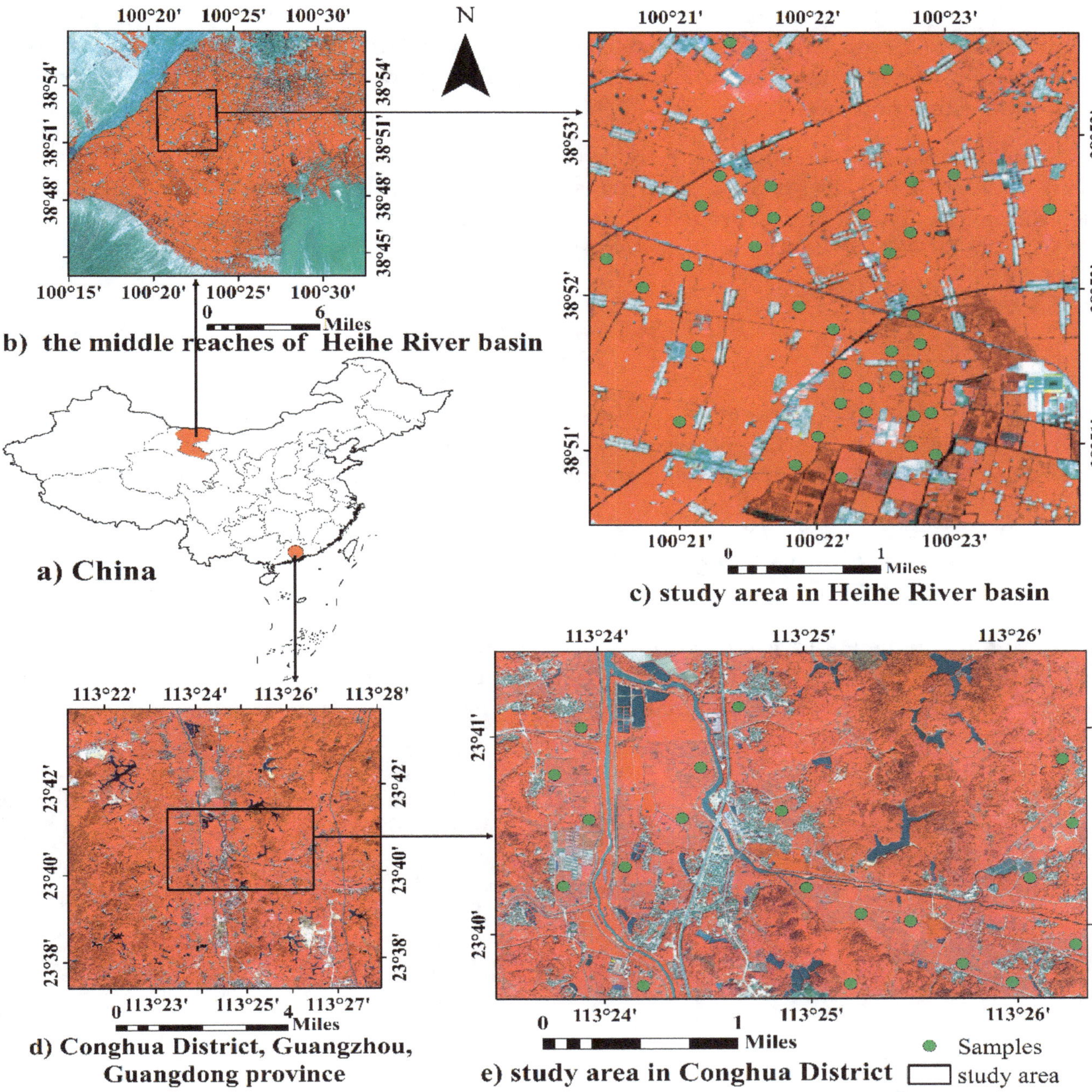

Figure 1. The location of the study areas shown using a false color composite image from ASTER data and GF-1 data: (**a**) China; (**b**) the middle reaches of Heihe River Basin; (**c**) study area in Heihe River Basin; (**d**) Conghua District, Guangzhou, Guangdong Province; (**e**) study area in Conghua District.

2.3. Methods

2.3.1. Leaf area index Reference Maps Generated Using the PROSAIL Model

To produce LAI reference maps (15 m ASTER and 8 m GF-1 LAI), it is critical to construct an estimation model. In this estimation model, the LAI is the dependent variable while canopy reflectance are the independent variables [24]. The empirical estimation model can be written as [25]:

$$\hat{y} = \mathrm{a} + \sum_{i=1}^{n} b_i \cdot x(\lambda_i) \tag{1}$$

where the dependent variable ($\hat{y}$) is the estimated value of LAI of reference maps; a and b represent constant values; $x(\lambda_i)$ represents the optimal relevant canopy reflectance at wavelength λ_i; and n is the total number of canopy reflectance variables. Multiple linear regression (MLR) was used to construct the empirical estimation model.

The field measured LAI in the study area were not sufficient for building an empirical model for deriving the ASTER and GF-1 LAI reference maps. Thus, we simulated ground LAI with the PROSAIL model, coupled by the leaf optical properties model PROSPECT [24] and the vegetation canopy model SAIL [25]. The PROSPECT model is expressed as [26]:

$$(\rho_1, \tau_1) = \mathrm{PROSPECT}(\mathrm{N}, \mathrm{Cab}, \mathrm{Cbrown}, \mathrm{Car}, \mathrm{Cw}, \mathrm{Cm}) \tag{2}$$

where N is a leaf structure parameter (unitless); Cab is the chlorophyll a and b content (μg/cm^2); Cbrown is the brown pigment content (μg/cm^2); Car is the carotenoid content (μg/cm^2); Cw is the water content(g/cm^2); and Cm is the dry matter content(μg/cm^2). The SAIL model is written as [27]:

$$\rho = \mathrm{SAIL}(\rho_1, \tau_1, \mathrm{LAI}, \mathrm{LAD}, \rho\mathrm{soil}, \mathrm{Diff}, \mathrm{Hspot}, \mathrm{SZA}, \mathrm{VZA}, \mathrm{RAA}) \tag{3}$$

where the input parameters are leaf reflectance ρ_1; transmittance τ_1; leaf area index LAI; leaf inclination distribution LAD; diffuse reflection coefficient Diff; soil coefficient ρsoil; hot spot parameter Hspot; atmosphere conditions and view-illumination geometry (solar zenith angle (SZA); view zenith angle (VZA); and relative azimuth angle (RAA).

In this study, qualitative and quantitative sensitivity analysis methods were driven to obtain sensitive parameters, as shown in Figures 2 and 3. The qualitative sensitivity analysis was performed by observing the change of the simulated top-of-the-canopy reflectance curves when varying one parameter while keeping the other parameters constant [28]. The quantitative sensitivity method used the sensitivity formula to obtain sensitivity parameters. The sensitivity (S) formula is as follows [28,29]:

$$\mathrm{S} = \frac{\sum_{j=1}^{n}\left(\rho_{Xo}^{(j)} - \rho_{Xo+\Delta X}^{(j)}\right)^2}{\rho_{Xo}^{(j)}} \tag{4}$$

where X_o is the original parameter value of the model; ΔX is the parameter step, which was determined through the PROSAIL model test according to relevant references [28]; $\rho_{Xo}^{(j)}$ is the original canopy reflectance of the jth added step; $\rho_{Xo+\Delta X}^{(j)}$ is the simulated canopy reflectance when the jth parameter value is $X + \Delta X$; and n is the number of added steps.

Figure 2 shows that the variation of the Cab curve was the largest in the range of 500 nm~700 nm, followed by LAD. In the range of 760 nm~1300 nm, the parameters LAI, LAD, SZA, VZA, Hspot, and Cm had a significant influence on canopy reflectance. In the range of 2080 nm~2350 nm, the curve variation of LAD, LAI SZA, and VZA were relatively larger. The parameters Car, Cbrown, ρsoil, Diff, and RAA had weak sensitivity in all wavebands.

Figure 3 shows that Cab was the most sensitive parameter in the range of 500 nm to 700 nm, the sensitivity of LAI was more significant than other parameters in the range of 720 nm to 1300 nm, and the LAD was also a sensitive parameter from 720 nm to 2500 nm. Therefore, Cab, LAI, and LAD were selected as sensitive parameters of the PROSAIL model. The specific ranges for the input parameters in the PROSAIL model set in this study are shown in Table 1.

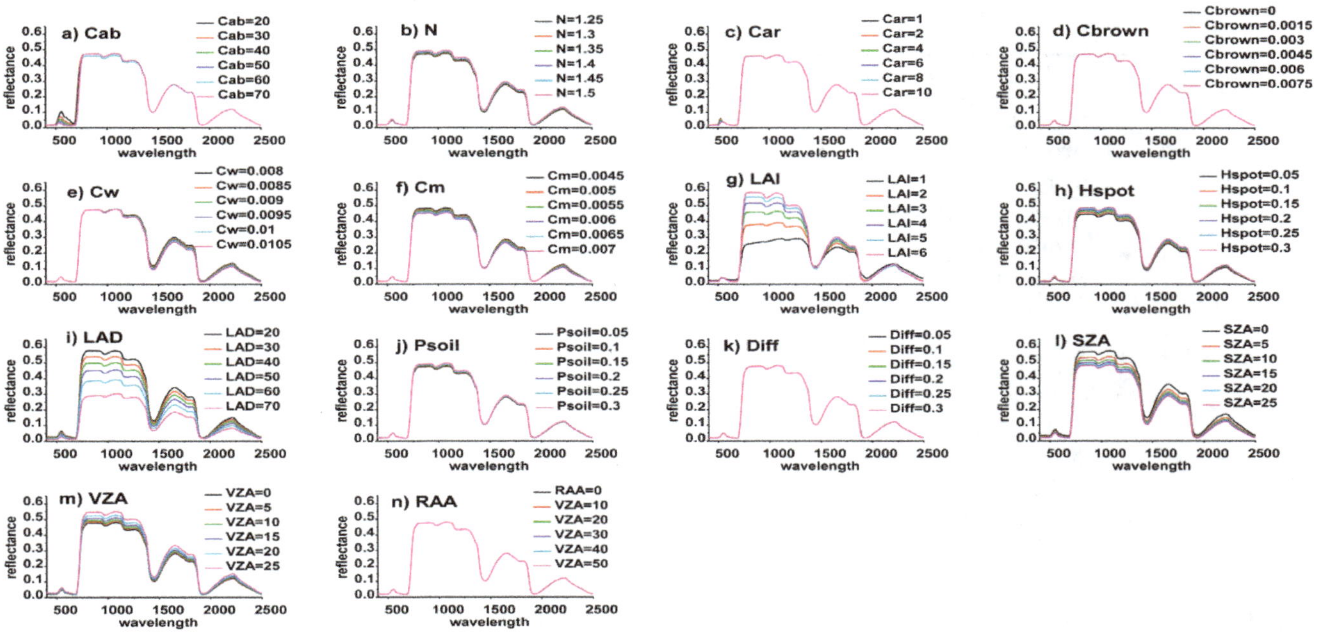

Figure 2. Influence of the input parameters on canopy reflectance in the PROSAIL model.

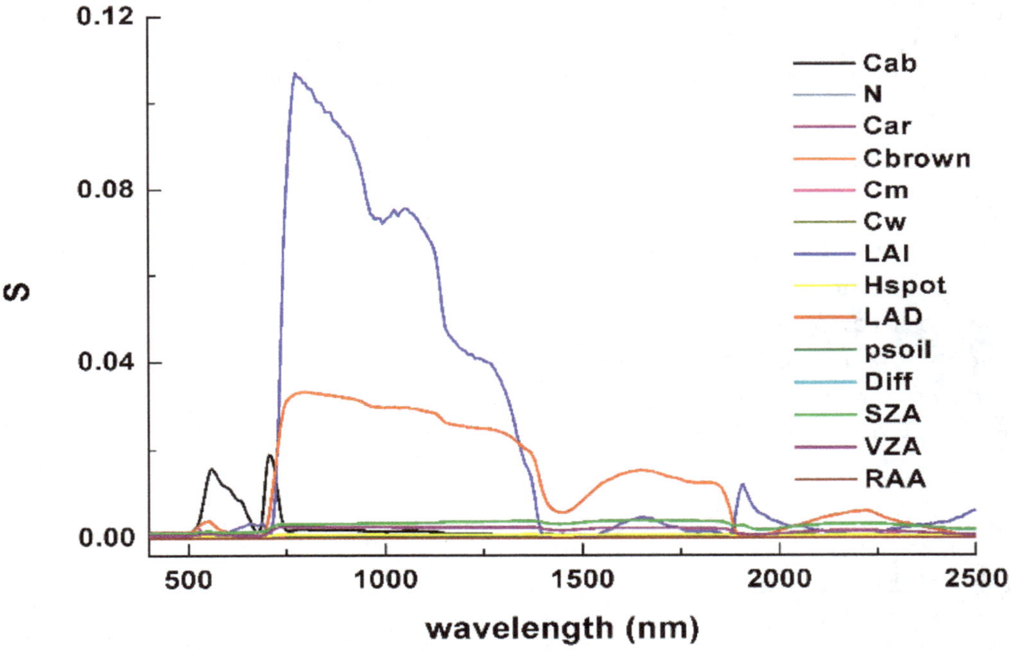

Figure 3. The sensitivities of the parameters in the PROSAIL model: the blue, red and black line indicate the sensitivities of LAI, LAD and Cab, respectively.

After deciding the input parameter values, the PROSAIL model was driven to generate the look-up-table (LUT) using the simulated LAI and corresponding canopy reflectance. In the LUT, one LAI value corresponded to 2101 canopy reflectance data from 2101 bands from 400 nm to 2500 nm. Therefore, in order to obtain an accurate empirical estimation model of LAI (Equation (1)), the appropriate canopy reflectance bands need to be determined from the 2101 bands, according to the Pearson correlation coefficient and significance level. In this study, the accurate empirical estimation model of LAI was constructed, taking LAI in the LUT as the dependent variable and the corresponding canopy reflectance of appropriate bands as independent variables.

Table 1. The input parameters setting in the PROSAIL model.

Parameters	Units	Min	Max	Step
leaf structure index N	unitless	1.3	1.3	-
leaf chlorophyll content a + b Cab	(μg/cm²)	20	70	5
carotenoid content Car	(μg/cm²)	8	8	-
brown pigment Cbrown	(μg/cm²)	0	0	-
water content Cw	(g/cm²)	0.0095	0.0095	-
dry matter content Cm	(μg/cm²)	0.0015	0.0015	-
leaf area index LAI	(m²/m²)	0.05	7	0.05
hot parameter Hspot	(m²/m²)	0.2	0.2	-
leaf angle distribution LAD	(°)	20	70	5
diffuse reflection coefficient Diff	(fraction)	0.2	0.2	-
soil coefficient ρsoil	unitless	0.5	0.5	-
Sun zenith angle SZA	(°)	32	32	-
view zenith angle VZA	(°)	0	0	-
Relative azimuth angle RAA	(°)	0	0	-

2.3.2. Validation of Multi-Scale MODIS LAI Products Based on the EBK Interpolation

The LAI products must be compared under the same spatial scale. Therefore, as a robust non-stationary algorithm for spatial interpolating geophysical corrections [30], the EBK interpolation in this study performed downscaling transformation on 500 m MODIS LAI products and upscaling transformation on the LAI reference maps to conduct the validation of multi-scale MODIS LAI products. In the EBK method, the expression of the predicted LAI values by Kriging $Z_{(x_0)}$ is as follows [31,32]:

$$Z_{(x_0)} = \sum_{i=1}^{n} \lambda_i \cdot Z_{(x_i)} + \sum_{i=1}^{n} s_i \cdot U_{(x_i)} \tag{5}$$

where Z_1, Z_2, Z_3 ... Z_n are the field measured LAI values; $\lambda_i (i = 1,2,3 \ldots \text{n})$ are kriging weights estimated using parameters of the cross-variograms; and $s_i (i = 1,2,3 \ldots \text{n})$ are kriging weights estimated on the basis of a cross-variogram between $Z_{(x_i)}$ and $U_{(x_i)}$. n denotes the total number of observations, where n represents 182 and 135,192 for 500 m MODIS LAI product and 15 m ASTER LAI in Heihe River Basin study area, respectively, and 140 and 276,219 for 500 m MODIS LAI product and 8 m GF-1 LAI in the Conghua study area. The variable $U_{(x)}$ was a standardized rank that was calculated as [33]:

$$U_{(x_i)} = \frac{R}{n} \tag{6}$$

where R is a rank of the Rth order statistic of LAI measured on the land surface at location x_i.

In this study, EBK interpolation was performed as an upscaling transformation of the 15 m ASTER and 8 m GF-1 LAIs to 500 m reference maps. The 500 m MODIS LAI products were then respectively downscaled to 15 m and 8 m. At the validation stage, the measured LAI data and satellite (MODIS, GF-1, and ASTER) LAI with the same resolution were compared to each other to validate the reliability of the 500 m MODIS LAI products at different spatial scales. Four statistical metrics (the determination coefficient R^2, mean relative error MRE, root-mean-square error RMSE and relative root mean square error RRMSE) were used to quantify the deviation between two datasets:

$$R^2 = 1 - \frac{\sum_{i=1}^{n} (y_i - \hat{y}_i)^2}{\sum_{i=1}^{n} (y_i - \overline{y_i})^2} \tag{7}$$

$$\text{MRE} = \frac{\sum_{i=1}^{n} |y_i - \hat{y}_i| / y_i}{n} \tag{8}$$

$$\mathrm{RMSE} = \sqrt{\frac{\sum_{i=1}^{n}(y_i - \hat{y}_i)^2}{n}} \quad (9)$$

$$\mathrm{RRMSE} = \frac{\mathrm{RMSE}}{\bar{y}} \times 100\% \quad (10)$$

where $\bar{y}$ is the mean value of the measured samples; y_i is the measured value of the ith sample; $\hat{y}_i$ is the estimated value of the ith sample; and n is the number of samples.

3. Results

3.1. Distribution Map of Multi-Scale ASTER, GF-1, and MODIS LAI over Crop during Its Growth Cycle

The Pearson correlation coefficients between the field simulated LAI and field measured spectral canopy reflectance are shown in Figure 4, which indicates that band 824 nm had the largest correlation with LAI compared to other spectral bands. Therefore, in this paper, the ASTER band 3, with a range of 760 nm–860 nm, was selected to create the ASTER LAI products at 15 m resolution.

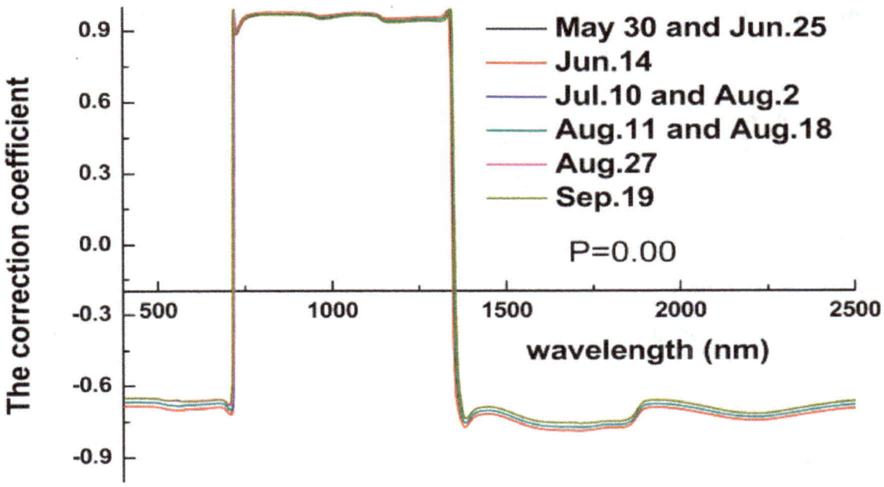

Figure 4. Pearson correlation coefficients between the field simulated leaf area index and field measured canopy reflectance.

The canopy reflectance data simulated by the PROSAIL model were spectrally resampled by the spectral resampling model in ENVI 5.3 to correspond to the spectral resolution (760–860 nm) of ASTER band 3. Then, the LUT was constructed by the resampled canopy reflectance data and LAI simulated by the PROSAIL model, which was used to solve the coefficients of Equation (1). Finally, the ASTER LAI and GF-1 LAI products with 15 m and 8 m spatial resolution were generated according to Equation (1), which is shown in the left figure of each phase in Figures 5 and 6. At the same time, the EBK interpolation method was driven to perform the downscaled transformation of the 500 m MODIS LAI products to generate MODIS LAI products at 15 m and 8 m resolution, which is shown in the right figure of each phase in Figures 5 and 6.

As shown in the left figure of each phase in Figures 5 and 6, LAI increased first and then decreased with time during the growth cycle of crops (corn and rice). LAI values were small at the early growing period (30 May and 14 June for corn; 3 August for rice), shown in red and orange shades in the graphs. LAI values began to increase at the middle stage (mid-to-late July and early August for corn and late September for rice) and peaked at the heading to flowering stage (mid-August for corn; 15 October for rice). Later, LAI values began to decrease at the maturity stage (mid-to-late September for corn and mid-to-late October for rice). During the crop growth cycle, the ASTER-derived LAI distribution maps were basically consistent with those of MODIS. While there was relatively larger difference in the early and late growth stage because 15 m MODIS products were acquired from 500 m MODIS products having more many mixed pixels than the other phases in the corn and rice fields.

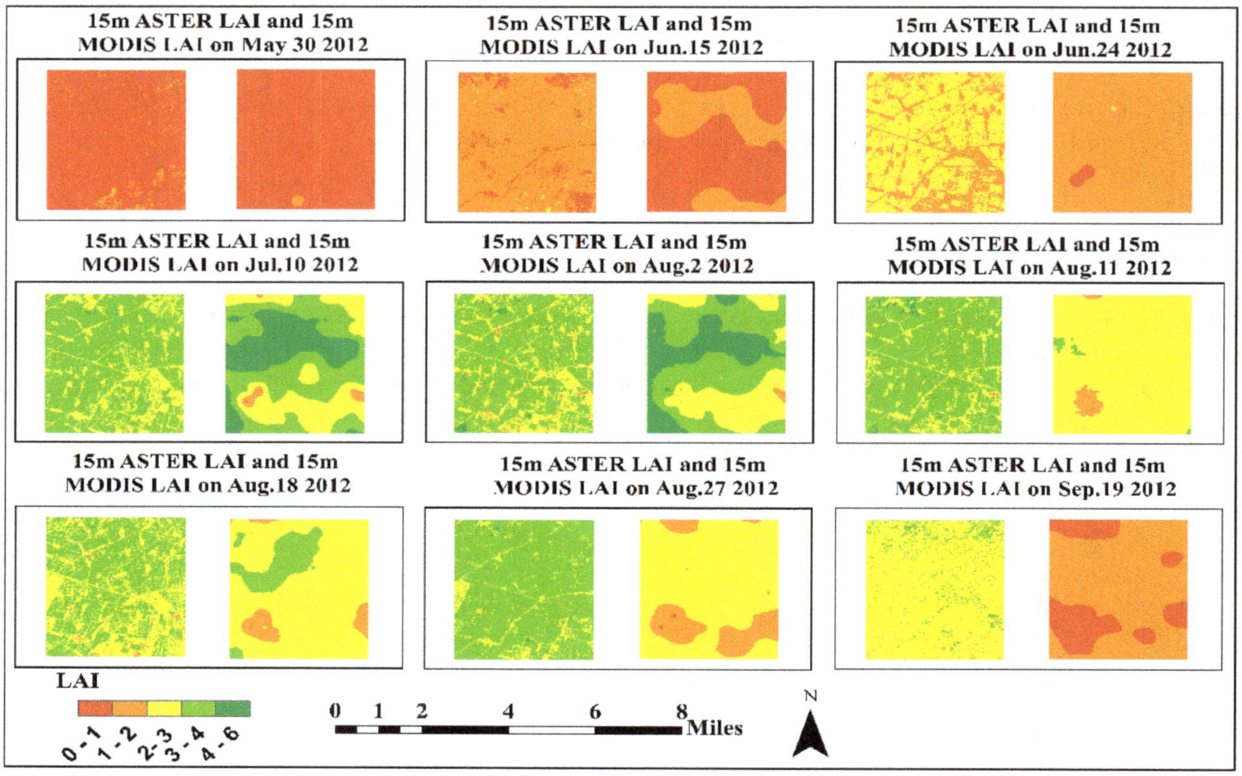

Figure 5. The time series distribution map of 15 m ASTER LAI (left figure of each period) and MODIS LAI with 15 m resolution interpolated by the EBK method (right figure of each period) during the growth cycle of corn.

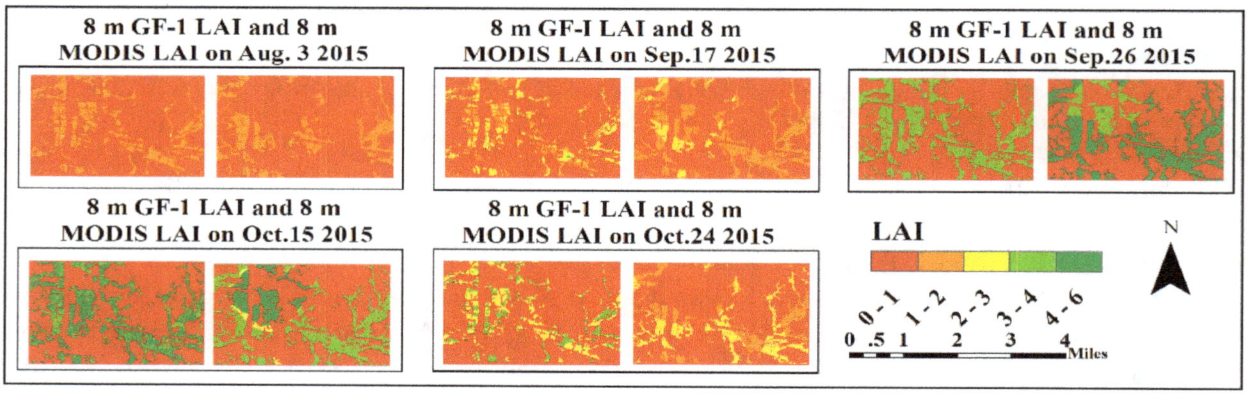

Figure 6. The time series distribution map of 8 m GF-1 (left figure of each period) and MODIS LAI with 8 m resolution interpolated by the EBK method (right figure of each period) during the growth cycle of rice.

The EBK interpolation method was used to perform the upscaling transformation of 15 m ASTER and 8 m GF-1 LAI to LAI products at 500 m resolution, respectively, which are shown in the left figure of each period in Figures 7 and 8. The 500 m LAI products obtained by the EBK method lost detailed information (such as roads and villages) as pixels became mixed. From the spatial distributions of LAI in Figures 5–8, the 500 m ASTER and GF-1 LAI were more consistent with the 500 m MODIS LAI products than other 15 m and 8 m LAI products with more detailed information. The frequency distribution (Figure 9) also indicates that more detailed spatial LAI patterns could be found in downscaled LAI, but were averaged out in the original MODIS LAI.

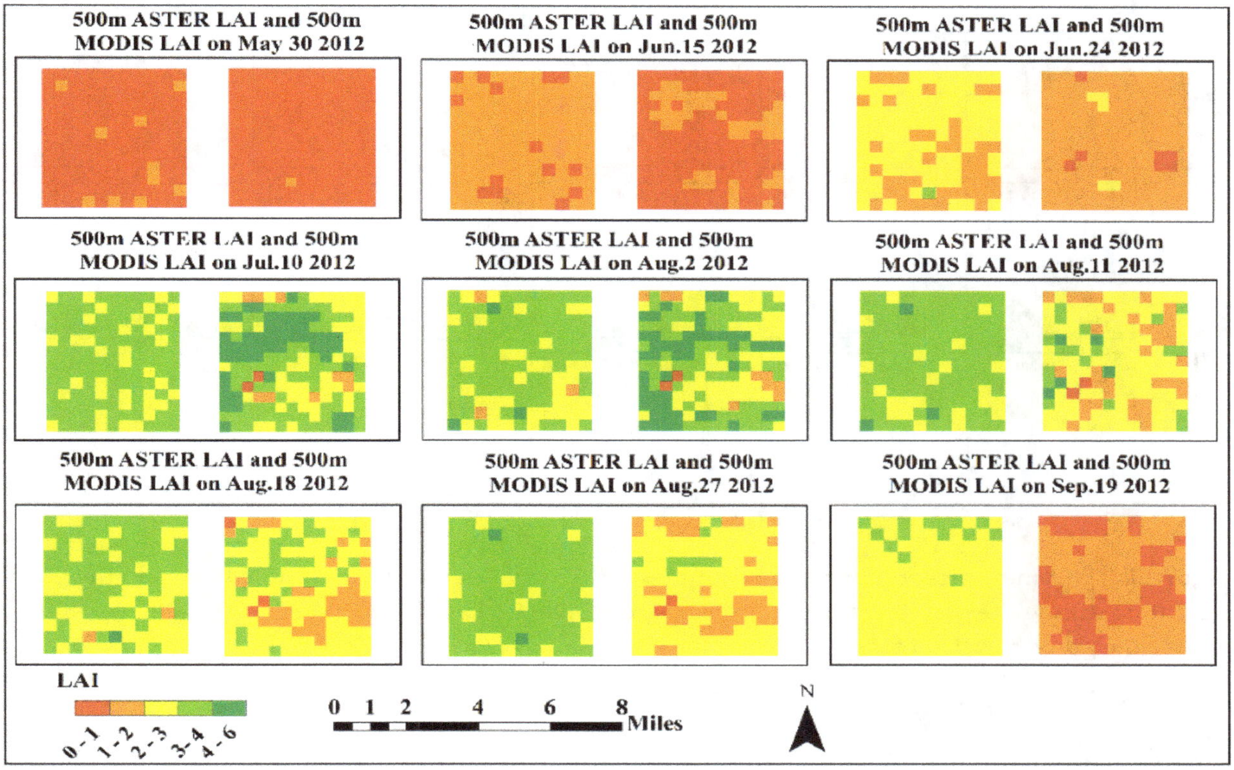

Figure 7. Time series distribution map of 500 m ASTER LAI (left figure of each period) interpolated by the EBK method and MODIS LAI with 500 m resolution (right figure of each period) during the corn growth cycle.

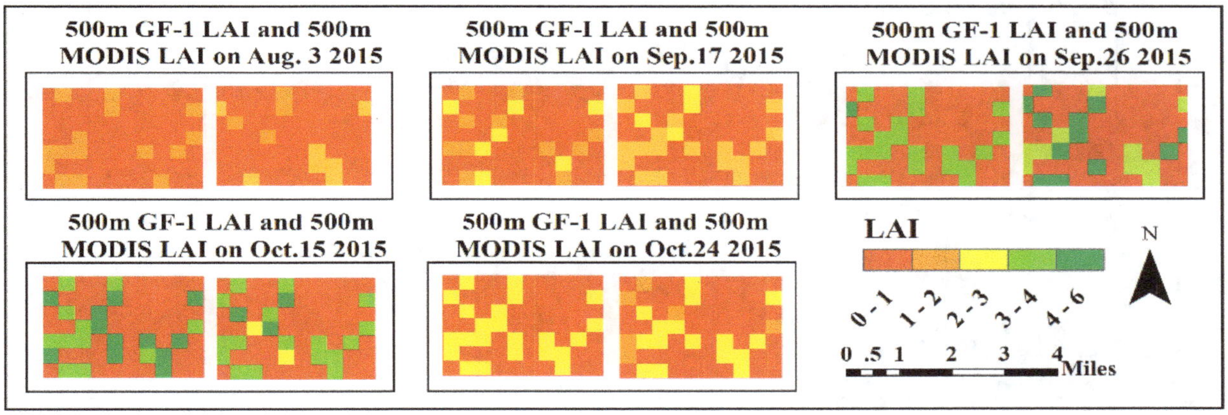

Figure 8. Time series distribution map of 500 m GF-1 LAI (left figure of each period) interpolated by the EBK method and MODIS LAI with 500 m resolution (right figure of each period) during the rice growth cycle.

3.2. Multi-Scale Validation of MODIS LAI Product

For this study, 121 and 100 in situ LAI measurements were collected in Heihe River Basin and Conghua District, respectively, during the crop growth cycle to validate the LAI products. The MODIS LAI products at 8 m, 15 m, and 500 m spatial resolution were compared with ASTER LAI products at 15 m and 500 m resolution, with GF-1 LAI products at 8 m and 500 m resolution and with the field measured LAI (shown in Figure 10). Figure 10 highlights the differences among MODIS LAI, ASTER LAI, GF-1 LAI, and measured LAI. In Figure 10, compared to the field measured LAI and ASTER LAI during the crop growth cycle, MODIS LAI tends to underestimate at low LAI and overestimate at high LAI. ASTER LAI values at 15 m resolution were closer to the MODIS LAI values than the field measured LAI values because of decreased spatial scale mismatch between the two pixel-based products.

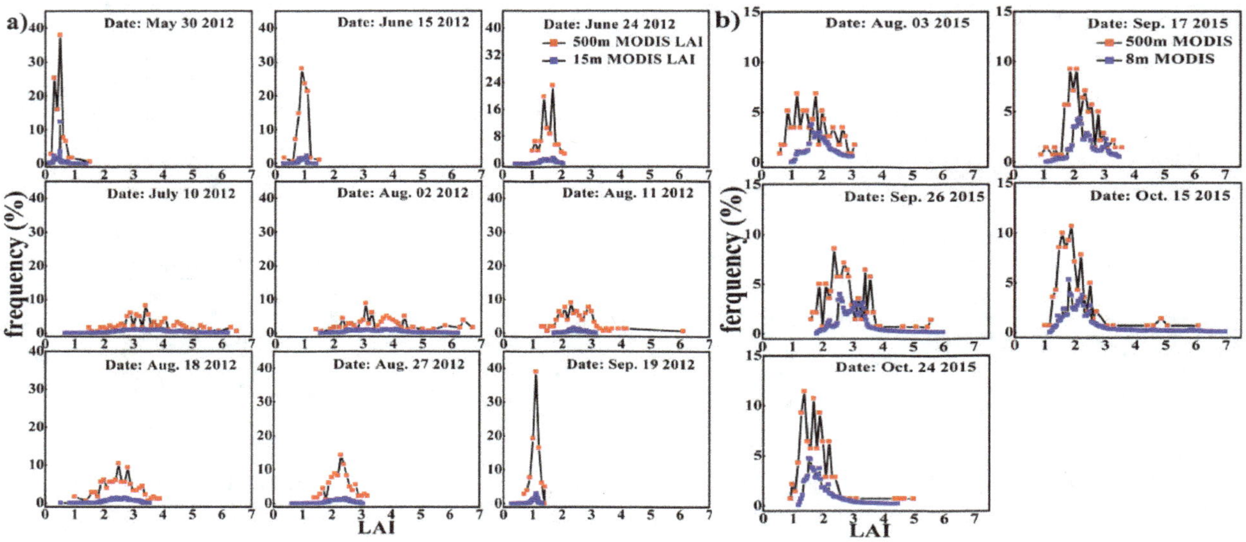

Figure 9. The time series frequency distribution: (**a**) 15 m MODIS LAI and MODIS LAI with 500 m resolution during the growth cycle of corn, (**b**) 8 m MODIS LAI and MODIS LAI with 500 m resolution during the growth cycle of rice.

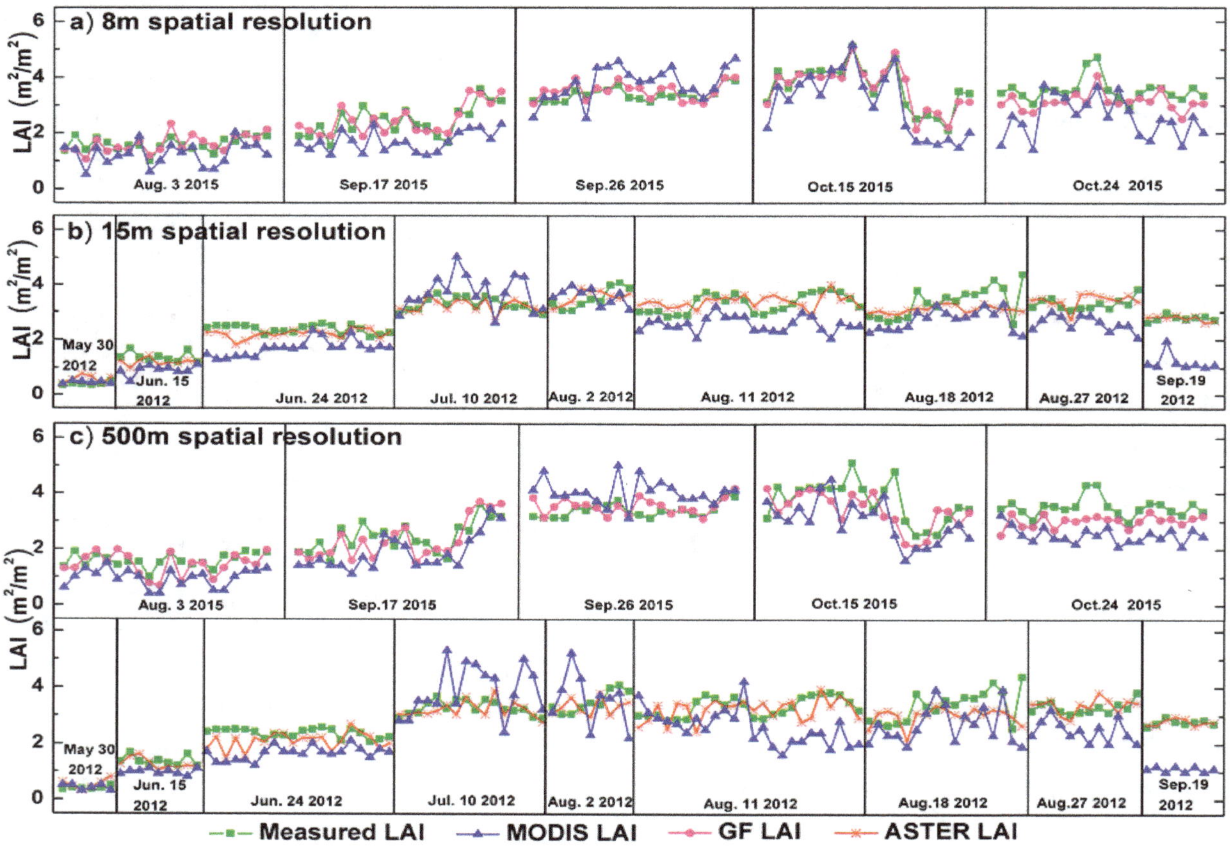

Figure 10. Sample comparison among MODIS LAI, ASTER LAI, GF-1 LAI, and field measured data during the crop growth cycle: (**a**) LAI products with 8 m resolution; (**b**) LAI products with 15 m resolution; (**c**) LAI products with 500 m resolution.

The MODIS LAI products, ASTER LAI products, GF-1 LAI products, and field measured LAI values from validation samples were compared by calculating the MRE, RMSE, and relative RMSE percentage (Figure 11). The highest R^2 was between MODIS LAI and ASTER LAI at 15 m resolution ($R^2 = 0.758$), followed by the correlation between MODIS LAI and measured LAI ($R^2 = 0.745$). The validation accuracy of MODIS LAI using ASTER LAI was higher than GF-1 LAI and field measured LAI, showing that the RRMSE was 28.73% for 15 m resolution, 29.00% for 8 m resolution, and 29.41%–30.80% for 500 m resolution, respectively, which suggests that the ASTER LAI products generated by the PROSAIL model were more suitable to validate the MODIS LAI products. The MODIS LAI products at 15 m resolution were closer to the field measured LAI than the 500 m and 8 m resolution MODIS LAI products, which implies that the MODIS LAI products at 15 m resolution would be more effective in monitoring surface vegetation cover and vegetation growth.

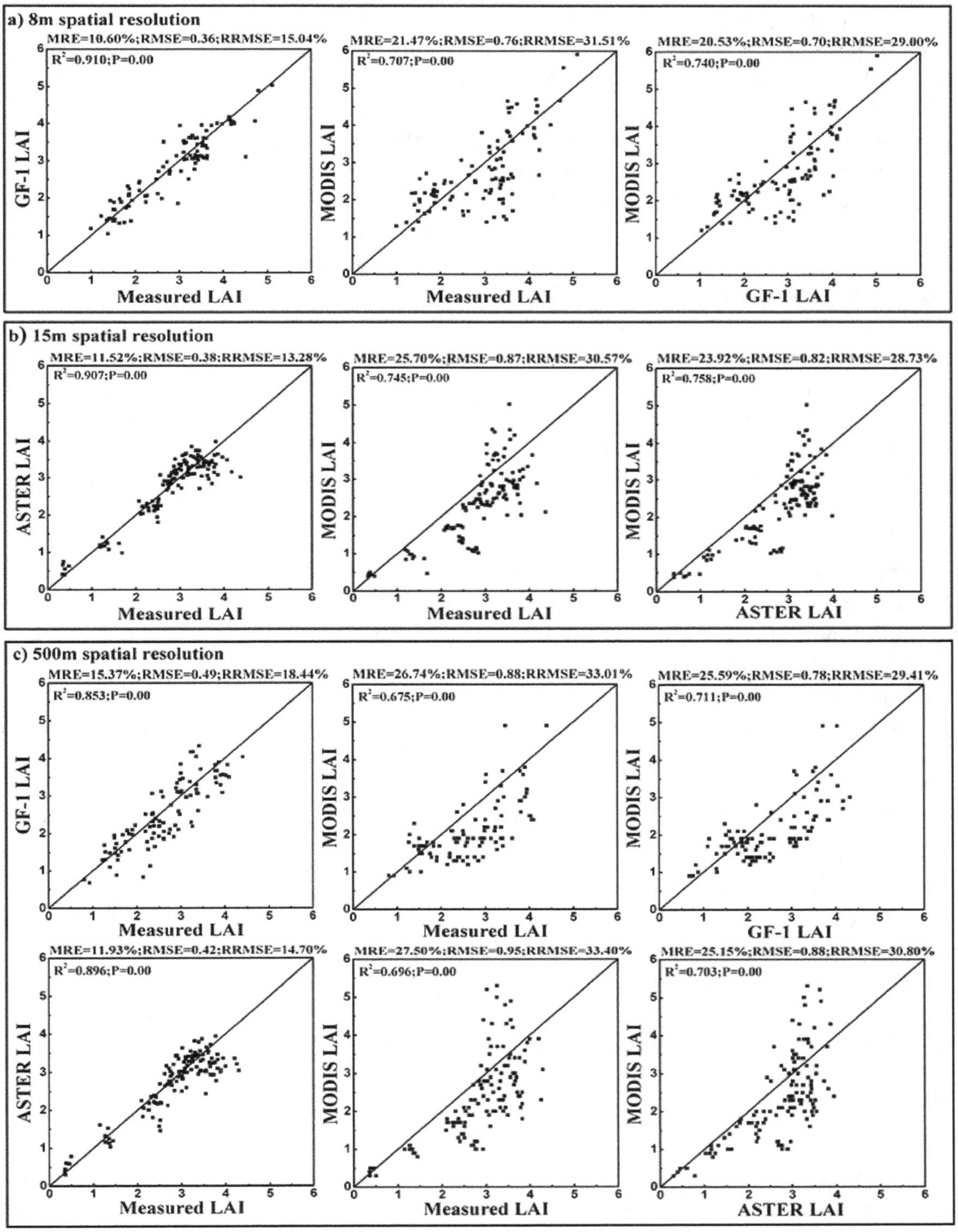

Figure 11. The LAI product versus LAI product scatter plots in the study areas using the corn and rice LAI values from the growth stages: (**a**) 8 m LAI product; (**b**) 15 m LAI product; (**c**) 500 m LAI product.

4. Discussion

4.1. Comparison with Other Similar Studies

Previous studies using direct and indirect methods to validate the accuracy of MODIS LAI products at a local or global scale [15,17,34–40], while providing valuable information, have some limitations.

First, previous validation methods [5,7] focused on validating MODIS LAI products at 500 m without downscaling the MODIS LAI to a finer resolution. Thus, these methods cannot evaluate the reliability of the 500 m MODIS LAI products at a fine local scale. The method proposed in this paper not only validated the accuracy of the MODIS LAI product with 500 m spatial resolution, but also validated the MODIS LAI with 15 m and 8 m spatial resolution generated by the EBK interpolation of the 500 m MODIS LAI. From Figure 11, the evaluation statistics derived from the comparison between the 15 m MODIS LAI and 15 m ASTER LAI were 0.758 of R^2, 23.92% of MRE, 0.82 of RMSE and 28.73% of RRMSE, which show improvements of 0.055, 1.23%, 0.06, and 2.07%, respectively, compared with the results calculated at 500 m spatial resolution. When compared to 8 m GF-1 LAI, the evaluation statistics of 8 m MODIS LAI products were 0.740 of R^2, 20.53% of MRE, 0.70 of RMSE, and 29.00% of RRMSE. These results show that the MODIS LAI with 15 m resolution obtained by EBK interpolation was more reliable than the MODIS LAI products with a 500 m resolution.

Second, previous validation studies [8,11] were based on a variety of vegetation types and rarely considered the crop growth cycle to validate the MODIS LAI product. In this study, the MODIS LAI products were validated based on the corn and rice growth cycles. Figure 10 indicates that the MODIS LAI product showed the mischaracterized corn and rice growth cycles except for the crop heading to the flower stage (August for corn and mid-October for rice), underestimating in low LAI regions and overestimating at LAI high regions. The results were in agreement with those of previous studies [17], where the inversion accuracy of the Collection 5 MODIS LAI (MOD15A2) products was examined for three typical vegetation types (i.e., alpine meadow, steppe, and subalpine shrub).

Third, though few downscaling techniques studies [19] have been able to downscale the lower spatial resolution satellite LAI products into the spatial resolution of the bridging LAI results and then conduct a comparison, some downscaling techniques have been applied to acquire other remote sensing data estimates [41,42]. Three machine learning approaches—random forest, boosted regression trees, and Cubist—were evaluated for a downscaling of the 25 km AMSR-E soil moisture data using 1 km MODIS products with a R^2 ranging from 70% to 75% [41]. Compared with the EBK downscaling method, the predictive accuracy of these downscaling methods showed no obvious difference. Thus, to improve the product validation accuracy, the optimal algorithm will be sieved in the future.

4.2. Prospects for Future Studies

The proposed method serves as a reference for future validation work. Due to the absence of other field vegetation type information, we cannot currently validate the MODIS LAI product uncertainties brought about by the type of vegetation. Thus, more field measured samples from different vegetation types need to be collected to further validate the reliability of the proposed validation method.

In this study, the scale-transformation method was performed between 500 m, 15 m, and 8 m spatial resolution pixels, which affects the accuracy of the validation due to the large difference in spatial scales. Future studies should consider introducing intermediate spatial scales of LAI to complete the MODIS LAI validation.

5. Conclusions

This study validated the MOD15A2H LAI products using a new multi-scale LAI product validation method based on the crop growth cycle. The experiment was conducted by long-term in situ observations during the corn and rice growth cycles. Results have shown that the MODIS LAI product with 500 m spatial resolution was overestimated during the medium growth stage (mid-to-late July and early August for corn and late September for rice) and underestimated during other growth stages (including tillering stage, jointing stage, heading to flowering stage, and maturity stage). The ASTER LAI values obtained by the PROSAIL model were more closely related to the MODIS LAI than GF-1 LAI and field measured LAI, with a R^2 of 0.758 for 15 m spatial resolution, and R^2 of 0.703 for 500 m. The relationship between MODIS LAI and field measured LAI had R^2 of 0.707 for 8 m, R^2 of 0.745 for 15 m, and R^2 of 0.675–0.696 for 500 m. Uncertainties in current satellite products still exist at different scales. Future studies should focus on considering the scale difference between field measurements and moderate resolution pixels utilizing higher resolution and pure pixel information of vegetation type observations.

Author Contributions: Z.L. and T.W. conceived and designed the experiments; T.W. analyzed the data, created the tables and figures, and finished the first version of the paper; Z.L., T.W., Y.Q., Ziqing Xia, and Y.P. contributed valuable opinions during the manuscript writing; Z.L. and Y.Q. revised the whole manuscript. All authors read and approved the final manuscript.

References

1. Moran, M.S.; Maas, S.J.; Pinter, P.J. Combining remote sensing and modeling for estimating surface evaporation and biomass production. *Remote Sens. Rev.* **1995**, *12*, 335–353. [CrossRef]
2. Norman, J.M.; Kustas, W.P.; Humes, K.S. Source approach for estimating soil and vegetation energy fluxes in observations of directional radiometric surface temperature. *Agric. For. Meteorol.* **1995**, *77*, 263–293. [CrossRef]
3. Anderson, M.C.; Norman, J.M.; Kustas, W.P.; Li, F.; Prueger, J.H.; Mecikalski, J.R. Effects of Vegetation Clumping on Two–Source Model Estimates of Surface Energy Fluxes from an Agricultural Landscape during SMACEX. *J. Hydrometeorol.* **2005**, *6*, 892–909. [CrossRef]
4. Doraiswamy, P. Crop condition and yield simulations using Landsat and MODIS imagery. *Remote Sens. Environ.* **2004**, *92*, 548–559. [CrossRef]
5. Sun, C.; Liu, L.; Guan, L.; Jiao, Q.; Peng, D. Validation and error analysis of the MODIS LAI product in Xilinhot grassland. *J. Remote Sens.* **2014**, *18*, 518–536.
6. Morisette, J.T.; Baret, F.; Privette, J.L.; Myneni, R.B.; Nickeson, J.E.; Garrigues, S.; Shabanov, N.V.; Fernandes, R.A.; Leblanc, S.G.; Kalacska, M.; et al. Validation of global moderate-resolution LAI products: A framework proposed within the CEOS land product validation subgroup. *IEEE Trans. Geosci. Remote Sens.* **2006**, *44*, 1804–1817. [CrossRef]
7. Garrigues, S.; Lacaze, R.; Baret, F.; Morisette, J.T.; Weiss, M.; Nickeson, J.E.; Fernandes, R.; Plummer, S.; Shabanov, N.V.; Myneni, R.B.; et al. Validation and intercomparison of global Leaf Area Index products derived from remote sensing data. *J. Geophys. Res. Biogeosciences* **2008**, *113*. [CrossRef]
8. Fang, H.; Wei, S.; Liang, S. Validation of MODIS and CYCLOPES LAI products using global field measurement data. *Remote Sens. Environ.* **2012**, *119*, 43–54. [CrossRef]
9. Yang, Y.; Li, A.; Jin, H.; Yin, G.; Zhao, W.; Lei, G.; Bian, J. Intercomparison Among GEOV1, GLASS and MODIS LAI Products over Mountainous Area in Southwestern China. *Remote Sens. Technol. Appl.* **2016**, *31*, 438–450.
10. Gessner, U.; Niklaus, M.; Kuenzer, C.; Dech, S. Intercomparison of leaf area index products for a gradient of sub-humid to arid environments in West Africa. *Remote Sens.* **2013**, *5*, 1235–1257. [CrossRef]
11. Jin, H.; Li, A.; Bian, J.; Nan, X.; Zhao, W.; Zhang, Z.; Yin, G. Intercomparison and validation of MODIS and GLASS leaf area index (LAI) products over mountain areas: A case study in southwestern China. *Int. J. Appl. Earth Obs. Geoinf.* **2017**, *55*, 52–67. [CrossRef]
12. Sea, W.B.; Choler, P.; Beringer, J.; Weinmann, R.A.; Hutley, L.B.; Leuning, R. Documenting improvement in leaf area index estimates from MODIS using hemispherical photos for Australian savannas. *Agric. For. Meteorol.* **2011**, *151*, 1453–1461. [CrossRef]
13. Zeng, Y.; Li, J.; Liu, Q. Review article: Global LAI ground validation dataset and product validation framework. *Adv. Earth Sci.* **2012**, *27*, 165–174.
14. Liu, Y.; Wang, J.; Zhou, H.; Xue, H. Upscaling approach for validation of LAI products derived from remote sensing observation. *J. Remote Sens.* **2014**, *18*, 1189–1198.

15. Yin, G.; Li, A.; Jin, H.; Zhao, W.; Bian, J.; Qu, Y.; Zeng, Y.; Xu, B. Derivation of temporally continuous LAI reference maps through combining the LAI Net observation system with CACAO. *Agric. For. Meteorol.* **2017**, *233*, 209–221. [CrossRef]
16. Liu, L. Simulation and correction of spatialscaling effects for leaf area index. *J. Remote Sens.* **2014**, *18*, 1158–1168.
17. Yang, J.; Chen, H.; Borjigin, N.; Zhao, M.; Zhou, Y.; Huang, Y. Validation of the MODIS LAI product in Qinghai Lake Basin combined with field measurements using Landsat 8 OLI data. *Acta Ecol. Sin.* **2017**, *37*, 322–331. [CrossRef]
18. Liu, Z.; Huang, R.; Hu, Y.; Fan, S.; Feng, P. Generating high spatiotemporal resolution LAI based on MODIS/GF-1 data and combined kriging-cressman interpolation. *Int. J. Agric. Biol. Eng.* **2016**, *9*, 120–131.
19. Houborg, R.; Mccabe, M.F.; Gao, F. A Spatio-Temporal Enhancement Method for medium resolution LAI (STEM-LAI). *Int. J. Appl. Earth Obs. Geoinf.* **2016**, *47*, 15–29. [CrossRef]
20. Qu, Y.; Zhu, Y.; Han, W.; Wang, J.; Ma, M. Crop leaf area index observations with a wireless sensor network and its potential for validating remote sensing products. *IEEE J. Sel. Top. Appl. Earth Obs. Remote Sens.* **2014**, *7*, 431–444. [CrossRef]
21. Li, X.; Liu, S.; Xiao, Q.; Ma, M.; Jin, R.; Che, T.; Wang, W.; Hu, X.; Xu, Z.; Wen, J.; et al. A multiscale dataset for understanding complex eco-hydrological processes in a heterogeneous oasis system. *Sci. Data* **2017**, *4*, 170083. [CrossRef] [PubMed]
22. Available online: http://www.heihedata.org (accessed on 10 October 2019).
23. LP DAAC - MOD15A2H. Available online: https://lpdaac.usgs.gov/products/mod15a2hv006/ (accessed on 20 October 2019).
24. Kancheva, R.; Georgiev, G. Assessing Cd-induced stress from plant spectral response. In Proceedings of the SPIE—The International Society for Optical Engineering, Amsterdam, The Netherlands, 22–25 September 2014; Volume 9239, pp. 1–12.
25. Krzywinski, M.; Altman, N. Multiple linear regression. *Nat. Methods* **2015**, *66*, 1103–1104. [CrossRef] [PubMed]
26. Feret, J.B.; François, C.; Asner, G.P.; Gitelsond, A.A.; Martin, R.E.; Bidel, L.P.R.; Ustin, S.L.; Maire, G.; Jacquemoud, S. PROSPECT-4 and 5: Advances in the leaf optical properties model separating photosynthetic pigments. *Remote Sens. Environ.* **2008**, *112*, 3030–3043. [CrossRef]
27. Jacquemoud, S.; Verhoef, W.; Baret, F.; Bacour, C.; Zarco-Tejada, P.J.; Asner, G.P.; François, C.; Ustin, S.L. PROSPECT + SAIL models: A review of use for vegetation characterization. *Remote Sens. Environ.* **2009**, *113*, S56–S66. [CrossRef]
28. Gu, C.; Du, H.; Zhou, G.; Han, N.; Xu, X.; Zhao, X.; Sun, X. Retrieval of leaf area index of Moso bamboo forest with Landsat Thematic Mapper image based on PROSAIL canopy radiative transfer model. *Chin. J. Appl. Ecol.* **2013**, *24*, 2248–2256.
29. Li, H. Leaf Area Index Retrieval Based on Prospect, Liberty and Geosail Models. *Sci. Silvae Sin.* **2011**, *47*, 75–81.
30. Krivoruchko, K.; Butler, K. *Unequal Probability-Based Spatial Sampling*; Esri: Redlands, CA, USA, 2013.
31. Goovaerts, P. Kriging and Semivariogram Deconvolution in the Presence of Irregular Geographical Units. *Math. Geosci.* **2008**, *40*, 101–128. [CrossRef]
32. Omre, H. Bayesian kriging-merging observations and qualified guesses in kriging. *Math. Geol.* **1987**, *19*, 25–39. [CrossRef]
33. Fabijańczyk, P.; Zawadzki, J.; Magiera, T. Magnetometric assessment of soil contamination in problematic area using empirical Bayesian and indicator kriging: A case study in Upper Silesia, Poland. *Geoderma* **2017**, *308*, 69–77. [CrossRef]
34. John, I.; Russell, C.; Timothy, L.; Andrew, P. Uncertainty Analysis in the Creation of a Fine-Resolution Leaf Area Index (LAI) Reference Map for Validation of Moderate Resolution LAI Products. *Remote Sens.* **2015**, *7*, 1397–1421.
35. Claverie, M.; Weiss, M.; Frédéric, B.; Hagolle, O.; Demarez, V. Validation of coarse spatial resolution LAI and FAPAR time series over cropland in southwest France. *Remote Sens. Environ.* **2013**, *139*, 216–230. [CrossRef]
36. Serbin, S.P.; Ahl, D.E.; Gower, S.T. Spatial and temporal validation of the MODIS LAI and FPAR products across a boreal forest wildfire chronosequence. *Remote Sens. Environ.* **2013**, *133*, 71–84. [CrossRef]

37. Jia, S.; Ma, M.; Yu, W. Validation of the LAI Product in Heihe River Basin. *Remote Sens. Technol. Appl.* **2014**, *29*, 1037–1045.
38. Qu, Y.; Han, W.; Ma, M. Retrieval of a Temporal High-Resolution Leaf Area Index (LAI) by Combining MODIS LAI and ASTER Reflectance Data. *Remote Sens.* **2015**, *7*, 195–210. [CrossRef]
39. Myneni, R.B.; Hoffman, S.; Knyazikhin, Y.; Privette, J.L.; Glassy, J.; Tian, Y.; Wang, Y.; Song, X.; Zhang, Y.; Smith, G.R.; et al. Global products of vegetation leaf area and fraction absorbed PAR from year one of MODIS data. *Remote Sens. Environ.* **2002**, *83*, 214–231. [CrossRef]
40. Liu, Y.; Liu, R.; Chen, J.; Cheng, X.; Zheng, G. Current Status and Perspectives of Leaf Area Index Retrieval from Optical Remote Sensing Data. *Geo-Inf. Sci.* **2013**, *15*, 734. [CrossRef]
41. Jing, W.; Yang, Y.; Yue, X.; Zhao, X. A Spatial Downscaling Algorithm for Satellite-Based Precipitation over the Tibetan Plateau Based on NDVI, DEM, and Land Surface Temperature. *Remote Sens.* **2016**, *8*, 655. [CrossRef]
42. Im, J.; Park, S.; Rhee, J.; Baik, J.; Choi, M. Downscaling of AMSR-E soil moisture with MODIS products using machine learning approaches. *Environ. Earth Sci.* **2016**, *75*, 1120. [CrossRef]

12

Disaster Mitigation in Urban Pakistan using Agent based Modeling with GIS

Ayesha Maqbool [1,*], Zain ul Abideen Usmani [2], Farkhanda Afzal [3] and Alia Razia [3]

1. Department of Computer Software Engineering, Military College of Signal, National University of Sciences and Technology, Islamabad 44000, Pakistan
2. Department of Information Security, Military College of Signal, National University of Sciences and Technology, Islamabad 44000, Pakistan; zusmani.msse3@students.mcs.edu.pk
3. Department of Humanities & Basic Sciences, Military College of Signal, National University of Sciences and Technology, Islamabad 44000, Pakistan; farkhanda@mcs.edu.pk (F.A.); aliarazia@mcs.edu.pk (A.R.)

* Correspondence: ayesha.maqbool@mcs.edu.pk.

Abstract: This study aims to propose an application of agent based modeling (ABM) and simulation for disaster mitigation in an urban region of Pakistan. Pakistan has been working over the past few decades to reduce the risk factor of disasters by using different disaster management approaches. However, these efforts are in an early stage. Although lack of planning and unchecked urbanization are the main hurdles, insufficient resources in terms of technology is also a major contributing factor that impedes achieving desired results. In this paper, we are proposing ABM and simulation of approaches using geographical information system (GIS) maps for disaster management in the urban locality of Pakistan. The conceptual model was implemented for analysis of resource allocation (RA) of first response units (ambulances, fire brigade, etc.). In the proposed model, we used two allocation algorithms; high severity level (HSL) and first come first serve (FCFS). These algorithms were simulated in NetLogo by creating a hypothetical disaster scenario in Rawalpindi city. In our experiments, the design was based on demand, resource agents, and their allocation behavior for disaster management. We analyzed the resource allocation mechanism using average wait time, overall number of demands, execution time, and unallocated demands as performance measures.

Keywords: agent based modeling; disaster management; resource allocation; high severity level; first come first serve; geographical information system

---

1. Introduction

Pakistan is one of the countries that has been badly affected by disasters in the last two decades. Pakistan is the sixth most populated country of the world. Hence, rapid urbanization has led to unchecked growth of small cities. These cities are growing without well-planned infrastructure, which is the root cause of urban disaster [1,2]. In populated cities, various incidents of road accidents and fire breakouts are causing hundreds of deaths and property damage at multiple levels. Modern countries around the world have worked very hard to manage the disasters while in developing countries, the efforts made are not up to the required standards. The degree of human suffering is the only thing that we have to control, and it is only possible with the significant decrease in risk and initial preparation for risk [3]. At present, the disaster management (DM) system in Pakistan does have a wide-ranging, well-connected arrangement at institutional level, but lacks an efficient, reliable, and speedy response plan. It is due to the reason that there is no effective resource allocation (RA) technique to deal with a disaster situation in accordance with the emerging technological trends. This results in a delayed response by administration. Therefore, whenever an emergency occurs in the region, it would result in huge loss of life, property, and capital.

Authors in [4] have presented a study of a tragedy of Baldia town factory, Karachi. The paper highlights the challenges faces by rescue services in Pakistan. The study is the motivation for our work as it shows the vulnerabilities of Urban Pakistan rescue and government services.

Considering the challenges of Urban Pakistan GIS study of current DM resources and infrastructures is an eminent challenge. Many existing studies have incorporated GIS for modeling and simulation of varied applications. In [5], the authors have established a technique that combines the use of GIS and remote sensing for achieving simulation and modeling of the landscape impact affected by construction. Taking a review of DM practices in Canada, the authors of [6] provided an in-depth investigation of reasons of failure of present DM practices, and suggested the application of knowledge and systems science to improve DM efforts.

Arain in [7] introduces a conceptual framework of knowledge-based methodology for instant and successful economical DM, while it is limited to post-disaster only. ABM has been employed to study human behavior and the complex adaptive effect on the overall system. In [8] the authors have presented a simulation of avoiding behaviors among social groups and synchronize behaviors among subgroups within the same social group.

The article [9] presents the effect of contextual decision making by authorities to reduce the impact of a disaster. The study presents the application of design science research to develop semi-structured DM plans. The research is focused on how the individual and social behavior affects the DM knowledge, which results in different approaches to disaster management. The complex interactions and information flows are modeled using ABM that signifies the application of ABM for DM.

Huang et al. [10] describe a problem under consideration of how humanitarian aid is distributed and delivered to the affected zone in an efficient way through a network of transportation to reduce the disaster effect. However, it is difficult for RA to determine which resources are helpful during allocation. Wise et al. [11] present precise work of several researchers and their opinion about urban framework elements. They are influenced by transportation concerning to ABM. Matveev et al. in [12] explain dependency and proficiency of emergency responses to incidents and other vehicle utilization during the emergency occurrence. In [13] the authors focus on how to deal with location and relocation, as well as dispatching decisions for emergency vehicles. The study highlights that using analytical methods and the inclusion of factors like dynamism and uncertainty makes a system computationally infeasible. In [14], Kong et al. proposed a model which is based on a backpropagation neural network composed of linear programming and RA requirement prediction, but resource management is always pre-assigned in the case of traffic accidents.

In [15], the researcher's consideration is centered on utilizing volunteer resources for mapping the damage. It expresses a spatial prototype that demonstrates an agent-based model using public utilization geographic information system (GIS) data.

Cavdur et al. in [16] has deliberated on a facility location problem for disaster operation management, and the proposed approach for solution is based on a deterministic model. Meng et al. determine the optimal location facilities for the terrorist attack in terms of emergency response in [17]. The study however contains the assumptions of boundless limitations, which are not practical.

In [18] authors address a maximum coverage location problem (MCLP) with partial coverage where distance and demand zones are rectilinear. Therefore, when addressing allocation in an attention zone, the coverage algorithm includes only the part that has relative demands. The authors developed a novel methodology for a specific facilities problem only having multiple objectives with the attention on an emergency location system [19].

Our presented work is aimed at establishing the use of agent-based modeling for disaster mitigation in Pakistan. The motivation behind the study is to provide soundness of using ABM in situations where the lack of data and dense urban maps of unstructured development poses a great challenge in making performance estimates of available resources in the case of emergencies. The rest of the paper is organized into four sections. Section 2 discusses the problem statement. In Section 3, the details of proposed model and two allocation algorithms first come first serve (FCFS) and high severity level (HSL) are provided. Section 4 presents results of our implemented simulation in NetLogo while Section 5 presents the conclusion and future work.

2. Problem Statement

Many developing countries have carried out extensive research to develop early warning systems, disaster relief distribution systems, and GIS to enhance their disaster response capacities to deal with adversity. In the field of disaster management, there has been a dire need to create and actualize RA models for improving the current RA techniques in Pakistan. This acute need to assess and devise a feasible RA model at national level will lead to better DM abilities. The problem statement is defined as:

"How to allocate resources for disaster management in an urban area by capitalizing on emerging technologies, to formulate and develop a system model using ABM for RA."

3. Proposed System Model

In this section, a system model is proposed to allocate resources in the case of disasters. The system model covers planning, strategy, and algorithm-based procedures RA. The model is based on ABM, which is a novel framework made up of self-governing agents and interaction between the agents. In our system, demand and resources are basic agents. An agent can be defined as self-governing, objective-oriented entity fit for watching and collaborating with other agents.

The objective of ABM for RA is to think about emergent complex behavior of the entire system as compared to a centralized decision-making agent the proposed system is a conceptual model utilized for resource allocation to the demands. The proposed model to implement ABM for RA is shown in Figure 1.

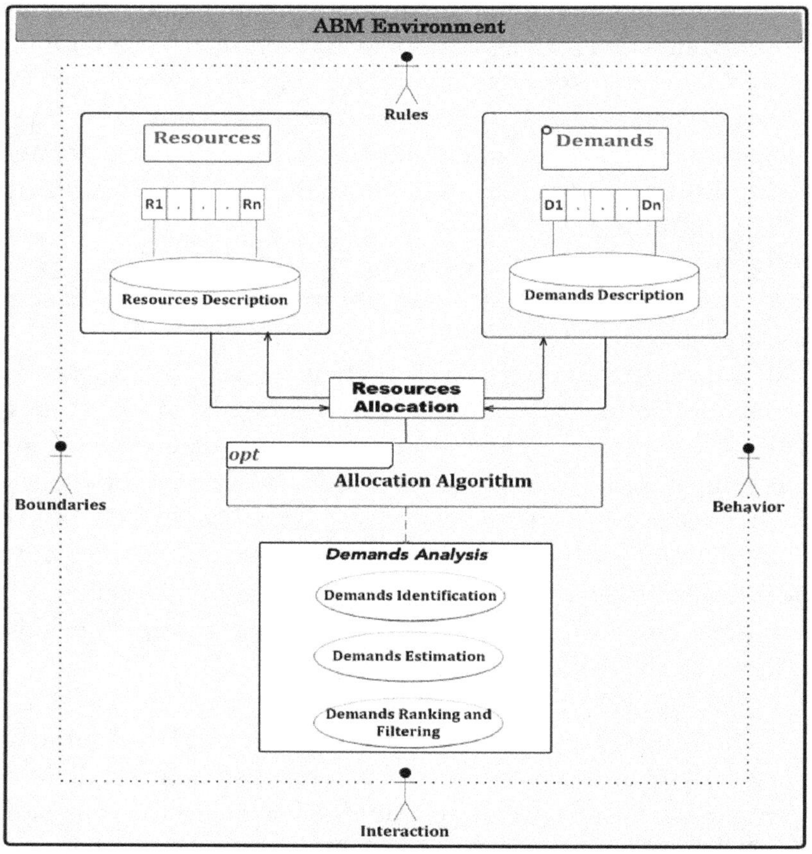

Figure 1. Conceptual model to implement agent based modeling (ABM) for resource allocation (RA).

3.1. Resources

Resources are organizations that provide emergency services (ambulance, fire brigade, etc.) to the demand site. Resources are required for all events including preventive action, early relief, emergency and disaster reduction.

An estimation of resources expected to achieve the target in terms of demand must be directed. A simple target may require simple resources, while the critical target will require numerous resources with critical abilities that are accessible in a short period. If resources are not available with the required quantities, then the requisite target may not be achieved. Other than recognizing explicit resources for the demand zone, the estimated requirements for resources should respond to different inquiries:

- What amount of resources is required?
- When a resource will be required?
- What ability does the resources need to fulfil? Are there any constraints?
- What is the timing for having resources accessible? Are there any liabilities related to the utilization of the resources?

In our simulation, we modeled the resource agent as an independent entity that responds and plans for appropriate resource allocation based on given selection criteria. Details of agent properties and behaviors are described in Table 1.

Table 1. Resource agent details.

Resource Agent Properties
1. Number of resources units R_n 2. Number of all allocated resource units *R_Alloc* 3. Number of all reserved resource units *R_Rese* 4. Following attribute of each unit *i* of resource, a. Allocated RA_i b. Reserved RRs_i c. Destination demand RD_i d. Selection criteria for measures (severity, distance) SC_i
Resource Agent Behavior
1. Resource receives demands from disaster sites 2. Demand details contain quantity needed, and selection criteria info. i.e., Distance and severity 3. While resources are available and the demands are active: a. The resource agent selects demand on the basis of selection criteria b. Sends confirmation and the reservation of available/required units of resources. Updates RRs_i, RD_i, and *R_Rese* 4. On reception of confirmation of reservation from demand site: a. The resource agent removes the resource unit from the Reservation list and add it to allocation RR_{si}, RD_i, RA_i, and *R_Rese*, *R_Alloc* b. Dispatches the resource to demand site. c. On completion of demand's need (measures in time duration of allocation). The resources are taken from allocation list and are added back to pool by clearing Rs_i, RD_i, RA_i, and *R_Rese*, *R_Alloc*.

Resource centers are capable of receiving demands from the disaster sites. Each resource center then bids to allocate the available units of resources to the demand site. Only on confirmation from the demand site the final allocation is performed. Once the resource's utility is completed at the demand site, the demand releases the allocated units and these units are added back to available pool. In the case where lack of available resources results in partial satisfaction of demand, in order to manage the complete demand, these new newly released units are then allocated to such sites.

3.2. Demands

Demand is considered as a disaster zone, a zone authoritatively announced to be the location of a crisis made by a disaster and subsequently qualified to get specific sorts of resources. There may be a hazardous situation is an area or a district, intensely harmed by common, innovative, or social perils. In our model, the demand agents, as described in Table 2, broadcast their requirement to all approachable resource agents. The resource agents then, depending upon their specific selection criteria, respond by reserving available units for the demand. This results in the demand site receiving multiple acknowledgments from multiple resource agents. The demand site then selects the nearest resource sites and confirms the reservations of that specific resource agent and declines the rest.

Table 2. Demand agent details.

Demand Agent Propertie

1. Number of demand units *Dm*
2. Number of all allocated demand units *D_Alloc*
3. Number of all reserved demand units *D_Rese*
4. Selection attribute of demand *D_Sa*
5. Following attribute of each unit j of demands,
    a. Allocated DA_j
    b. Reserved DRs_j
    c. Resource agent DRa_j
    d. Arrival time Dt_j

Resource Agent Behavior

Till the demands are met or demand expires:
1. Demand broadcast the request to all approachable demand sites.
2. Demand details contain quantity needed, and selection criteria info. i.e., distance and severity, *D_Sa*, Dn_n
3. Wait of response from resources
4. On reception of responses from resource agents *D_Rese*
    a. Select the nearest resource agent and send the confirmation of allocation updates *D_Alloc*, DRs_j
    b. Declines the rest of the reservations
5. Wait for the resource arrival
6. On arrival of resource, it holds the resource until requirement is met update DA_j
7. At completion of request, releases the resources updates D_*Alloc*, DA_j, DRa_j

The demand site then waits for resource units. At arrival of the units, the demand site retains the resource for the required duration. After completion of the task the resources are released back. In the case of partial allocation, the demand sites keep on broadcasting the remaining requirements to all resource centers until the demand is met or expires. In our model, the demand expiration depicts the situation where the demand's initial requested quantity is never met but the demand's requirements are met without these unallocated units. For example, for a certain incidence, initially 10 ambulances were requested, but only two vehicles were available at the time of request. The rest of the support is provided by self-help basis from citizens and injured are taken to medical facilities in other vehicles.

3.3. *Resource Allocation*

In general, RA is an arrangement for utilizing accessible resources to accomplish objectives. RA within the umbrella of ABM involves deployment of a large numbers of rescue services within an urban environment in the context of disaster. It is fundamental to emergency response planning. One of the challenges faced in emergency response planning is the optimal or near to optimal allocation of resources to specific emergencies. In the context of RA, the problem may be stated as resource R being allocated to demand D as shown in Figure 2a.

The RA process (Figure 2b) tracks which resource is held by what type of demand and which demand is waiting for specific kind of resource. Each cycle comprises of demand information and at least one allocation of demand.

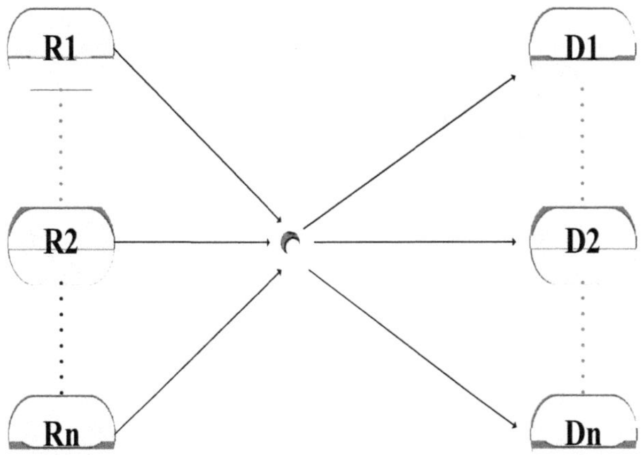

(a)

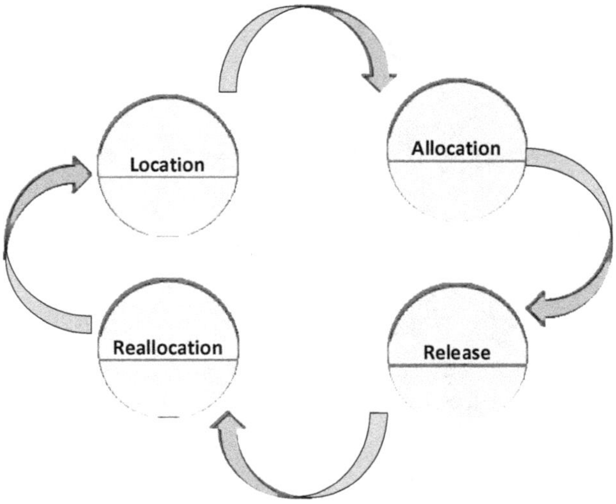

(b)

Figure 2. General Resource Allocation Process (**a**) Overall Resource Allocation Cycle with release and reallocation (**b**).

3.4. *Allocation Algorithm*

The allocation algorithm defines RA selection criteria for utilizing accessible resources to achieve demand objectives. The allocation algorithm for the demand zone is a major DM activity that takes into consideration procedure execution.

In the allocation algorithm, the main focus is demand analysis that relies on a procedure by which RA decides on the creation, assignment, utility of resources, and estimation. How much a resource site produces resources depends on the capacity of creation that is based on the potential demand for the zone. Demand analysis is fundamental to conceivable linkage as well as in making an extension to basic RA. Demand analysis is vital to create satisfactory and efficient strategies for implementing the RA in a crisis or emergency situation. In our recent work, we used two selection algorithms, high severity level (HSL) and first come first serve (FCFS).

HSL is a technique for planning and utilizing resources that is dependent on severity of demand. In this technique, the demand zone utilizes the resource zone to fulfil its objectives according to their severity. In the severity algorithm, every demand is rigorous, but the demand with higher severity is allocated first, while demands of equal severity are allocated based on FCFS. Resources are progressively allocated in the demand zone when they are required. If the resources are not available, at that point the demand site within their time frame waits for the release of resources for utilization. The steps of HSL based RA are shown in Algorithm I.

Algorithm I: HSL Algorithm

Given Resource R_i with net units R_n, number of already allocated units *R_Alloc*, and number of reserved units *R_Rese* **and** Demand D_j with demand of units D_m, and severity level *D_Sa*.

1. Given broadcast of quantity D_m from Demand site D_j is received by Resource R_i.
2. D_j has Severity level D_Sa_j and maximum time Dt_total_j
3. While there is a demand j with quantity $(D_m > 0)$ and highest severity $D\_Sa_j > D\_Sa_k$ for all received demands k.
4. If "x" demands with similar severity arrives select one with

$$min(\text{TravellingDist}) \quad (\text{for } k = 0, \ldots x)$$

5. If resources are available reserve "w" Resource to Demand and update Reserved resources

$$w = \begin{cases} 0 & (R\_Rese^{(old)} + R\_Alloc) \geq R_n \\ (R\_Rese^{(old)} + R\_Alloc) - R_n & otherwise \end{cases}$$

$$R\_Rese^{(new)} = R\_Rese^{(old)} + w$$

6. If acknowledgement from demand site arrives for t reserved resources allocate resources to demand and update allocated and reserved resources

$$R\_Alloc^{(new)} = R\_Alloc^{(old)} + t$$

$$R\_Rese^{(new)} = R\_Rese^{(old)} - t$$

7. After time Dt_total_j the released z and are added back to available pool

$$R\_Alloc^{(new)} = R\_Alloc^{(old)} - z$$

8. end while

In the FCFS allocation algorithm, as the name suggests, the demand which appears first is granted the resources foremost. The demand site releases the resources after complete utilization of resources within its time frame. The steps of FCFS based RA are shown in Algorithm II. In the presented model demand location, number units needed by demands and allocation time are generated randomly Such random behavior is simulated using agents and the allocations are performed by individual policy of resource rather than system wide allocation. The conventional analytical models for RA perform optimization for entire system. In our work all resources and demands are independent agents lacking a centralized optimization authority.

Algorithm II: FCFS Algorithm

Given Resource R_i with net units R_n, number of already allocated units *R_Alloc*, and number of reserved units *R_Rese* **and** Demand D_j with demand of units D_m maximum time Dt_total_j

1. Given broadcast of quantity D_m from Demand site D_j is received by Resource R_i.

   Location of *ith* Resource is *Rxi, Ryi*
   Location of *jth* Demand is *Dxj, Dyj*

2. While there is a demand *j* with quantity ($D_m > 0$)
3. Select demand D_j with min (*TravellingDist*)
4. If resources are available reserve "*w*" Resource to Demand and update reserved resources.

$$w = \begin{cases} 0 & (R\_Rese^{(\text{old})} + R\_Alloc) \geq R_n \\ (R\_Rese^{(\text{old})} + R\_Alloc) - R_n & otherwise \end{cases}$$

$$R\_Rese^{(\text{new})} = R\_Rese^{(\text{old})} + w$$

5. If acknowledgement from demand site arrives for ***t*** reserved resources allocate resources to demand and update allocated & reserved resources

$$R\_Alloc^{(\text{new})} = R\_Alloc^{(\text{old})} + t$$

$$R\_Rese^{(\text{new})} = R\_Rese^{(\text{old})} - t$$

6. After time Dt_total_j the released z units are added back to availed pool

$$R\_Alloc^{(\text{new})} = R\_Alloc^{(\text{old})} - z$$

7. end while

4. Simulation and Results

In our case, we used the NetLogo environment that is a good simulator of ABM with libraries to add GIS data. ABM captures emerging phenomena and provides a natural environment for the analysis of certain systems. It is a versatile technique especially in relation to geospatial model development.

4.1. RA in GIS Space

GIS is one of the most advanced technologies that have changed the manners in which RA has been used previously. These technologies have affected RA procedures explicit to the manner by which resources are utilized by the demand zone. In GIS space, it allows agents to be utilized as resources and demands into geospatial condition characterized by a GIS map. It includes a function to set and get the present agent space, allocate the resources to demands space according to RA procedure, to execute activity upon appearance, to execute the agent in its predefined environment, to set up interaction dependent on agents, design, and another attribute.

By using the proposed ABM in the previous section, we simulated a scenario for disaster in the Rawalpindi city in Pakistan. We took 65 union councils of Rawalpindi district. A disaster is created on a random building, which is our demand zone. It requires several resources to reduce the disaster effect in the zone. Figure 3 shows the steps that are involved in the creation of a scenario.

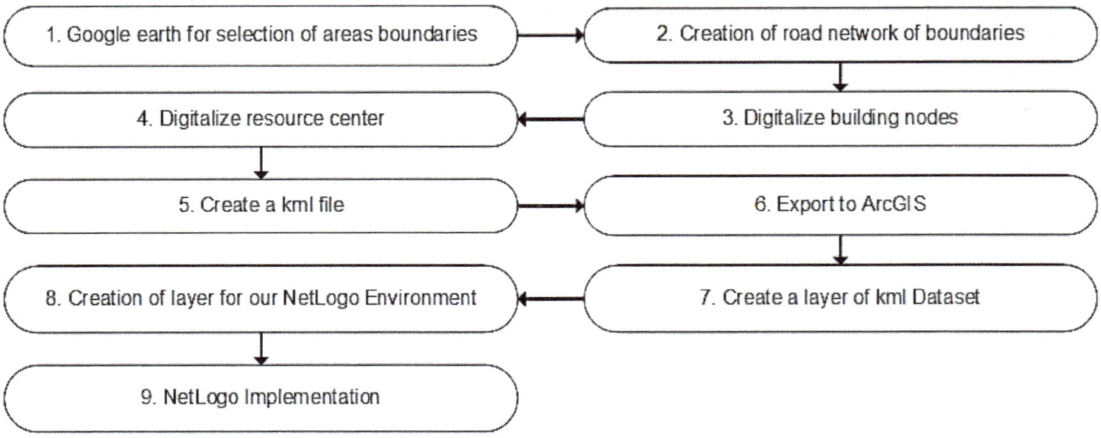

Figure 3. Steps for creation of scenario.

In Figure 4a, boundaries of the selected area are shown, and their total area is 122,982,339.2 m^2. Whereas, Figure 4b shows road network of these selected areas. The roads network is created near to the resource station and building. We took four resource/rescue stations and 186 buildings for the scenario creation, which are shown in Figure 4c,d. A screenshot of the ArcGIS layer structure for NetLogo implementation is shown in Figure 5. Different experiments were simulated according to factors given in Table 3 in NetLogo environments.

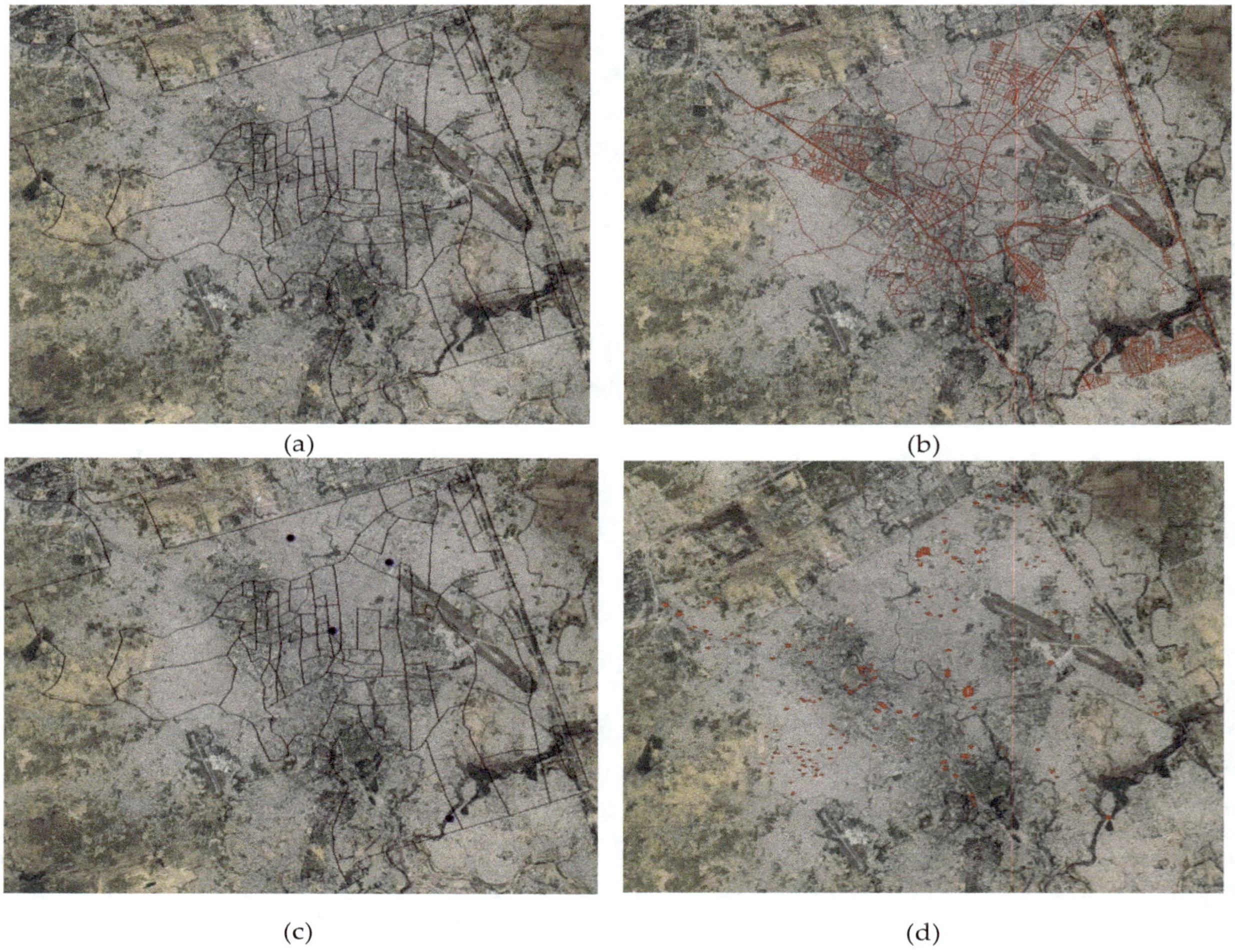

Figure 4. Area boundaries (**a**), road networks (**b**), resources center (**c**), and digitalized building node (**d**).

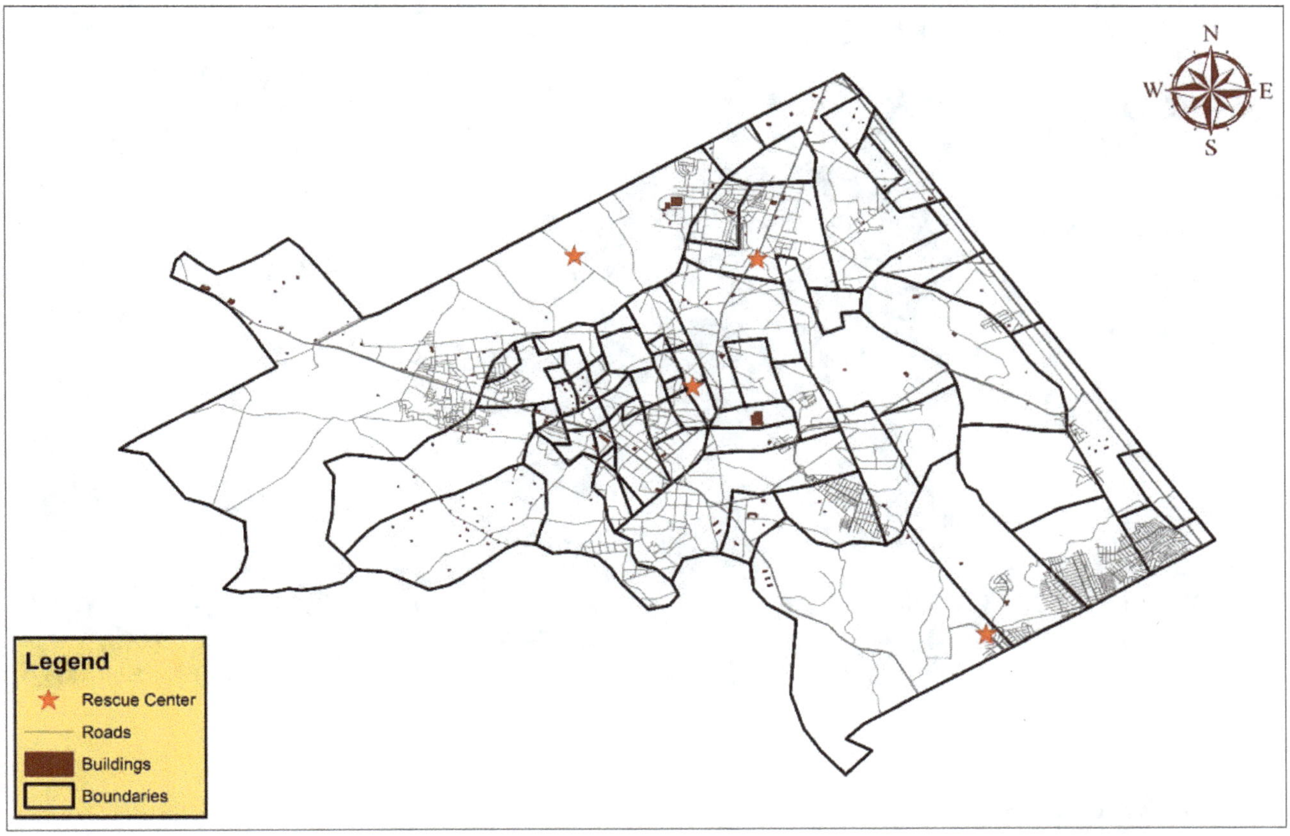

Figure 5. A screenshot of the NetLogo environment for experiment 1 given in Table 3.

Table 3. Simulation design.

Allocation Algorithm	Experiment	Demand Creation on Building Node	Demand Quantity	Fixed Resource Centre: Four with Quantity	Max Demand Duration (Unit Time)	Traffic Delay
	Experiment 1	5	1 ≤ random ≤ 10	1 ≤ random ≤ 10		
	Experiment 2	10	"	"	<20	
	Experiment 3	15	"	"	<40	
HSL and FCFS	Experiment 4	20	"	"	<60	With and without traffic
	Experiment 5	25	"	"	<80	
	Experiment 6	30	"	"	<100	
	Experiment 7	35	"	"		

4.2. Experiment Design

An experiment scenario was designed for the city of Rawalpindi in Pakistan, we took 65 union councils of Rawalpindi district and their total area is 122,982,339.2 m^2. We used NetLogo environment for simulations in this region as shown in Figure 6.

In the simulations, we divided agents into two categories; demands and resources. Resource agents are fixed according to the actual location of the resource center while demand agents are created randomly on 186 building nodes according to simulation. Furthermore, travelling distance was estimated using GIS road information instead of Euclidian distance. Each scenario was simulated with and without randomizing traffic delay of maximum 50% of travelling time. These scenarios were simulated 150 times for each experiment according to given factors in Table 3. ANOVA test results are presented in Table 4.

Figure 6. RA in the geographic information system (GIS) Space NetLogo implementation.

Table 4. ANOVA test summary.

	Factor	Effect	F Value	Pr (>F)
HSL	Duration of commitment	Net allocated	6.45	0.0112
	Number of demand sites	Unallocated demands	7841	$<2 \times 10^{-16}$
	Traffic vs. without traffic		3957	$<2 \times 10^{-16}$
FCFS	Duration of commitment	Net allocated	5.939	0.0149
	Number of demand sites	Unallocated demands	7962	$<2 \times 10^{-16}$
	Traffic vs. without traffic		3417	$<2 \times 10^{-16}$

4.3. Result Analysis

Simulation results of each algorithm were analyzed in R studio with several experiments that are given in Table 3. We performed analysis on the following data to evaluate simulation:

(1) Effects of selection algorithm HSL vs. FCFS;
(2) Effects of max demand duration on demand allocation;
(3) Effect of traffic on demand allocation.

The two algorithms, FCFS and HSL were analyzed for significantly different behavior. As shown in Figure 7a, the HSL behaves slightly better than FCFS. The unallocated demands in HSL are slightly less than that of FCFS allocation. The reason behind this performance is selective allocation of HSL. In HSL, the demand with high severity is allocated at first thus a certain number of demands get served early. Secondly, the effects of duration of demand i.e., short or long jobs on allocation algorithm are observed. It is observed (Figure 7b) that HSL is more sensitive to demand duration as compared to FCFS. As the simulation depicts non-preemptive allocation, the low priority waiting jobs with longer duration stay in HSL wait queues and eventually receive the required resource, and this improves the percentage of allocated jobs. Here each job size is simulated 150 times for numbers of demand sites 5, 10, 15, 20, 25, 30, and 35 respectively. Their requested quantity at each site lies between $1 \leq \text{random} \leq 10$.

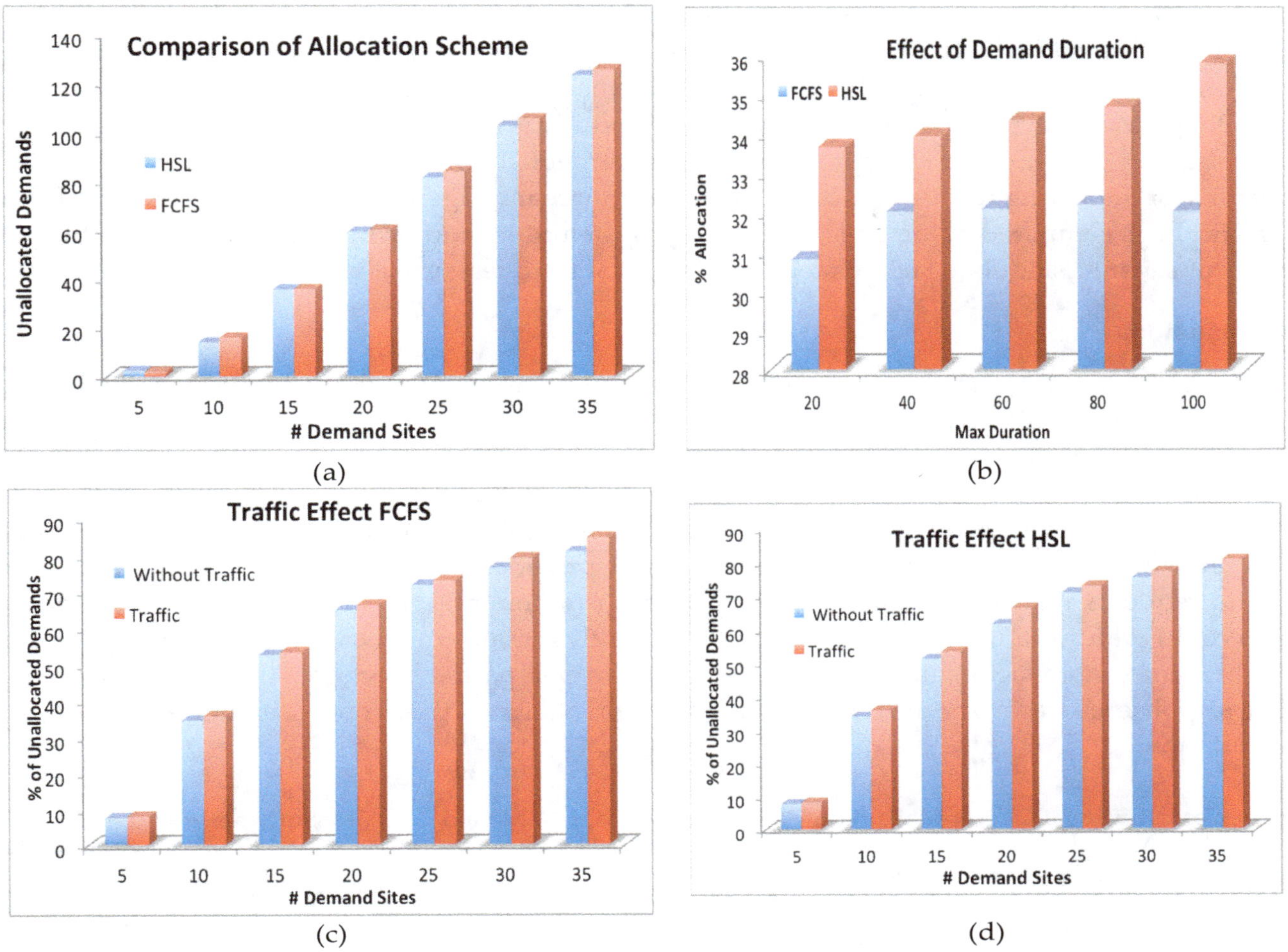

Figure 7. Effect of number of demand sites on total unallocated demands. (**a**). Percentage of allocated demands w.r.t duration of commitment demands (**b**). Effect of traffic on % of unallocated demands (first come first serve—FCFS) (**c**). Effect of traffic on % of unallocated demands (high severity level—HSL) (**d**).

The effects of traffic delay on resource allocation are presented in Figure 7c,d. A random delay of maximum 50% of the travelling time was introduced for both FCFS and HSL. The increase in unallocated demands is observed due to traffic delay as these delays increase the duration of the commitment of the resources.

The average wait time will be less if we increase the number of resource quantities that fulfils the demand quantities within its time frame. Average wait time also depends on the RA cycle; if it releases resources as soon as possible before the request of other demands, it will, in turn, also affect overall average wait time.

Increase in the number of demands affects the overall resource allocation. If number of demands is less than or equal to number of resources, then its average time is almost near to zero. For all experimental data shown in Table 3, we increased the demand node while resources remained constant with same quantity between $1 \leq$ random ≤ 10 for both algorithms. From all experiments, we analyzed that when we have more demands as compared to the available resources then more demands stay unallocated.

4.4. Discussion of Results

Our work focuses on unallocated demands, as these are a number of deficient resources required in cases of emergencies and disaster. It is seen that even with equal number of demands and resource sites, there is a chance that up to 10% of demands will not be entertained. Pakistan has population to health facility ratio of 1:11,413. Our model shows that in the case of disaster, the resulting resource to demands ratio is of 1:7. All allocation schemes result in up to 90% deficiency. It is seen that in Pakistan only 4% of patients [20] arriving to the hospital use the emergency services like ambulances. It is due to chronic lack of resources that has resulted in mistrust in the system, thus, most of the emergencies are handled by self-help basis with no medical expert on site. Our study is the first step towards establishing effects of several factors that result in poor disaster management. In our work, we showed the policies of managing or allocating resources does affect the resource management. However, that difference is insignificant as compared to the deficiencies in the system. In future work, we will highlight the areas that are at severe risk due to lack of facilities and distance to resources centers.

5. Conclusions

Our study is first of its kind to address disaster mitigation in general and resource allocation, in particular, using ABM for analysis of disaster management strategies in urban locality of Pakistan. We used two different algorithms; HSL and FCFS, both have slightly different performance based on allocation. We tested our proposed system on multiple union councils of Rawalpindi city. This scenario shows a significant result for both allocation algorithms. While comparing the results of the FCFS algorithm with the HSL algorithm, we can see that HSL provides better RA results. The degree to which demand requests and resources react to wait time is fundamental. If number of demands is higher, then overall wait time increases, and a lot of demands remain unallocated during their execution time. In future, more factors like damage due to late arrival, partial allocation, traffic, damage caused, public response, etc., are to be introduced in the implementation of this model to enhance the realization of simulation.

Author Contributions: Conceptualization, A.M.; methodology, A.M. and Z.u.A.U.; Software A.M. and Z.u.A.U.; validation, A.M. and F.A.; formal analysis, A.M. and Z.u.A.U.; investigation, A.M. and Z.u.A.U.; data curating, A.M., F.A. and A.R.; writing original draft, A.M. and Z.u.A.U.; writing review and editing, A.M., F.A. and A.R.; visualization, F.A. and A.R.; supervision, A.M.; project administration, A.M. and F.A.; funding acquisition, A.M., Z.u.A.U., F.A. and A.R. All authors have read and agreed to the published version of the manuscript.

Acknowledgments: We thanks to anonymous referees for improving our paper.

References

1. Haider, M.; Badami, M.G. Urbanization and local governance challenges in Pakistan. *Environ. Urban. ASIA* **2010**, *1*, 81–96. [CrossRef]
2. Made for Minds. Available online: https://www.dw.com/en/pakistans-urbanization-a-challenge-of-great-proportions/a-18163731 (accessed on 8 September 2019).
3. Rafiq, L.; Blaschke, T. Disaster risk and vulnerability in Pakistan at a district level. *Geomat. Nat. Hazards Risk* **2012**, *3*, 324–341. [CrossRef]
4. Waheeda, M.A.A. Approach to fire-related disaster management in high density urban-area. *Procedia Eng.* **2014**, *77*, 61–69. [CrossRef]
5. Martínez-Graña, A.M.; Rodríguez, V.V. Remote Sensing and GIS Applied to the Landscape for the Environmental Restoration of Urbanizations by Means of 3D Virtual Reconstruction and Visualization (Salamanca, Spain). *ISPRS Int. J. Geo-Inf.* **2016**, *5*, 2. [CrossRef]
6. Simonovic, S.P. Systems Approach to Management of Disasters—A Missed Opportunity? *IDRiM J.* **2015**, *5*, 70–81. [CrossRef]
7. Arain, F. Knowledge-based approach for sustainable disaster management: Empowering emergency response management team. *Procedia Eng.* **2015**, *118*, 232–239. [CrossRef]
8. Huang, L.; Gong, J.; Li, W.; Xu, T.; Shen, S.; Liang, J.; Feng, Q.; Zhang, D.; Sun, J. Social force model-based group behavior simulation in virtual geographic environments. *ISPRS Int. J. Geo-Inf.* **2018**, *7*, 79. [CrossRef]
9. Inan, D.I.; Beydoun, G. Disaster knowledge management analysis framework utilizing agent-based models: Design science research approach. *Procedia Comput. Sci.* **2017**, *124*, 116–124. [CrossRef]

10. Huang, K.; Jiang, Y.; Yuan, Y.; Zhao, L. Modeling multiple humanitarian objectives in emergency response to large-scale disasters. *Transp. Res. E Logist. Transp. Rev.* **2015**, *75*, 1–17. [CrossRef]
11. Wise, S.; Crooks, A.; Batty, M. Transportation in agent-based urban modeling. In Proceedings of the International Workshop on Agent Based Modelling of Urban Systems, Singapore, 10 May 2016; Springer: Cham, Switzerland, 2016; pp. 129–148. [CrossRef]
12. Matveev, A.; Maksimov, A.; Vodnev, S. Methods improving the availability of emergency-rescue services for emergency response to transport accidents. *Transp. Res. Procedia* **2018**, *36*, 507–513. [CrossRef]
13. Bélanger, V.; Ruiz, A.; Soriano, P. Recent optimization models and trends in location, relocation, and dispatching of emergency medical vehicles. *Eur. J. Oper. Res.* **2019**, *272*, 1–23. [CrossRef]
14. Kong, Y.; Pan, S. Intelligent Prediction Method for Transport Resource Allocation. *Sens. Mater.* **2019**, *31*, 1917–1925. [CrossRef]
15. Crooks, A.T.; Wise, S. GIS and agent-based models for humanitarian assistance. *Comput. Environ. Urban Syst.* **2013**, *41*, 100–111. [CrossRef]
16. Cavdur, F.; Kose-Kucuk, M.; Sebatli, A. Allocation of temporary disaster response facilities under demand uncertainty: An earthquake case study. *Int. J. Disaster Risk Reduct.* **2016**, *19*, 159–166. [CrossRef]
17. Meng, L.; Kang, Q.; Han, C.; Zhou, M. Determining the optimal location of terror response facilities under the risk of disruption. *IEEE Trans. Intell. Transp. Syst.* **2017**, *19*, 476–486. [CrossRef]
18. Bansal, M.; Kianfar, K. Planar maximum coverage location problem with partial coverage and rectangular demand and service zones. *INFORMS J. Comput.* **2017**, *29*, 152–169. [CrossRef]
19. Karatas, M.; Yakıcı, E. An iterative solution approach to a multi-objective facility location problem. *Appl. Soft Comput.* **2018**, *62*, 272–287. [CrossRef]
20. Zia, N.; Shahzad, H.; Baqir, S.M.; Shaukat, S.; Ahmad, H.; Robinson, C.; Hyder, A.A.; Razzak, J. Ambulance use in Pakistan: An analysis of surveillance data from emergency departments in Pakistan. *BMC Emerg. Med.* **2015**, *15*, S9. [CrossRef] [PubMed]

13

GIS-Based Mapping of Seismic Parameters for the Pyrenees

José Lázaro Amaro-Mellado [1,2,*] and Dieu Tien Bui [3]

1 Department of Graphic Engineering, University of Seville, 41092 Seville, Spain
2 National Geographic Institute of Spain, Andalusia Division, 41013 Seville, Spain
3 Geographic Information System Group, Department of Business and IT, University of Southeast Norway, Gulbringvegen 36, N-3800 Bø i Telemark, Norway; dieu.t.bui@usn.no
* Correspondence: jamaro@us.es

Abstract: In the present paper, three of the main seismic parameters, maximum magnitude -M_{max}, *b-value*, and annual rate -*AR*, have been studied for the Pyrenees range in southwest Europe by a Geographic Information System (GIS). The main aim of this work is to calculate, represent continuously, and analyze some of the most crucial seismic indicators for this belt. To this end, an updated and homogenized Poissonian earthquake catalog has been generated, where the National Geographic Institute of Spain earthquake catalog has been considered as a starting point. Herein, the details about the catalog compilation, the magnitude homogenization, the declustering of the catalog, and the analysis of the completeness, are exposed. When the catalog has been produced, a GIS tool has been used to drive the parameters' calculations and representations properly. Different grids ($0.5 \times 0.5°$ and $1 \times 1°$) have been created to depict a continuous map of these parameters. The *b-value* and *AR* have been obtained that take into account different pairs of magnitude–year of completeness. M_{max} has been discretely obtained (by cells). The analysis of the results shows that the Central Pyrenees (mainly from Arudy to Bagnères de Bigorre) present the most pronounced seismicity in the range.

Keywords: seismic parameters; GIS; seismicity; spatial analysis; *b-value*; earthquake catalog

---

1. Introduction

In seismicity studies, some parameters have a particularly important role. Among others, these are the maximum magnitude (recorded, possible, expected), the *b-value* of the Gutenberg–Richter (GR) frequency-magnitude (FMD) relation, and a parameter related to the seismic activity (mean seismic activity rate or the *a-value* of GR) [1–13]. Some of these studies are based on seismic zonations [1,2], and others are found in a purely geographical grid division [7,11,14,15]. In both cases, the usage of a system capable of adequately integrating different sources of geographic data is advisable. Thus, a Geographic Information System (GIS) can provide the rigor, flexibility, and the power to calculate, show, and analyze the parameters. This fact has been demonstrated previously for natural hazards works [7,16–19].

Upon Gutenberg and Ritcher [20], the GR relation establishes that the number of earthquakes, *N*, with magnitude larger than or equal to a given magnitude *M* (cut-off magnitude), can be expressed as

$$\log_{10} N(M) = a - bM \quad (1)$$

where *a-value* estimates the seismic productivity, whereas the slope (known as *b-value*) measures the size–distribution parameter. The latter expresses the relationship between small and large events. It is usually considered as the essential seismic parameter, which varies from 0.6 to 1.5 for global seismicity [21]. Values between 0.9 and 1.0 can be established as normal *b-value* [22–24]. In many cases, the *b-value* is considered as a stress-meter, where the higher its value is, the lower the stress is held [25–27]. However, other studies [23,28] state that its value must be 1.0, and any variation is due to some issues such as improper calculation, insufficient data, or inhomogeneous detection network.

The main aim of this work is to calculate, represent continuously, and analyze some of the most crucial seismic indicators for the Pyrenees range. Previously, an updated, homogeneous, and extensive catalog has been generated.

The importance of earthquake catalogs in seismic studies is crucial [8,10,11,29–31]. Due to the different content of seismic information for both historical and instrumental epochs, a critical point in the catalog analysis is to generate a reliable, updated, extensive, and homogeneous catalog for the studied region to make the results comparable. In areas where historical seismicity data are available, they must be included in the working catalog to conduct a complete and robust analysis [1–3,10,32–34]. Besides, the size of the (non-small) earthquakes is usually given in moment magnitude (M_w) [35], as it has a direct relation with the released energy through scalar seismic moments and does not get saturated for larger events. To this end, both global and regional parameters can be found in the literature to convert both intensities and different magnitude types into M_w [35–37]. Currently, the evolution and improvement of the seismic networks allows for dealing with a considerable amount of data with very high precision in both location and magnitude.

In this work, after preparing the working catalog, the information has been integrated into a GIS. The calculations and representations have been carried out in its environment as it offers some remarkable advantages, such as combining alphanumeric and geographic information or generating high-quality maps.

The rest of the paper has been organized as follows. In Section 2, the study area is presented. Section 3 contains the data and methodology employed. Results and analysis are shown in Section 4. Finally, Section 5 presents the conclusions of the work.

2. Study Area

2.1. Geological Settings

The Pyrenees is a range of mountains located in the France–Spain border, which spans 450 km long (E–W) and 150 km wide (N–S) [27]. It was formed from the convergence between the Eurasian (to the north) and Iberian (to the south) plates. It happened after an extensive period related to the opening of the Bay of Biscay in the Lower Cretaceous [38–40]. Nowadays, the Iberian Peninsula is located in the southwest of the Eurasian plate, which presents a collision movement with the African plate with an estimated rate of between 2 and 5 mm/year in the NW–SE to WNW–ESE direction [2]. Currently, this rate is considerably lower for the Pyrenees [41], despite being the second most seismically active area in the Iberian Peninsula after the Betics.

A general view of this belt's geological structure can be found in [39,42–45]. Figure 1 presents a general sketch of the geological context.

The main units can be summarized as follows [39]:

- The North Pyrenean Zone (NPZ), which was mainly formed by Mesozoic (Cretaceous) deposits, rides northward over the Aquitaine basin along with the North Pyrenean Frontal Trust (NPFT). The NPZ contains Paleozoic outcrops.
- The Paleozoic Axial Zone, which is located at the central part and presents the highest peaks of the Pyrenees, is composed of Hercynian structures that were reactivated during the Alpine orogeny.
- The South Pyrenean Zone (SPZ), which slid southward down from the Axial zone when it rose, presents Mesozoic (Cretaceous) and Cenozoic (Oligocene) sediments in its composition.

Besides, the so-called North Pyrenean Fault (NPF) is a major east–west tectonic suture, which separates the Axial Zone of the NPZ. This fault is frequently considered as the limit between the

Iberian and Eurasian plates. It coincides with an important vertical shift of the Moho discontinuity, where the Iberian side presents a thicker crust.

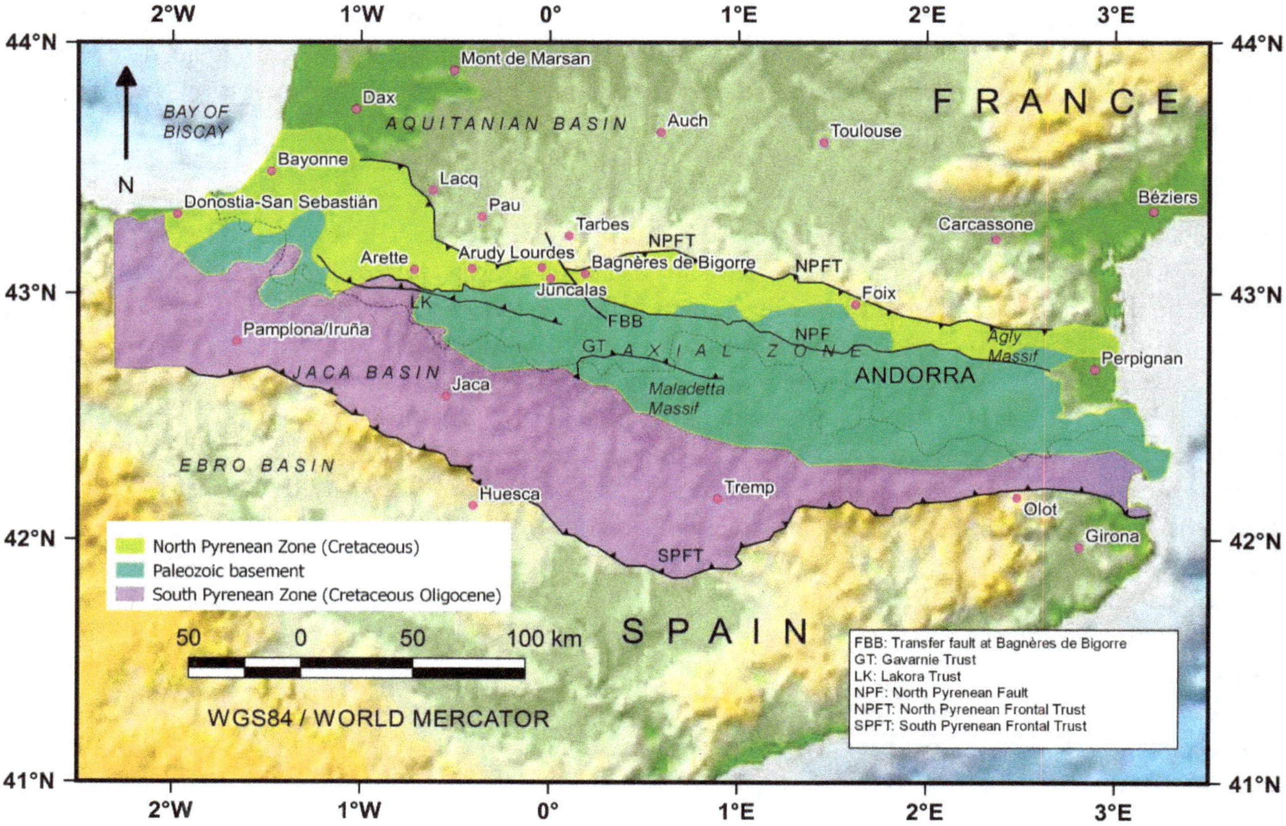

Figure 1. Sketch map with the main structural units of the Pyrenees. Redrawn after [43–45].

2.2. Seismicity

The spatial distribution of earthquakes in this belt is complex with different mechanisms, such as denudation, and gravitational collapse, or lithospheric flexure [43]. Therefore, it is necessary to study it carefully when performing a Probabilistic Seismic Hazard Analysis (PSHA). Although the Pyrenees seismicity can be considered as low to moderate; however, some shocks still produced significant damages in the historical period, especially with events either with magnitudes exceeding 6 (M6+), or MSK intensities equal to or larger than VIII [41,46]. For example, on the Spain side, the 1373 earthquake (I_o = VIII-IX) occurred in the Ribagorça County (by Maladetta Massif) in the south part of the Central Pyrenees [47]; the Catalan seismic crisis during 1427 and 1428 reached I_o = IX, destroyed Olot and Queralbs and caused more than 1000 fatalities [47–49]. On the France side, the 1660 Bigorre Earthquake (I_o = IX), near Lourdes in the Central Pyrenees [46,50,51], and more recently, the 1750 Juncalas (I_o = VIII) in a nearby area [41]. Besides, in the instrumental period, some strong earthquakes have occurred. Among others, 1967 Arette (M5.3) [46,51], in the Atlantic Pyrenees, with I_o = VIII, or others in the eastern Pyrenees, in the Agly Massif, such as St-Paul de Fenouillet (M5.0) in 1996 [39,52].

The seismicity takes part mainly in the western part of the NPZ, while in the east, it is lower and more dispersed. Regarding the focal depth, it has to be pointed out that it is shallow, mainly less than 20 km. Besides, the seismogenic crust could correspond to the first 15–20 km of the crust [49].

There is a lack of in-depth and extensive knowledge of every fault that is capable of producing an earthquake. This fact, and the seismicity patterns, means that the PSHA in this area has been derived mainly from seismogenic zones rather than from specific faults [1,2,53].

2.3. *Related Works on the Pyrenees Seismicity*

Various studies have explored the variation in the *b-value* in the Pyrenees. Gallart et al. [54] analyzed it for the Arette–Arudy region in the western Pyrenees. Kijke-Kassala et al. [38] calculated the *b-value* from a regionalization (nine zones). In other works, the Arudy region was analyzed by Sylvander et al. [55], whereas Secanell et al. [56] conducted a PSHA, where they calculated the *b-value*. Afterward, Secanell et al. [57] divided the Pyrenees into ten seismogenic zones from the ISARD project earthquake catalog and tectonic models, and then *b-values* were obtained. In addition, in the frame of the PSHA models for Spain and their analysis, some studies have been addressed [1,2,6]. In recent research, Amaro-Mellado et al. [7] produced a continuous *b-value* map for the whole Iberian Peninsula and adjacent areas; however, an in-depth specific study for the Pyrenees has not been conducted. Rigo et al. [27] analyzed the *b-value* variation with the depth relating to the differential stress, considering only *b-values* from a seismic zonation proposed by Rigo et al. [58]. Nevertheless, to the best of our knowledge, no continuous map has been deployed for this specific region, taking into account its seismic circumstances.

3. Data and Methods

In this section, the input data and the methodology conducted to produce the maps of the seismic parameters are described.

3.1. *Catalog generation*

As has been pointed out above, the importance of working with a high quality and reliable catalog is a key issue. The steps required to this end are shown in Figure 2.

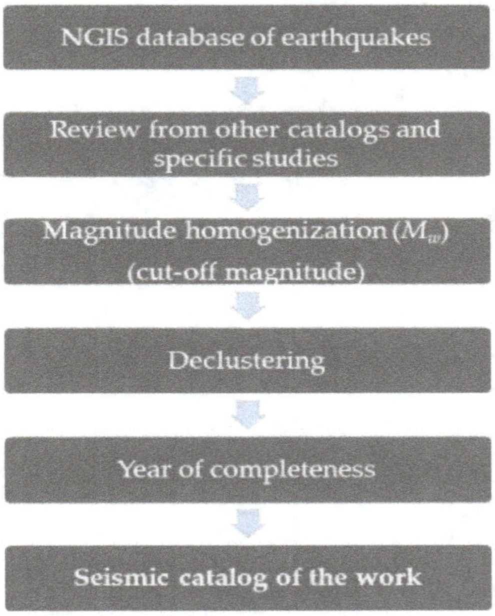

Figure 2. Seismic catalog generation workflow.

3.1.1. The National Geographic Institute of Spain Earthquake Catalog

The earthquake catalog employed in this work takes as a starting point the earthquake catalog of the National Geographic Institute of Spain (hereinafter, NGIS), which can be downloaded from [59]. It has more than 135,000 events between 1373 and the end of 2019, corresponding to the Iberian Peninsula and the Canary Islands and their adjacent areas. Through the years, the sizes of the events have been recorded in different ways (macroseismic intensity or several magnitude types, as will be shown later). The fields recorded in the database include elements such as an ID, date, hour,

3D location, intensity, magnitude, or magnitude type. A detailed study of the NGIS catalog can be found in González [60], where the seismic network evolution is shown.

Regarding the precision in earthquakes' location, for the whole catalog, shocks from 1997 have a better location than 3 km (4 km in 1985; 7 km in 1983–1984, and it is worse for previous events) [60], but, in regions like the Pyrenees, they are lower thanks to the data contributed to the catalog by other networks [43,60]. Currently, this precision is significatly better, as can be checked in [59].

For this work, the geographical extent considered has been limited by 41° N and 44° N parallels and 2.5° W and 3.5° E meridians. The earthquake distribution of the 24,282 resulting events is shown in Figure 3, where 248 (mostly aftershocks from pre-instrumental and historical periods) have no size assigned.

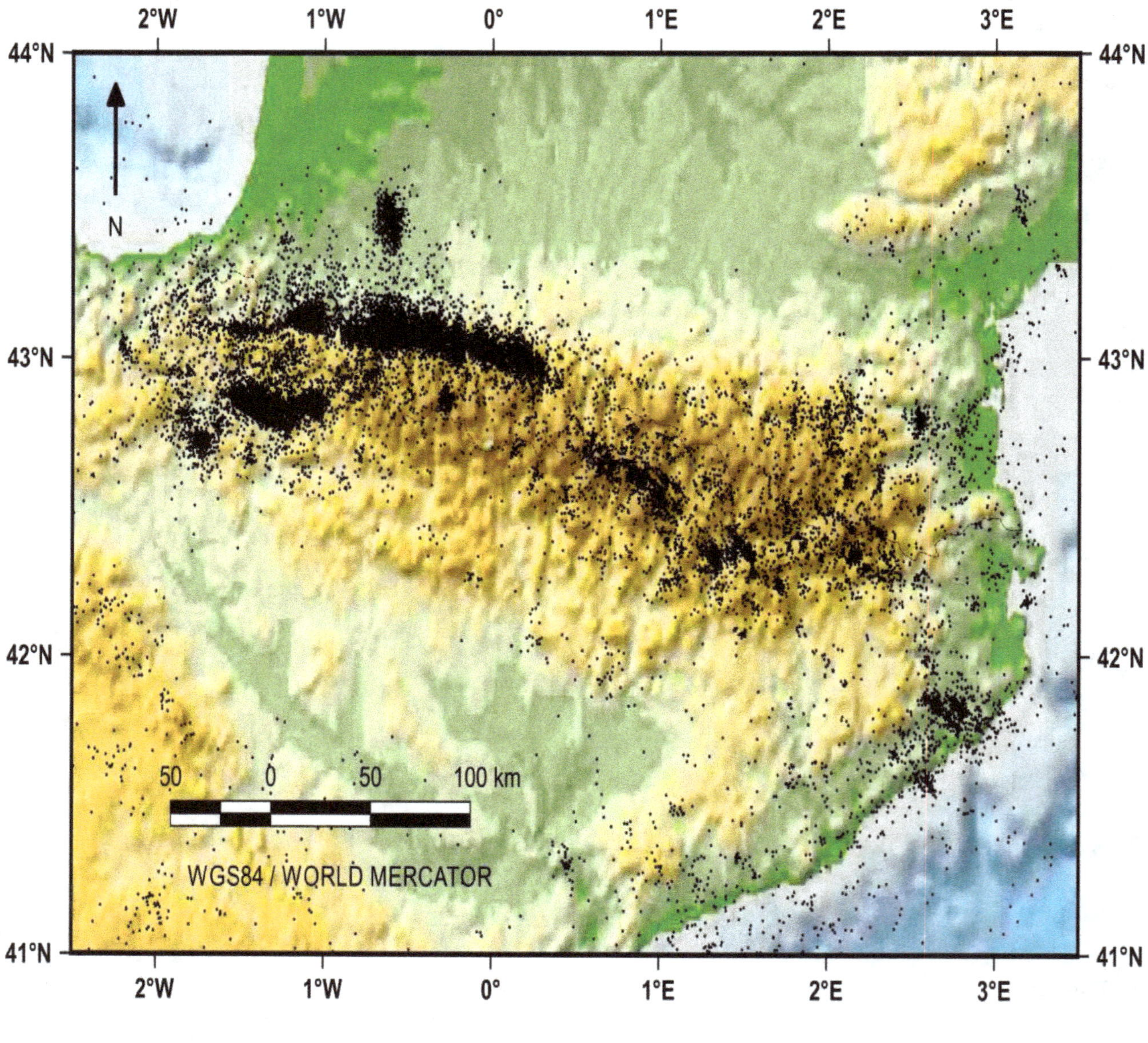

Figure 3. The National Geographic Institute of Spain (NGIS) catalog for the working area (from 1373 to 2019).

Additionally, for the instrumental period, a map showing the originally recorded magnitudes in the NGIS catalog is presented in Figure 4.

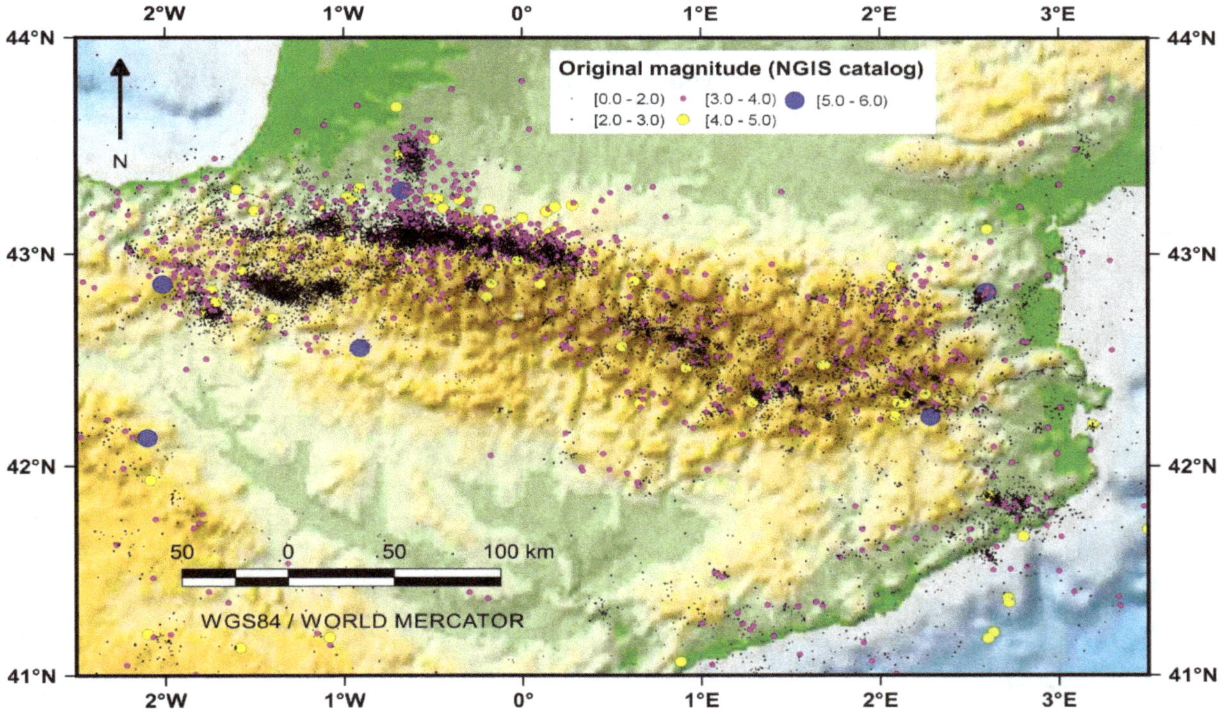

Figure 4. The NGIS catalog (original magnitudes for the instrumental period).

3.1.2. Review from Other Catalogs and Specific Studies

The seismic catalog must be improved with data from other sources, such as from other catalogs or from specific publications (mainly, journal papers) where individual events have been re-evaluated, as has been conducted in similar works such as in [2,6]. Therefore, a thorough revision has been developed, and the size of 31 events has been reviewed, particularly to assign a reliable M_w to the largest events. Special mention is deserved for some earthquakes with a marine epicenter, and with only macroseismic information. Their intensities recorded in the catalog could not be the epicentral ones, due to attenuation, and should be revised. The works that have supported the review are found in [6,39,46,47,49,50,52,55,61–67].

3.1.3. Magnitude Homogenization and Cut-off Magnitude

Over time, the size of the earthquakes has been calculated in line with the evolution of the seismic network. Besides, it has been stored in different ways in the NGIS catalog, for historical, pre-instrumental, and instrumental periods. It can be found in [2], and it is summarized here (with its uncertainty):

1. Epicentral or maximum intensity (until 1923) ($0.5 \leq \sigma \leq 1.5$).
2. Pre-instrumental 1924–February 1962: Duration magnitude [68], m_D(MMS) ($\sigma = 0.4$).
3. March 1962–February 1998: Surface–wave magnitude [68], m_{bLg}(MMS) ($\sigma = 0.3$) (<1985); ($\sigma = 0.2$) (>1985).
4. March 1998–2002: Body–wave magnitude [69], m_b(VC) ($\sigma = 0.2$) and m_{bLg}(MMS) ($\sigma = 0.2$).
5. March 2002–onward: Surface–wave magnitude [70], m_{bLg}(L) ($\sigma = 0.2$) or m_b(VC) ($\sigma = 0.2$), or, for the most significant events, M_w ($\sigma = 0.1$).

To work with a homogeneous catalog, all the events must possess the same kind of magnitude, and the preferred scale has been M_w for the advantages mentioned earlier. Different authors have proposed global and regional parameters to convert the original size into M_w [35–37,71–74]. Given that both independent and dependent variables have errors, a reduced major axis (RMA) is currently preferred to a least-square ordinary regression [36,71]. In this work, the conversion from the original size of every event to M_w has been conducted from the parameters suggested by Cabañas et al. [36], as they are specific for the Iberian Peninsula and adjacent areas.

After those conversions, a proper *b-value* requires estimating the magnitude of completeness (M_c), which is an essential and mandatory step to conduct a seismic analysis. M_c can be defined as the lowest magnitude at which 100% of the earthquakes in a space–time volume are detected [10,75]. There are multiple methods to estimate this value, and in-depth analysis can be found in Mignan and Woessner [75].

The method employed in this paper is based on the earthquake catalog, and consists of plotting the non-cumulative FMD in addition to the standard (cumulative) FMD. Assuming self-similarity, M_c is simply the magnitude increment at which the FMD departs from the linear trend in the log–lin plot [75]. The estimate of M_c can be assessed from Figure 5 at 1.8 ($M_c = 1.8$).

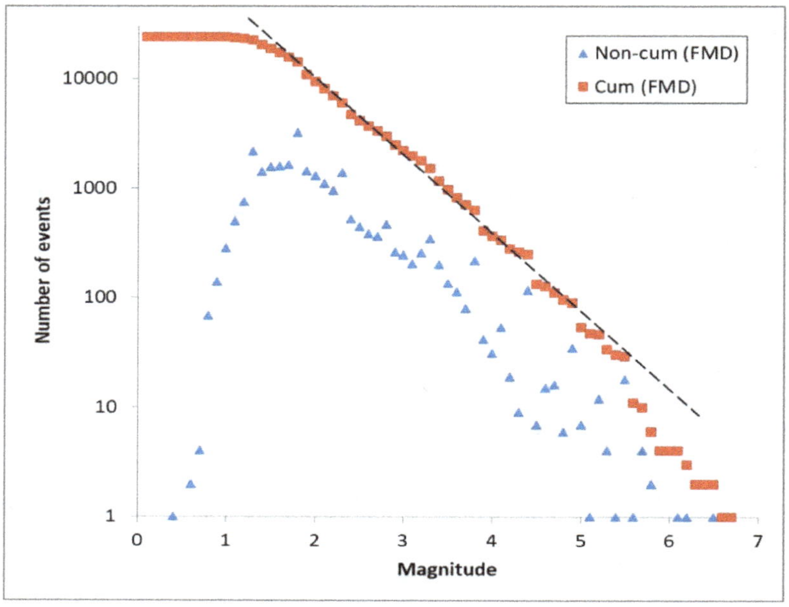

Figure 5. Frequency-magnitude (FMD) of the catalog for M_c estimation.

After determining the M_c, regarding the *b-value* calculation, the cut-off magnitude ($M_{co} \geq M_c$) for this work has been set as 2.0 (M2), to have a margin and, besides, consider a standard value.

3.1.4. Declustering

Although the studies of the aftershocks are necessary to some applications [76,77], the computation of the PSHA is mainly based on a Poisson distribution. Herein, these events must be independent; thus, foreshocks, aftershocks, and seismic swarms must be deleted. Consequently, the resulting catalog contains mainshocks only. This is a tricky process, known as declustering, where two principal methodologies, proposed by [78,79], used to be applied, and a thorough study on it can be found in [80]. In this work, the method suggested by Gardner and Knopoff [78] was selected due to its clarity, simplicity, and stability [2,3,32]. Besides, this method has proven its efficiency in most of the recent research [8,10,31,81,82], and particularly for the studied area [2,7]. As the more energetic an earthquake is, the more temporal and geographical extension is involved, this method defines temporal and spatial windows depending on the main earthquake. There are different values to establish these windows. In this research, the proposal by Uhrhammer [83] has been chosen as it is a conservative solution and widely used [29,81,82,84,85].

The results of declustering, conducted by the ZMAP application [86], show that there are 1434 clusters, whose seismic moment released is about 4.4% of the total seismic moment of the catalog. Finally, 19,625 events remain in the declustered catalog.

3.1.5. Year of Completeness

The NGIS catalog spans more than 600 years in its records, and this study has taken advantage of this sizeable temporal extent. As has been pointed out, the magnitude of completeness must be related to a reference date, due to the seismic network heterogeneities and evolution, and it is crucial to estimate a *b-value* properly. The larger an event is, the more likely it is to be recorded in distant times.

The best method to calculate the different M_c–year of completeness pairs is a much-discussed subject. In this research, the cumulative method [87–89] has been used to estimate the completeness periods for different levels of magnitude, as in other recent works [10,81]. Thus, plotting the cumulative number of earthquakes and time, the year of completeness can be estimated. The catalog is considered complete for a threshold magnitude concerning the time where there is approximately a straight line of the plotted date, so the completeness period varies with time [10]. In this study, the year of completeness has been determined for M2, M3, M4, and M5, as can be seen in Figure 6.

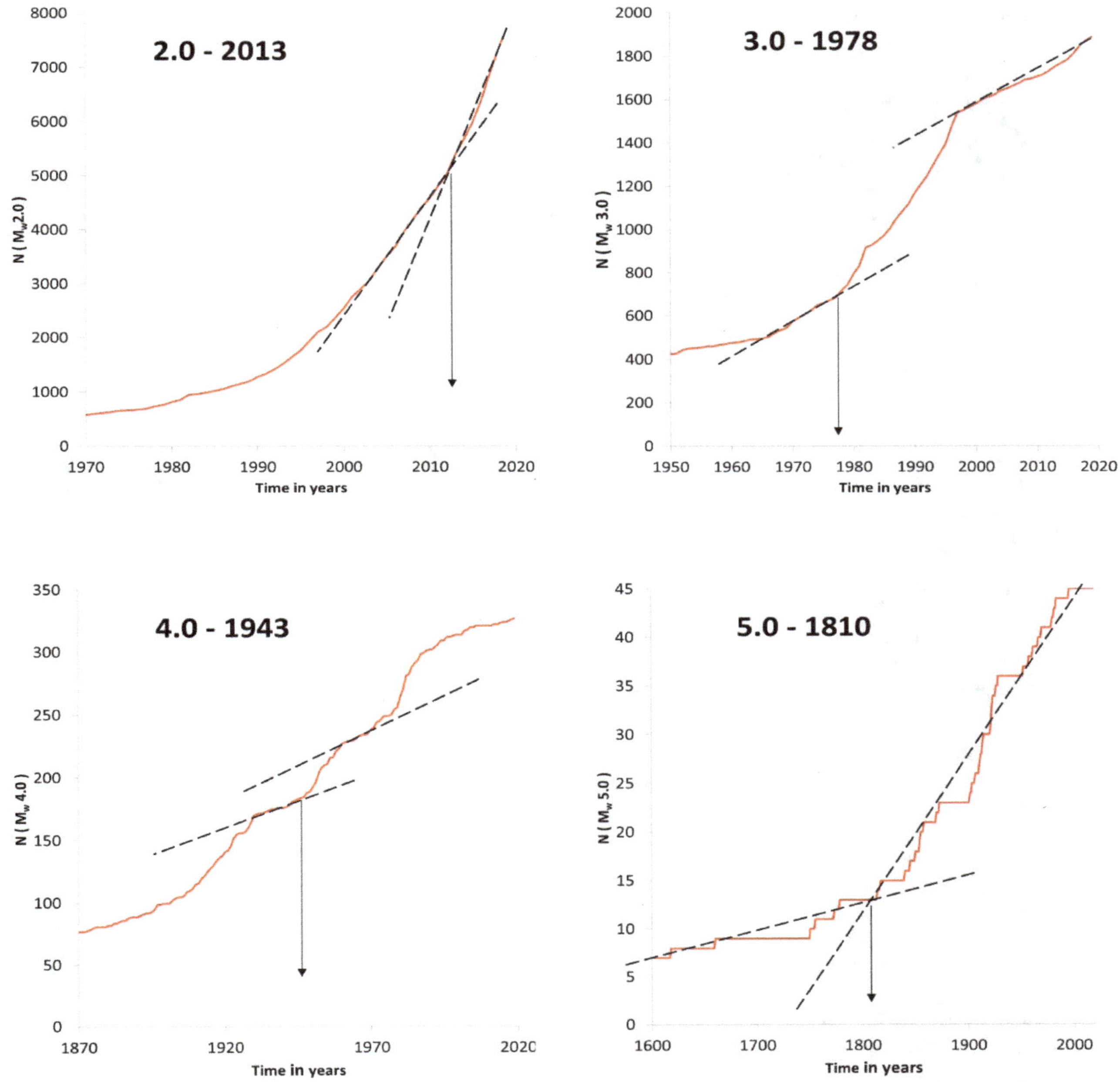

Figure 6. Year of completeness determination from the cumulative method for different values of magnitude.

The result is shown in Table 1. They are in line with those obtained for other authors for an area in which the Pyrenees are included [1,2].

Table 1. Year of completeness vs. magnitude.

Magnitude	Year
$M_w \geq 2.0$	2013
$M_w \geq 3.0$	1978
$M_w \geq 4.0$	1943
$M_w \geq 5.0$	1810

3.1.6. Seismic Catalog of the Work

After all the steps described previously, the events deeper than 65 km have been removed, as they are not relevant for seismic hazard in the area [2].

The final catalog consists of 7706 events with M_w from 1373 to 2019 is shown in Figure 7.

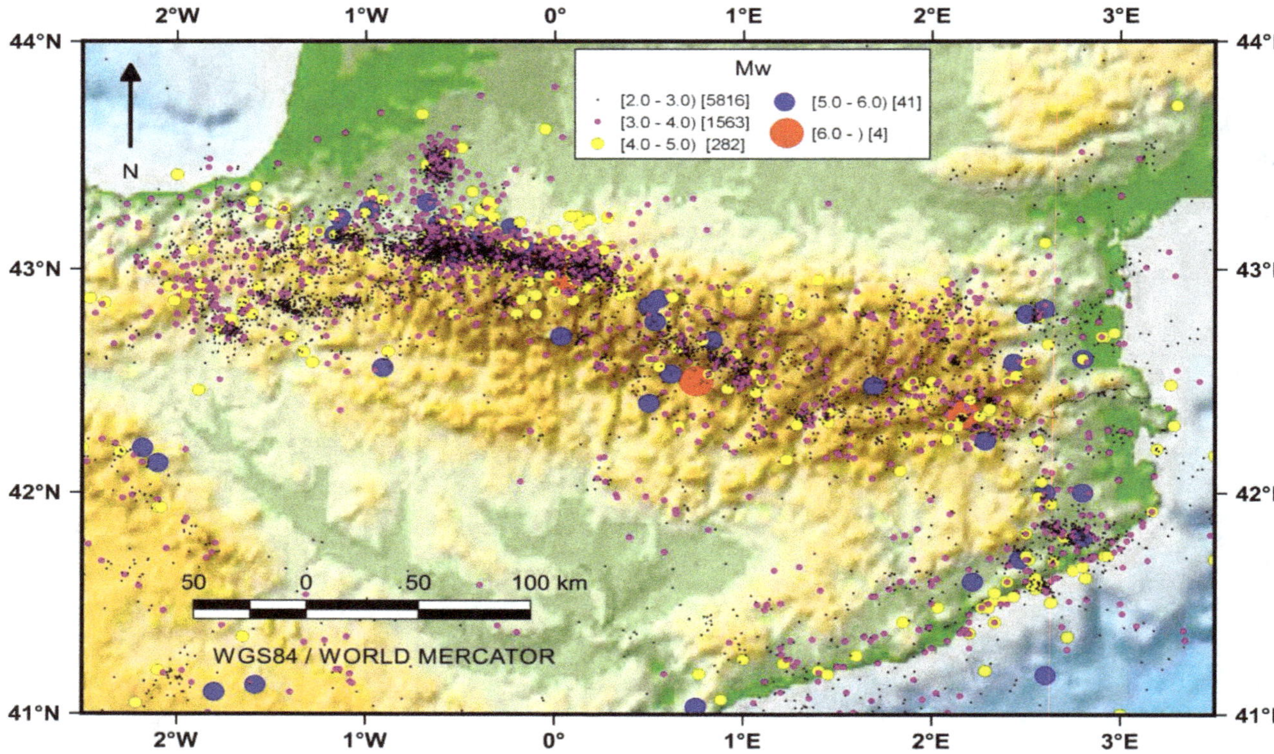

Figure 7. Seismic catalog of the work ($M_w \geq 2.0$). Temporal extent: 1373–2019.

3.2. Seismic Parameters

In this subsection, several relevant seismic parameters will be introduced as they are indicators of the seismicity of an area. These parameters are the maximum magnitude, the *b-value*, and the normalized annual rate for events exceeding M3.

3.2.1. The Maximum Magnitude

The maximum magnitude (M_{max}) is closely related to the seismic hazard for an area. Therefore, to appreciate its distribution, a $0.5 \times 0.5°$ grid has been deployed using a GIS tool and covering the whole working area, in a GIS environment. There are different concepts related to the maximum magnitude, such as the largest recorded event, *a-value/b-value*, the largest physically possible. Besides, there are different approaches (parametric, non-parametric) to calculate its value. Thorough works can be found in [90,91]. In this work, as the temporal extent of the catalog is long, the maximum magnitude of the working catalog for every grid's cell has been chosen, being aware of the limitations.

3.2.2. The *b-value*

The *b-value* is the parameter most studied in seismology and corresponds to the slope of the FMD in a log–log plot. The majority of the authors stated its relationship with the physics of the studied area. It has been deeply studied and is usually analyzed in PSHA works, in prediction, in locating asperities, periodical tidal loading, and energetic characterization [4,8,11,16,25,92–95].

Although firstly, the least square solution was usually employed to calculate its value, currently, there is a consensus in considering the Maximum-Likelihood-Estimate (MLE) as the best approach to obtain it. This is due to the fact that it does not present interdependency between variables [2]. Over time, a considerable amount of different methods have been suggested for its calculations, such as those found in [13,96–100]. One of the most employed formulae for MLE was proposed by [98,99] for binned data.

$$b = \frac{\log_{10} e}{\overline{M} - M_c - \frac{\Delta M}{2}} \tag{2}$$

where M_c is the cut-off magnitude, $\overline{M}$ is the average magnitude of the earthquakes whose magnitude is larger than or equal to M_c, and ΔM is the binning interval of the magnitude.

In this research, the bin interval is 0.1. The solution proposed by Kijko and Smit [13], through the exposed MLE expression, has been applied. It permits taking into account a longer temporal extent and different magnitude–year of completeness pairs. Moreover, it is said to be simple, manageable, and it is not based on iterations [13].

The method suggested by Kijko and Smit [13] is based on dividing the catalog into more coherent s sub-catalogs, each of a different level of completeness, and with its corresponding year of completeness. It is particularly indicated for incomplete or inhomogeneous catalogs. For every sub-catalog, the MLE proposed by [98,99] is used. Later, *b-value* is estimated as a weighted solution as:

$$\hat{b} = \frac{n}{\frac{n_1}{b_1} + \frac{n_2}{b_2} + \ldots + \frac{n_s}{b_s}} \tag{3}$$

where b_i is the *b-value* of each of the s sub-catalogs, n_i is the sample size of the sub-catalogs, and n is the total number of events considered ($n = n_1 + n_2 + \ldots + n_s$).

Each sub-catalog has a known but different level of completeness, M^1_{min}, M^2_{min}, …, M^s_{min}, and it spans $t_1, t_2, \ldots, t_s$ years.

Finally, from now on, the obtained *b-value* after the correction proposed by [101], to minimize the overestimation produced for small samples, will be noted as *b-value*, or only *b*.

$$\hat{b}_{Og} = \frac{(n-1)\hat{b}}{n} \tag{4}$$

Besides, despite the method proposed by Kijko and Smit [13] gives the expression to calculate the standard deviation, that suggested by Shi and Bolt [102] has been preferred as it considers the real dispersion of the sample:

$$\delta b = 2.3b^2 \sqrt{\frac{\Sigma_i^n \left(M_i - \overline{M}\right)^2}{n(n-1)}} \tag{5}$$

where n is the total samples.

As stated above, the method requires pairs of values to estimate the *b-value*. These have been previously calculated and are shown in Table 2.

Table 2. Sub-catalog division (M_C-Years).

Magnitude	Year
$M_w \geq 2.0$	2013–2019
$M_w \geq 3.0$	1978–2012
$M_w \geq 4.0$	1943–1977
$M_w \geq 5.0$	1810–1942

Once the formulae have been defined, the minimum number of events to generate a representative *b-value* must be established. It is a very controversial issue, and there is no general academic agreement: extraordinarily, Dominique and Andre [103] considered only six events: Bender [98] or Skordas and Kulhánek [104] chose 25 events; Mousavi [11] and Amorèse et al. [28] proposed 50; González [60] suggested 60, and Roberts et al. [105] established 200. A thorough study regarding the relationship between error and number of events for different *b-values* can be found in Nava et al. [24].

When a continuous representation is prepared, another crucial parameter is how the geographical space is divided. This split can lead to completeness issues, as the number of events for every cell may be small, particularly when a grid division is employed [106], as in [7,11,107]. The minimum number of events is a trade-off between accuracy and coverage, whereas cell size is a trade-off between coverage and resolution [11]. In this work, using a GIS tool, two different grid sizes have been established as in [7,11]. Considering these trade-offs, a $0.5 \times 0.5°$ grid was selected with 100 events as a minimum value for the most active area; and a $1 \times 1°$ grid has been considered for the whole area, with a minimum of 50 events. Besides, to reduce the border effect, four overlapped grids have been defined (the original; one shifted half of cell size to the south; another moved half to the east; and finally, displaced in south and east), as was done previously in [7,15].

For every cell, the average geographical coordinates of the epicenters have been estimated. Thus, seismic parameters have been assigned to this location. Later, the method proposed by Kijko and Smit [13] has been adopted to compute the *b-value*. This has been for every cell of every grid. Finally, to conduct the spatial analysis, an interpolation by the Inverse Distance Weighting (IDW) algorithm has been applied. It must be noted that where the minimum number of events has not been reached, its associated *b-value* has not been considered in the interpolation, so caution should be taken when analyzing these areas.

Different color maps have been produced with the conditions above exposed to represent the *b-value* distribution. The location of the points used to generate the maps is shown.

3.2.3. The Annual Rate (Normalized)

Finally, the annual rate (the number of events exceeding a threshold in a unit of time–year) is calculated and, besides, it is easily derived from the *b-value*. [13] proposed its calculation from:

$$\lambda(M_{min}) = \frac{n}{\sum_{i=1}^{s} t_i 10^{-b(M_{min}^i - M_{min})}} \tag{6}$$

where M_{min} is the minimum magnitude considered (2.0), and the rest of the parameters have been defined above.

The annual rate is a usual parameter in PSHA studies and, in this case, given that M2.0 earthquakes are not particularly relevant for seismic hazards, it has been obtained straightforwardly for M3.0 from:

$$\lambda(3) = \lambda(2)10^{-b(3-2)} \tag{7}$$

This value is more comparable if it is related to a unit area (km^2), resulting in the normalized, *AR*, (for M3), so:

$$AR = \frac{\lambda(3)}{Area\left(in\ \mathrm{km}^2\right)} \tag{8}$$

As this value is very low, it has been multiplied by 10^4 for representation.

4. Results and Discussion

In this section, the spatial distribution of several of the chosen seismic parameters is shown by maps, after the processes conducted into a GIS environment. The maximum magnitude, the *b-value*, and the normalized mean seismic activity rate (annual rate) are presented here.

4.1. The Maximum Magnitude

The maximum magnitude in the working catalog for every cell of the 0.5 × 0.5° grid is presented in Figure 8. It must be highlighted that it corresponds to a temporal extent of more than 600 years, and the moment magnitude has been chosen.

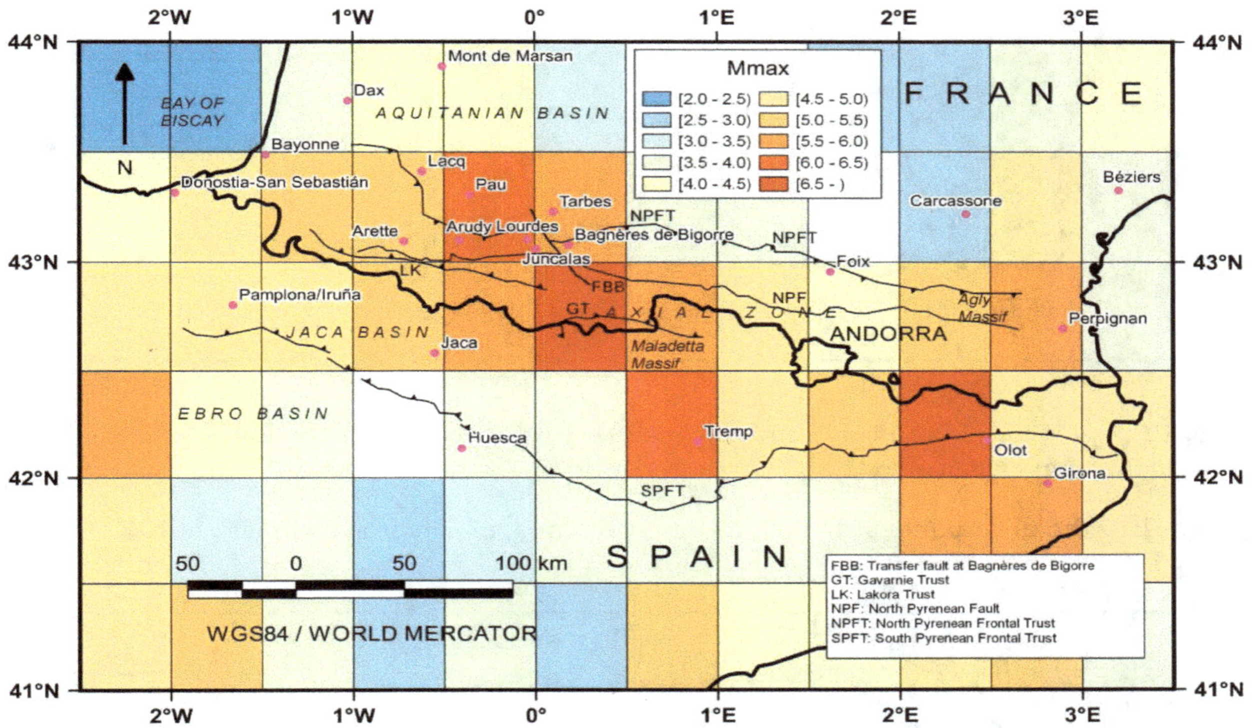

Figure 8. Maximum magnitude in the working catalog (1373–2019).

This M_{max} map is a representation of the maximum magnitude from a deterministic point of view. It shows that more significant events have occurred in the Central Pyrenees in the Bagnères de Bigorre area, reaching M6.7, and on the Spanish side of the Eastern Pyrenees, reaching M6.5, near Olot. The Arudy region reaches M6+, the Maladetta Massif, and the Agly Massif (in the Eastern Pyrenees) are hot spots with M5.5+. Along with the border limit, M_{max} exceeding 5.0 is frequent, except for its eastern extreme. At the north and the south of the belt, M_{max} is notably lower, showing several areas M_{max} below 4.0, mainly on the French side.

In general, the highest values are located in the Central Pyrenees more than in the extremes.

4.2. The b-value

As has been stated, *b-value* is the most studied parameter because it is related to the physics of the source. Lower values could mean that additional stress is being held and might be released abruptly later; contrariwise, higher values indicate continuous energy released. Figure 9 shows the *b-value* map for the Pyrenees region.

The calculations show that the extreme values are 1.42 (by 1.6° E, 41.4° N) and 0.95 in the Bagnères de Bigorre. A *b-value* of 1.14 is found in the western extreme and 1.18 in the eastern extreme, and, for

Agly Massif, it is 1.14; for the central southern and the north part, there are not enough data. A general value can be set as 1.10, and higher values are located in the Central Pyrenees.

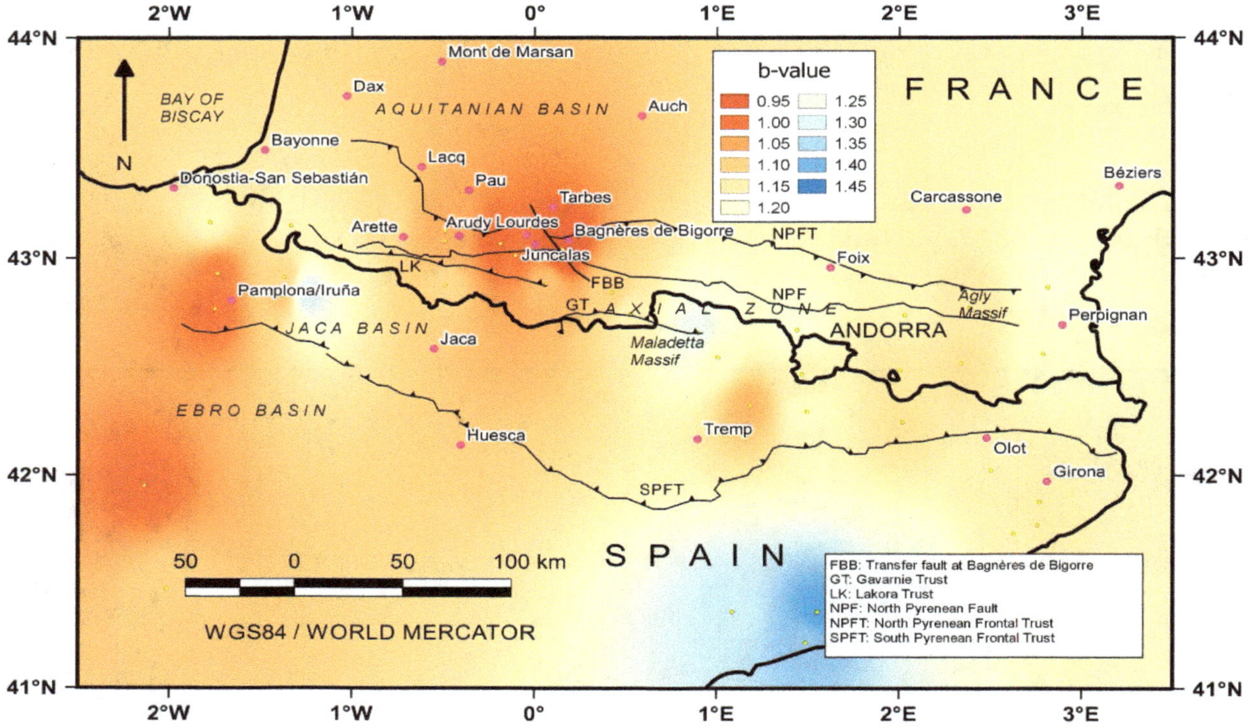

Figure 9. *b-value* map from a 1 × 1° grid and N ≥ 50 events. A yellow point indicates that it has been used in the interpolation process.

Regarding the uncertainties, a map has been prepared, as can be seen in Figure 10.

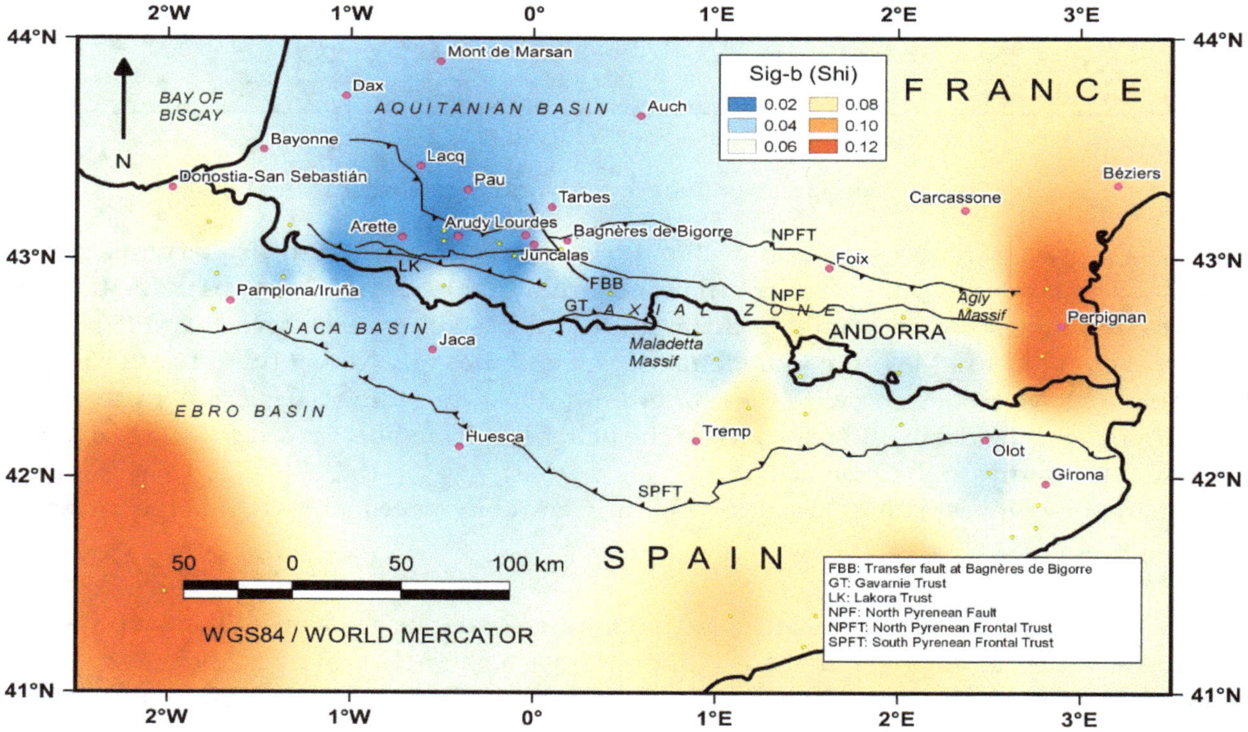

Figure 10. Uncertainty in *b-value* (1 sigma) from [102]. A yellow point indicates that it has been used in the interpolation process.

As expected in the Arette–Arudy-Bagnères de Bigorre environment, lower values are found (0.02). Besides, while in the south of the Ebro Basin and the eastern extreme, higher values are shown (0.10–0.12). Moreover, a reference value can be established at 0.05.

As can be stated, for the central area, a denser grid has been considered. Besides, the minimum number of events required for calculations has been set as 100 for this grid. The result is shown in Figure 11.

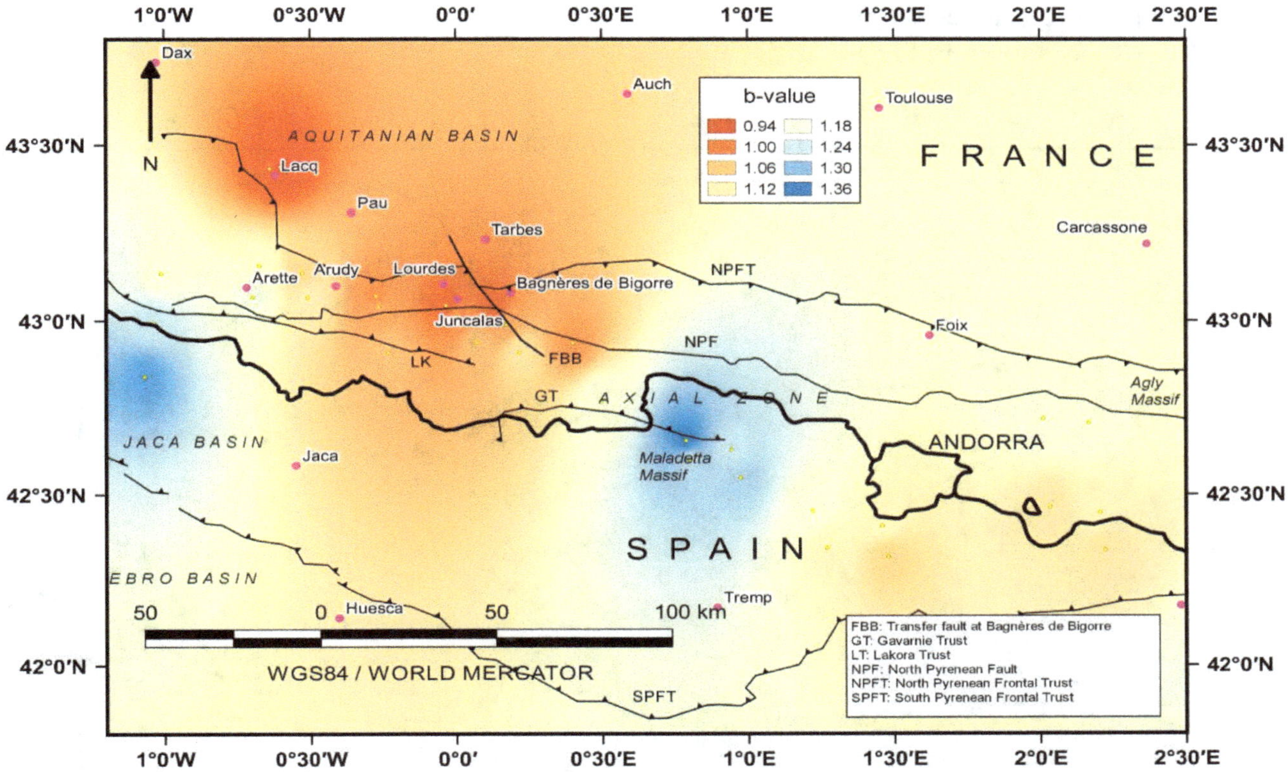

Figure 11. *b-value* map from a 0.5 × 0.5° grid and N ≥ 100 events. A yellow point indicates that it has been used in the interpolation process.

The specific map illustrates that the lower *b-value* is found in the Lacq environment, due to the seismicity induced by the gas extraction in that area [108]. Another peak value (0.97) is located in Bagnères de Bigorre. From there and toward the west, the value increases to approximately 1.05 to reach Arudy and 1.17 in Arette.

The obtained results are in line with other researchers who have studied the *b-value* variation in the Pyrenees. Gallart et al. [54] studied the *b-value* for the Arette–Arudy region in the western Pyrenees, and they concluded that the existence of a lateral variation, from 1.09 ± 0.06 for the west (W) and 0.91 ± 0.09 for the east (E), is possibly related to a depth variation 16 km to 4 km. Kijke-Kassala et al. [38] obtained smaller *b-values* (1.04 ± 0.11) for the central Pyrenees than for the extremes (1.15 ± 0.10 W and 1.19 ± 0.08 E), and a variation with depth for the Arette region 1.01 ± 0.04 for the first 5 km, 0.82 ± 0.04 for 5–10 km, and 0.82 ± 0.07 when deeper than 10 km. The Arudy region was also studied by Sylvander et al. [55], where they found an area with a low *b-value*, understood as an asperity. Secanell et al. [56] conducted a PSHA to calculate the *b-value* and afterward, Secanell et al. [57] divided the Pyrenees into ten seismogenic zones from tectonic models, and they found a wide *b-value* range from 0.91 to 1.64.

Contrariwise, they are not so consistent with Rigo et al. [27] who analyzed the *b-value* variation with the depth relating to the differential stress. These authors took into account only *b-values* with an error better than 0.15, and a magnitude of completeness of 1.5 ± 0.1. Thus, they calculated a *b-value* of 0.80 ± 0.01, as the overall value and considered the seismic zonation proposed by Rigo et al. [58].

The obtained values ranged from 0.71 for the southern zone to 0.99 for the northeasternmost coastal zone. Besides, all the values are between 0.59 and 0.99, but varying with the depth. The *b-value* decreases from 0.92 at 1 km to 0.75 at 11 km, and it is 0.85 ± 0.06 at 19 km. It presents no representative values when deeper than 21 km. The discrepancy between both works can be seen as a result of such a low M_c value, which could have led to underestimating the *b-value* in that work.

4.3. The Annual Rate (Normalized)

Finally, the third seismic parameter to be represented has been normalized by the surface annual rate for M3 (*AR*), as shown in Figure 12.

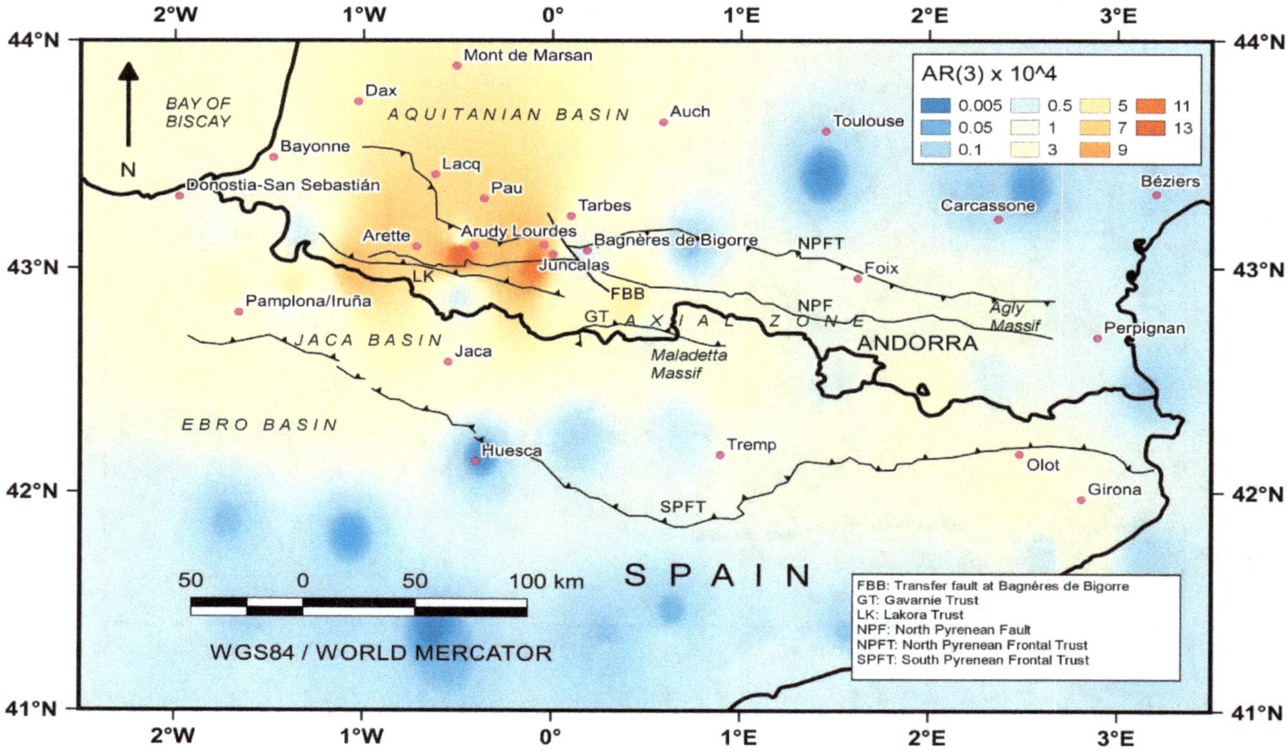

Figure 12. Annual rate (normalized) for M3+.

The analysis of the map indicates that the Arudy region is the most active area in the Pyrenees. In addition, it is one of the highest in the whole Iberian Peninsula and its surroundings, compared with [6,7] (for M3). An *AR* value exceeding 12 events every year × 10^{-4}/km^2 is higher than any other, as it is about 7–9 near Granada in the Betics. Other regions with higher values are in the central-western area (six or more) and particularly near Lourdes (greater than 10). On the contrary, in the east is unlikely to reach three. In a considerable extension of south territory, almost no zones get 1, as in the northeast.

These data are consistent with those calculated by IGN-UPM-WorkingGroup [2], where both Z16 for GM12 zonation and Z15 for A12 are among the most active zones in the Iberian Peninsula and adjacent areas.

After these results, the environment of Arudy and Bagnères de Bigorre has the most pronounced seismicity, as M_{max} and *AR* are high, and the *b-value* is lower than in the rest of the Pyrenees.

5. Conclusions

This paper presents the calculation, continuous representation, and analysis of some of the most relevant seismic parameters for the Pyrenees range. These parameters are the maximum magnitude,

M_{max}, the *b-value* of the GR relation, and the annual rate *AR*. The information has been integrated into the GIS for calculation and visualization, which allows proper handling of the data.

For this purpose, a reliable, homogeneous, extensive, reviewed, and updated catalog has been compiled. First of all, the NGIS earthquake database (with reliable records form 1373) has been chosen as a starting point. After that, it has been reviewed with other catalog and specific studies. Then, the size of all the events has been converted to moment magnitude (M_w), as the state-of-the-art establishes, through Cabañas et al. [36]. Besides, the completeness has been analyzed, leading to set the cut-off magnitude as 2.0. Next, the non-main events have been removed by the declustering process proposed by Gardner and Knopoff [78] with the Uhrhammer [83] parameters. Later, the pairs magnitude–year of completeness have been obtained. Finally, the seismic catalog of the work is ready to be integrated into the GIS. In its environment, different grids, of $0.5 \times 0.5°$ and $1.0 \times 1.0°$, have been established to produce continuous representation.

Concerning the results for the seismic parameters, some findings can be stated. The M_{max} value ranges from 6.7 to no data (or $M < 2$) in some grids' cells. Through the belt, near the Spain–France border, most of the zones have M5+. The maximum values are located in the Bagnères de Bigorre region, and to the north of Olot (Eastern Pyrenees), in Spain.

The *b-value* calculations have employed more than 200 years to include historical events. Furthermore, the method proposed by Kijko and Smit [13], which considers different pairs of magnitude–year of completeness, has been used. The lower values (so, higher stress held) have been found in the Central area, near Bagnères de Bigorre (0.97) and to the west to reach 1.05 in the Arudy region, and 1.17 in the Arette region. In both extremes of east and west, the values are approximately 1.15–1.18.

The *AR* for M3 shows that the Arudy region is the most active area in the belt. Besides, the Bagnères de Bigorre area shows high *AR* values too, with *AR* values higher than 10^{-4} earthquakes/km^2 in a year.

Finally, from the analysis of the parameters, conducted by the GIS, it arises that the Central part, mainly that from Arudy to Bagnères de Bigorre, presents the most pronounced seismicity of the region, as it has the highest M_{max} and *AR*, and the lowest *b-value*. Therefore, this would be the area where the PSHA should be focused on, and a GIS tool would be very handy for it.

Author Contributions: Conceptualization, J.L.A.-M. and D.T.B.; methodology, J.L.A.-M.; software, J.L.A.-M.; validation, J.L.A.-M. and D.T.B.; formal analysis, J.L.A.-M. and D.T.B.; investigation, J.L.A.-M.; resources, J.L.A.-M.; data curation, J.L.A.-M.; writing—original draft preparation, J.L.A.-M.; writing—review and editing, J.L.A.-M. and D.T.B.; visualization, J.L.A.-M.; supervision, J.L.A.-M. and D.T.B.; project administration, J.L.A.-M. All authors have read and agreed to the published version of the manuscript.

Acknowledgments: The authors would like to thank researchers from NGIS for the information, and particularly, José Manuel Martínez-Solares for his comments that have helped to enrich this document.

References

1. Mezcua, J.; Rueda, J.; García Blanco, R.M. A new probabilistic seismic hazard study of Spain. *Nat. Hazards* **2011**, *59*, 1087–1108. [CrossRef]
2. IGN-UPM-Working Group. *Actualización de Mapas de Peligrosidad Sísmica 2012*; Instituto Geográfico Nacional: Madrid, Spain, 2013; ISBN 9788441626850.
3. Hamdache, M.; Peláez, J.A.; Kijko, A.; Smit, A.; Sawires, R. Seismic Hazard Parameters of Main Seismogenic Source Zone Model in the Algeria-Morocco Region. In Proceedings of the Geo-Risks in the Mediterranean and their Mitigation, Valletta, Malta, 20–21 July 2015; pp. 99–101.
4. Hamdache, M.; Peláez, J.A.; Kijko, A.; Smit, A. Energetic and spatial characterization of seismicity in the Algeria–Morocco region. *Nat. Hazards* **2017**, *86*, 273–293. [CrossRef]
5. Hong, T.-K.; Park, S.; Lee, J.; Kim, W. Spatiotemporal Seismicity Evolution and Seismic Hazard Potentials in the Western East Sea (Sea of Japan). *Pure Appl. Geophys.* **2020**, 1–14. [CrossRef]
6. Amaro-Mellado, J.L.; Morales-Esteban, A.; Asencio-Cortés, G.; Martínez-Álvarez, F. Comparing seismic parameters for different source zone models in the Iberian Peninsula. *Tectonophysics* **2017**, *717*, 449–472. [CrossRef]

7. Amaro-Mellado, J.L.; Morales-Esteban, A.; Martínez-Álvarez, F. Mapping of seismic parameters of the Iberian Peninsula by means of a geographic information system. *Cent. Eur. J. Oper. Res.* **2018**, *26*, 739–758. [CrossRef]
8. Hossain, B.; Hossain, S.S. Probabilistic estimation of seismic parameters for Bangladesh. *Arab. J. Geosci.* **2020**, *13*. [CrossRef]
9. Popandopoulos, G.A.; Chatziioannou, E. Gutenberg-Richter Law Parameters Analysis Using the Hellenic Unified Seismic Network Data Through FastBee Technique. *Earth Sci.* **2014**, *3*, 122. [CrossRef]
10. Sawires, R.; Santoyo, M.A.; Peláez, J.A.; Corona Fernández, R.D. An updated and unified earthquake catalog from 1787 to 2018 for seismic hazard assessment studies in Mexico. *Sci. data* **2019**, *6*, 241. [CrossRef]
11. Mousavi, S.M. Mapping seismic moment and b-value within the continental-collision orogenic-belt region of the Iranian Plateau. *J. Geodyn.* **2017**, *103*, 26–41. [CrossRef]
12. Beirlant, J.; Kijko, A.; Reynkens, T.; Einmahl, J.H.J. Estimating the maximum possible earthquake magnitude using extreme value methodology: The Groningen case. *Nat. Hazards* **2019**, *98*, 1091–1113. [CrossRef]
13. Kijko, A.; Smit, A. Extension of the Aki-Utsu b-Value Estimator for Incomplete Catalogs. *Bull. Seismol. Soc. Am.* **2012**, *102*, 1283–1287. [CrossRef]
14. Ghosh, A.; Newman, A.V.; Thomas, A.M.; Farmer, G.T. Interface locking along the subduction megathrust from b-value mapping near Nicoya Peninsula, Costa Rica. *Geophys. Res. Lett.* **2008**, *35*, 1–6. [CrossRef]
15. Mapa_Sismotectónico_WG. *Análisis Sismotectónico de la Península Ibérica, Baleares y Canarias*; Instituto Geográfico Nacional: Madrid, Spain, 1992; Volume 26.
16. Deligiannakis, G.; Papanikolaou, I.D.; Roberts, G. Fault specific GIS based seismic hazard maps for the Attica region, Greece. *Geomorphology* **2018**, *306*, 264–282. [CrossRef]
17. De Donatis, M.; Pappafico, G.; Romeo, R. A Field Data Acquisition Method and Tools for Hazard Evaluation of Earthquake-Induced Landslides with Open Source Mobile GIS. *ISPRS Int. J. Geo-Information* **2019**, *8*, 91. [CrossRef]
18. Tien Bui, D.; Shahabi, H.; Omidvar, E.; Shirzadi, A.; Geertsema, M.; Clague, J.; Khosravi, K.; Pradhan, B.; Pham, B.; Chapi, K.; et al. Shallow Landslide Prediction Using a Novel Hybrid Functional Machine Learning Algorithm. *Remote Sens.* **2019**, *11*, 931. [CrossRef]
19. Naegeli, T.J.; Laura, J. Back-projecting secondary craters using a cone of uncertainty. *Comput. Geosci.* **2019**, *123*, 1–9. [CrossRef]
20. Gutenberg, B.; Richter, C.F. *Seismicity of the Earth*; Princeton University: Princeton, NJ, USA, 1954.
21. Udías, A.; Mezcua, J. *Fundamentos de Geofísica*; Alianza Universidad Textos; Alianza Editorial: Madrid, Spain, 1997.
22. Hiemer, S.; Woessner, J.; Basili, R.; Danciu, L.; Giardini, D.; Wiemer, S. A smoothed stochastic earthquake rate model considering seismicity and fault moment release for Europe. *Geophys. J. Int.* **2014**, *198*, 1159–1172. [CrossRef]
23. Frohlich, C.; Davis, S.D. Teleseismic b values; Or, much ado about 1.0. *J. Geophys. Res.* **1993**, *98*, 631–644. [CrossRef]
24. Nava, F.A.; Ávila-Barrientos, L.; Márquez-Ramírez, V.H.; Torres, I.; Zúñiga, F.R. Sampling uncertainties and source b likelihood for the Gutenberg-Richter b value from the Aki-Utsu method. *J. Seismol.* **2018**, *22*, 315–324. [CrossRef]
25. Tormann, T.; Wiemer, S.; Mignan, A. Systematic survey of high-resolution b value imaging along Californian faults: Inference on asperities. *J. Geophys. Res. Solid Earth* **2014**, *119*, 2029–2054. [CrossRef]
26. Morales-Esteban, A.; Martínez-Álvarez, F.; Reyes, J. Earthquake prediction in seismogenic areas of the Iberian Peninsula based on computational intelligence. *Tectonophysics* **2013**, *593*, 121–134. [CrossRef]
27. Rigo, A.; Souriau, A.; Sylvander, M. Spatial variations of b-value and crustal stress in the Pyrenees. *J. Seismol.* **2018**, *22*, 337–352. [CrossRef]
28. Amorèse, D.; Grasso, J.R.; Rydelek, P.A. On varying b-values with depth: Results from computer-intensive tests for Southern California. *Geophys. J. Int.* **2010**, *180*, 347–360. [CrossRef]
29. Pandey, A.K.; Chingtham, P.; Roy, P.N.S. Homogeneous earthquake catalogue for Northeast region of India using robust statistical approaches. *Geomatics, Nat. Hazards Risk* **2017**, 1–15. [CrossRef]
30. Haque, D.M.E.; Chowdhury, S.H.; Khan, N.W. Intensity Attenuation Model Evaluation for Bangladesh with Respect to the Surrounding Potential Seismotectonic Regimes. *Pure Appl. Geophys.* **2020**, *177*, 157–179. [CrossRef]

31. Waseem, M.; Khan, S.; Asif Khan, M. Probabilistic Seismic Hazard Assessment of Pakistan Territory Using an Areal Source Model. *Pure Appl. Geophys.* **2020**. [CrossRef]
32. Peláez, J.A.; Chourak, M.; Tadili, B.A.; Aït Brahim, L.; Hamdache, M.; López Casado, C.; Martínez Solares, J.M. A catalog of main Moroccan earthquakes from 1045 to 2005. *Seismol. Res. Lett.* **2007**, *78*, 614–621. [CrossRef]
33. Hamdache, M.; Pelaez, J.A.; Talbi, A.; Casado, C.L. A Unified Catalog of Main Earthquakes for Northern Algeria from A.D. 856 to 2008. *Seismol. Res. Lett.* **2010**, *81*, 732–739. [CrossRef]
34. Zare, M.; Amini, H.; Yazdi, P.; Sesetyan, K.; Demircioglu, M.B.; Kalafat, D.; Erdik, M.; Giardini, D.; Khan, M.A.; Tsereteli, N. Recent developments of the Middle East catalog. *J. Seismol.* **2014**, *18*, 749–772. [CrossRef]
35. Das, R.; Sharma, M.L.; Wason, H.R.; Choudhury, D.; Gonzalez, G. A seismic moment magnitude scale. *Bull. Seismol. Soc. Am.* **2019**, *109*, 1542–1555. [CrossRef]
36. Cabañas, L.; Rivas-Medina, A.; Martínez-Solares, J.M.; Gaspar-Escribano, J.M.; Benito, B.; Antón, R.; Ruiz-Barajas, S. Relationships Between M w and Other Earthquake Size Parameters in the Spanish IGN Seismic Catalog. *Pure Appl. Geophys.* **2015**, *172*, 2397–2410. [CrossRef]
37. Scordilis, E.M. Empirical Global Relations Converting MS and mb to Moment Magnitude. *J. Seismol.* **2006**, *10*, 225–236. [CrossRef]
38. Njike-Kassala, J.-D.; Souriau, A.; Gagnepain-Beyneix, J.; Martel, L.; Vadell, M. Frequency-magnitude relationship and Poisson's ratio in the Pyrenees, in relation to earthquake distribution. *Tectonophysics* **1992**, *215*, 363–369. [CrossRef]
39. Rigo, A.; Pauchet, H.; Souriau, A.; Grésillaud, A.; Nicolas, M.; Olivera, C.; Figueras, S. The February 1996 earthquake sequence in the eastern Pyrenees: First results. *J. Seismol.* **1997**, *1*, 3–14. [CrossRef]
40. Vissers, R.L.M.; Meijer, P.T. Iberian plate kinematics and Alpine collision in the Pyrenees. *Earth-Science Rev.* **2012**, *114*, 61–83. [CrossRef]
41. Lacan, P.; Ortuño, M. Active Tectonics of the Pyrenees: A review. *J. Iber. Geol.* **2012**, *38*, 9–30. [CrossRef]
42. Singh, C. Spatial variation of seismic b-values across the NW Himalaya. *Geomat. Nat. Hazards Risk* **2016**, *7*, 522–530. [CrossRef]
43. Souriau, A.; Rigo, A.; Sylvander, M.; Benahmed, S.; Grimaud, F. Seismicity in central-western Pyrenees (France): A consequence of the subsidence of dense exhumed bodies. *Tectonophysics* **2014**, *621*, 123–131. [CrossRef]
44. Teixell, A.; Labaume, P.; Lagabrielle, Y. The crustal evolution of the west-central Pyrenees revisited: Inferences from a new kinematic scenario. *Comptes Rendus-Geosci.* **2016**, *348*, 257–267. [CrossRef]
45. Roigé, M.; Gómez-Gras, D.; Remacha, E.; Daza, R.; Boya, S. Tectonic control on sediment sources in the Jaca basin (Middle and Upper Eocene of the South-Central Pyrenees). *Comptes Rendus-Geosci.* **2016**, *348*, 236–245. [CrossRef]
46. Baize, S.; Cushing, E.M.; Lemeille, F.; Jomard, H. Updated seismotectonic zoning scheme of Metropolitan France, with reference to geologic and seismotectonic data. *Bull. la Soc. Geol. Fr.* **2013**, *184*, 225–259. [CrossRef]
47. Olivera, C.; Redondo, E.; Lambert, J.; Riera-Melis, A.; Roca, A. *Els terratrèmols de segles XIV i XV a Catalunya*; Institut Cartogràfic de Catalunya: Barcelona, Spain, 2006; ISBN 8439369611.
48. Insitituto Geográfico Nacional Terremotos más Importantes. Available online: https://www.ign.es/web/ign/portal/terremotos-importantes (accessed on 29 April 2020).
49. Perea, H. The Catalan seismic crisis (1427 and 1428; NE Iberian Peninsula): Geological sources and earthquake triggering. *J. Geodyn.* **2009**, *47*, 259–270. [CrossRef]
50. Honoré, L.; Courboulex, F.; Souriau, A. Ground motion simulations of a major historical earthquake (1660) in the French Pyrenees using recent moderate size earthquakes. *Geophys. J. Int.* **2011**, *187*, 1001–1018. [CrossRef]
51. Cara, M.; Alasset, P.J.; Sira, C. Magnitude of Historical Earthquakes, from Macroseismic Data to Seismic Waveform Modelling: Application to the Pyrenees and a 1905 Earthquake in the Alps. In *Historical Seismology. Modern Approaches in Solid Earth Sciences*; Fréchet, J., Meghraoui, M., Stucchi, M., Eds.; Springer: Cham, Switzerland, 2008; Volume 2, pp. 369–384.
52. Rigo, A.; Massonnet, D. Investigating the 1996 Pyrenean Earthquake (France) with SAR interferograms heavily distorted by atmosphere. *Geophys. Res. Lett.* **1999**, *26*, 3217–3220. [CrossRef]
53. Ministerio de Fomento (Gobierno de España). *Norma de la Construcción Sismorresistente Española (NCSE-02)*; Boletín Oficial del Estado: Madrid, Spain, 2002.

54. Gallart, J.; Dainières, M.; Banda, E.; Suriñach, E.; Hirn, A. The eastern Pyrenean domain: Lateral variations at crust-mantle level. *Ann. Geophys.* **1980**, *36*, 141–158.
55. Sylvander, M.; Souriau, A.; Rigo, A.; Tocheport, A.; Toutain, J.-P.; Ponsolles, C.; Benahmed, S. The 2006 November, M L = 5.0 earthquake near Lourdes (France): New evidence for NS extension across the Pyrenees. *Geophys. J. Int.* **2008**, *175*, 649–664. [CrossRef]
56. Secanell, R.; Martin, C.; Goula, X.; Susagna, T.; Tapia, M.; Bertil, D. Evaluacion probabilista de la peligrosidad sísmica de la región pirenaica. In 3° *Congreso Nacional de Ingeniería Sísmica*; Asociación Española de Ingeniería Sísmica: Madrid, Spain, 2007; pp. 1–17.
57. Secanell, R.; Bertil, D.; Martin, C.; Goula, X.; Susagna, T.; Tapia, M.; Dominique, P.; Carbon, D.; Fleta, J. Probabilistic seismic hazard assessment of the Pyrenean region. *J. Seismol.* **2008**, *12*, 323–341. [CrossRef]
58. Rigo, A.; Vernant, P.; Feigl, K.L.; Goula, X.; Khazaradze, G.; Talaya, J.; Morel, L.; Nicolas, J.; Baize, S.; Chery, J.; et al. Present-day deformation of the Pyrenees revealed by GPS surveying and earthquake focal mechanisms until 2011. *Geophys. J. Int.* **2015**, *201*, 947–964. [CrossRef]
59. Insitituto Geográfico Nacional Catálogo de Terremotos. Available online: http://www.ign.es/web/ign/portal/sis-catalogo-terremotos (accessed on 29 April 2020).
60. González, Á. The Spanish National Earthquake Catalogue: Evolution, precision and completeness. *J. Seismol.* **2017**, *21*, 435–471. [CrossRef]
61. Mezcua, J.; Rueda, J.; Blanco, R.M.G. Reevaluation of historic Earthquakes in Spain. *Seismol. Res. Lett.* **2004**, *75*, 75–81. [CrossRef]
62. Martínez-Solares, J.M.; Mezcua, J. *Catálogo Sísmico de la Península Ibérica (800 a. C.-1900)*; Instituto Geográfico Nacional: Madrid, Spain, 2002.
63. Susagna, T.; Roca, A.; Goula1, X.; Batlló, J. Analysis of macroseismic and instrumental data for the study of the 19 November 1923 earthquake in the Aran Valley (Central Pyrenees). *Nat. Hazards* **1994**, *10*, 7–17. [CrossRef]
64. Martín, R.; Stich, D.; Morales, J.; Mancilla, F. Moment tensor solutions for the Iberian-Maghreb region during the IberArray deployment (2009–2013). *Tectonophysics* **2015**, *663*, 261–274. [CrossRef]
65. García-Mayordomo, J.; Insua-Arévalo, J.M. Seismic hazard assessment for the Itoiz dam site (Western Pyrenees, Spain). *Soil Dyn. Earthq. Eng.* **2011**, *31*, 1051–1063. [CrossRef]
66. Stich, D.; Ammon, C.J.; Morales, J. Moment tensor solutions for small and moderate earthquakes in the Ibero-Maghreb region. *J. Geophys. Res. Solid Earth* **2003**, *108*. [CrossRef]
67. Stich, D.; Martín, R.; Batlló, J.; Macià, R.; Mancilla, F.d.L.; Morales, J. Normal Faulting in the 1923 Berdún Earthquake and Postorogenic Extension in the Pyrenees. *Geophys. Res. Lett.* **2018**, *45*, 3026–3034. [CrossRef]
68. Mezcua, J.; Martínez-Solares, J.M. *Sismicidad del área Íbero-Magrebí*; Instituto Geográfico Nacional: Madrid, Spain, 1983; Volume 203.
69. Veith, K.F.; Clawson, G.E. Magnitude from short period P-wave data. *Bull. Seismol. Soc. Am.* **1972**, *62*, 435–452.
70. López, C. *Nuevas fórmulas de Magnitud para la Península Ibérica y su Entorno*; Universidad Complutense de: Madrid, Spain, 2008.
71. Das, R.; Wason, H.R.; Gonzalez, G.; Sharma, M.L.; Choudhury, D.; Lindholm, C.; Roy, N.; Salazar, P. Earthquake magnitude conversion problem. *Bull. Seismol. Soc. Am.* **2018**, *108*, 1995–2007. [CrossRef]
72. Johnston, A.C. Seismic moment assessment of earthquakes in stable continental regions-I. Instrumental seismicity. *Geophys. J. Int.* **1996**, *124*, 381–414. [CrossRef]
73. Johnston, A.C. Seismic moment assessment of earthquakes in stable continental regions-II. Historical seismicity. *Geophys. J. Int.* **1996**, *125*, 639–678. [CrossRef]
74. Castellaro, S.; Mulargia, F.; Kagan, Y.Y. Regression problems for magnitudes. *Geophys. J. Int.* **2006**, *165*, 913–930. [CrossRef]
75. Mignan, A.; Woessner, J. Estimating the Magnitude of Completeness for Earthquake Catalogs. CORSSA (Community Online Resource for Statistical Seismicity Analysis). 2012, pp. 1–45. Available online: https://tcs.ah-epos.eu/eprints/1587/ (accessed on 29 April 2020).
76. Page, M.T.; van der Elst, N.; Hardebeck, J.; Felzer, K.; Michael, A.J. Three Ingredients for Improved Global Aftershock Forecasts: Tectonic Region, Time-Dependent Catalog Incompleteness, and Intersequence Variability. *Bull. Seismol. Soc. Am.* **2016**, *106*, 2290–2301. [CrossRef]

77. Karimzadeh, S.; Matsuoka, M.; Kuang, J.; Ge, L. Spatial prediction of aftershocks triggered by a major earthquake: A binary machine learning perspective. *ISPRS Int. J. Geo-Inf.* **2019**, *8*, 462. [CrossRef]
78. Gardner, J.K.; Knopoff, L. Is the sequence of earthquakes in Southern California, with aftershocks removed, Poissonian? *Bull. Seismol. Soc. Am.* **1974**, *64*, 1363–1367.
79. Reasenberg, P. Second-order moment of Central California Seismicity, 1960–1982. *J. Geophys. Res.* **1985**, *90*, 5479–5495. [CrossRef]
80. Van Stiphout, T.; Zhuang, J.; Marsan, D. *Seismicity Declustering, Community Online Resource for Statistical Seismicity Analysis*; ETH—Swiss Federal Institute of Technology: Zurich, Switzerland, 2012. [CrossRef]
81. Khan, M.M.; Munaga, T.; Kiran, D.N.; Kumar, G.K. Seismic hazard curves for Warangal city in Peninsular India. *Asian J. Civ. Eng.* **2020**, *21*, 543–554. [CrossRef]
82. Pandey, A.K.; Chingtham, P.; Prajapati, S.K.; Roy, P.N.S.; Gupta, A.K. Recent seismicity rate forecast for North East India: An approach based on rate state friction law. *J. Asian Earth Sci.* **2019**, *174*, 167–176. [CrossRef]
83. Uhrhammer, R. Characteristics of Northern and Central California Seismicity. *Earthq. Notes* **1986**, *57*, 21.
84. Khan, M.M.; Kalyan Kumar, G. Statistical Completeness Analysis of Seismic Data. *J. Geol. Soc. India* **2018**, *91*, 749–753. [CrossRef]
85. Papadakis, G.; Vallianatos, F.; Sammonds, P. Evidence of Nonextensive Statistical Physics behavior of the Hellenic Subduction Zone seismicity. *Tectonophysics* **2013**, *608*, 1037–1048. [CrossRef]
86. Wiemer, S. A Software Package to Analyze Seismicity: ZMAP. *Seismol. Res. Lett.* **2001**, *72*, 373–382. [CrossRef]
87. Mulargia, F.; Tinti, S. Seismic sample areas defined from incomplete catalogues: An application to the Italian territory. *Phys. Earth Planet. Int.* **1985**, *40*, 273–300. [CrossRef]
88. Tinti, S.; Mulargia, F. Confidence intervals of b values for grouped magnitudes. *Bull. Seismol. Soc. Am.* **1987**, *77*, 2125–2134.
89. Wiemer, S.; Wyss, M. Minimum magnitude of completeness in earthquake catalogs: Examples from Alaska, the Western United States, and Japan. *Bull. Seismol. Soc. Am.* **2000**, *90*, 859–869. [CrossRef]
90. Wesseloo, J. Addressing misconceptions regarding seismic hazard assessment in mines: B-value, Mmax, and space-time normalization. *J. South. Afr. Inst. Min. Metall.* **2020**, *120*, 67–80. [CrossRef]
91. Kijko, A.; Singh, M. Statistical tools for maximum possible earthquake magnitude estimation. *Acta Geophys.* **2011**, *59*, 674–700. [CrossRef]
92. Asim, K.M.; Moustafa, S.S.; Niaz, I.A.; Elawadi, E.A.; Iqbal, T.; Martínez-Álvarez, F. Seismicity analysis and machine learning models for short-term low magnitude seismic activity predictions in Cyprus. *Soil Dyn. Earthq. Eng.* **2020**, *130*, 105932. [CrossRef]
93. Tan, Y.J.; Waldhauser, F.; Tolstoy, M.; Wilcock, W.S.D. Axial Seamount: Periodic tidal loading reveals stress dependence of the earthquake size distribution (b value). *Earth Planet. Sci. Lett.* **2019**, *512*, 39–45. [CrossRef]
94. Arvanitakis, K.; Gounaropoulos, C.; Avlonitis, M. Locating Asperities by Means of Stochastic Analysis of Seismic Catalogs. *Bull. Geol. Soc. Greece* **2017**, *50*, 1293. [CrossRef]
95. Martínez-Álvarez, F.; Reyes, J.; Morales-Esteban, A.; Rubio-Escudero, C. Determining the best set of seismicity indicators to predict earthquakes. Two case studies: Chile and the Iberian Peninsula. *Knowl.-Based Syst.* **2013**, *50*, 198–210. [CrossRef]
96. Weitcher, D.H. Estimation of the earthquake recurrence parameters for unequal observation periods for different magnitudes. *Bull. Seismol. Soc. Am.* **1980**, *70*, 1337–1346.
97. Kamer, Y.; Hiemer, S. Data-driven spatial b value estimation with applications to California seismicity: To b or not to b. *J. Geophys. Res. Solid Earth* **2015**, *120*, 5191–5214. [CrossRef]
98. Bender, B. Maximum likelihood estimation of b values for magnitude grouped data. *Bull. Seismol. Soc. Am.* **1983**, *73*, 831–851.
99. Aki, K. Maximum likelihood estimate of b in the formula logN = a-bM and its confidence limits. *Bull. Earthq. Res. Inst.* **1965**, *43*, 237–239.
100. Utsu, T. A method for determining the value of b in a formula log n = a-bM showing the magnitude-frequency relation for earthquakes. *Geophys. Bull. Hokkaido Univ.* **1965**, *13*, 99–103.
101. Ogata, Y.; Yamashina, K. Unbiased estimate for b-value of magnitude frequency. *J. Phys. Earth* **1986**, *34*, 187–194. [CrossRef]
102. Shi, Y.; Bolt, B.A. The standard error of the magnitude-frequency b value. *Bull. Seismol. Soc. Am.* **1982**, *72*, 1677–1687.
103. Dominique, P.; Andre, E. Probabilistic seismic hazard map on the French national territory. In Proceedings of

the 12th World Conference on Earthquake Engineering, Auckland, New Zealand, 30 January–4 February 2000; pp. 1–8.
104. Skordas, E.; Kulhánek, O. Spatial and temporal variations of Fennoscandian seismicity. *Geophys. J. Int.* **1992**, *111*, 577–588. [CrossRef]
105. Roberts, N.S.; Bell, A.F.; Main, I.G. Are volcanic seismic b-values high, and if so when? *J. Volcanol. Geotherm. Res.* **2015**, *308*, 127–141. [CrossRef]
106. Marzocchi, W.; Spassiani, I.; Stallone, A.; Taroni, M. How to be fooled searching for significant variations of the b-value. *Geophys. J. Int.* **2020**, *220*, 1845–1856. [CrossRef]
107. Wiemer, S.; Wyss, M. Mapping spatial variability of the frequency-magnitude distribution of earthquakes. *Adv. Geophys.* **2002**, *45*, 259–302. [CrossRef]
108. Bardainne, T.; Dubos-Sallée, N.; Sénéchal, G.; Gaillot, P.; Perroud, H. Analysis of the induced seismicity of the Lacq gas field (Southwestern France) and model of deformation. *Geophys. J. Int.* **2008**, *172*, 1151–1162. [CrossRef]

14

Abandoned Farmland Location in Areas Affected by Rapid Urbanization using Textural Characterization of High Resolution Aerial Imagery

Juan José Ruiz-Lendínez

Departamento de Ingeniería Cartográfica, Geodésica y Fotogrametría. Escuela Politécnica Superior de Jaén, Universidad de Jaén, 23071 Jaén, Spain; lendinez@ujaen.es.

Abstract: Several studies have demonstrated that farmland abandonment occurs not only in rural areas, but is also closely interlinked with urbanization processes. Therefore, the location of abandoned land and the registration of the spatial information referring to it play important roles in urban land management. However, mapping abandoned land or land in the process of abandonment is not an easy task because the limits between the different land uses are not clear and precise. It is therefore necessary to develop methods that allow estimating and mapping this type of land as accurately as possible. As an alternative to other geomatics methods such as satellite remote sensing, our approach proposes a framework for automatically locating abandoned farmland in urban landscapes using the textural characterization and segmentation of aerial imagery. Using the city of Poznań (Poland) as a case study, results demonstrated the feasibility of applying our approach, reducing processing time and workforce resources. Specifically and by comparing the results obtained with the data provided by CORINE Land Cover, 2275 ha (40.3%) of arable land within the city limits were abandoned, and the area of abandoned arable land was almost 9.2% of the city's area. Finally, the reliability of the proposed methodology was assessed from two different focuses: (i) the accuracy of the segmentation results (from a positional point of view) and (ii) the efficiency of locating abandoned land (as a specific type of land use) in urban areas particularly affected by rapid urbanization.

Keywords: farmland abandonment mapping; textural segmentation; aerial imagery; land use; Poznań

---

1. Introduction

Farmland abandonment (FLA) can be defined as the cessation of agricultural activities on a given surface of land [1]. FLA has significant environmental consequences, and it is often associated with social and economic problems in rural areas. FLA occurs in areas where agricultural activities are no longer viable, but is also found to occur in areas that are well connected to transport networks and with intense economic activity [2]. According to Grădinaru et al. [2], this latter type of abandonment is interlinked with rapid urbanization processes. Thus, new economic opportunities, urbanization, real estate preferences, and changes in demographic structure are considered the most important drivers of this loss of interest in agricultural activities [3]. Consequently, the land becomes underused, which is considered actual abandonment. The term rapid urbanization refers to scattered, land-consuming, and unorganized urban expansion, developed because of permissive legislation or even non-compliance with planning regulations. In the last few decades, construction has been the main driving force for the economy of many European countries, so peri-urban agricultural land has been especially under pressure, suffering processes of degradation and abandonment in waiting for urban development [4].

The location of abandoned land and registration of the spatial information referring to it play important roles in urban land management. Thus, mapping abandoned land may bring to light future trajectories of land change. This can enable local authorities to evaluate the effectiveness of land use regulations and adapt urban planning for a more sustainable use of land [2]. The location of abandoned land or land in the process of abandonment is difficult because the limits between the different land uses are not clear and precise. According to Pointereau et al. [1], the frontier between utilized agricultural land, non-utilized agricultural land, and forest may be established as shown in Figure 1. Thus, non-utilized agricultural areas could be defined as wooded areas that are not identifiable as forest, rough grassland, or fallow lands. Utilized agricultural areas (UAA) comprise arable lands, temporary meadows, and permanent meadows.

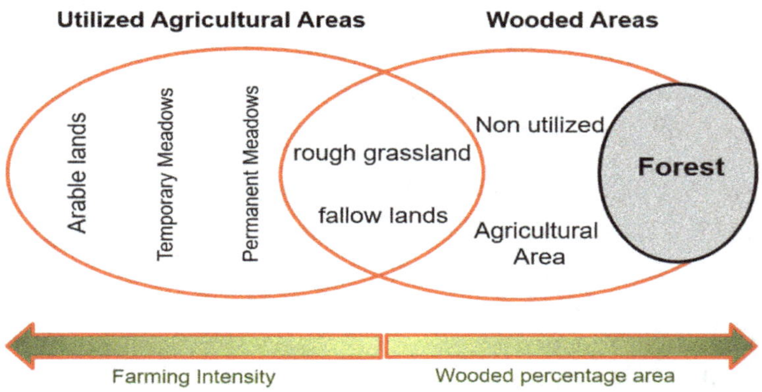

Figure 1. The frontier between utilized agricultural land, non-utilized agricultural land, and forest.

On the basis of the above, it is necessary to develop methods that allow us to estimate and map FLA as accurately as possible. Thus, as an effective alternative to traditional methods based on direct observations that can be time consuming, expensive, and require extensive field work, new forms of estimation are emerging as a result of the on-going and rapid development of geomatics technologies. One of these geomatics technologies employed for the study of landscapes and for mapping land use over large areas is satellite remote sensing (SRS) [5–8]. There are several studies related to the application of SRS to FLA mapping, and in all of them, SRS has proven to be an effective and valuable tool. Among these studies, we highlight the works of Estel et al., Liu et al., Grădinaru et al., and Stryjakiewicz et al. [2,9–11]. Liu et al. [10] investigated the utility of high resolution airborne images (digital multi spectral imagery (DMSI)) obtained over two seasons to estimate carbon mitigation in the revegetation of abandoned agricultural land. Their results indicated that the proposed methodology can potentially be applied to the large-scale mapping of abandoned land in mountain areas. In their paper, Grădinaru et al. [2] analyzed how time series of seasonal images influence classification accuracy in heterogeneous urban landscapes characterized by high fragmentation and high crop variety. In order to do this, they proposed a method for the rapid assessment of FLA by using seasonal time series of Landsat data. Specifically, they used high resolution Landsat imagery (30 m) that allowed them to fine-scale map abandonment rates and identify spatial patterns. However, Landsat data have been used on many earlier occasions in order to: (i) determine driving forces and abandonment rates in agricultural areas [12,13], (ii) analyze the influence of Landsat ETM/ETM+ availability and image acquisition dates on the detection of abandoned land, and (iii) demonstrate the importance of multi-seasonal datasets in obtaining high classification accuracy [14]. On the other hand, Stryjakiewicz et al. [11] assessed the degree of abandonment of agricultural land using Sentinel imagery. Although this type of very high resolution imagery (10 m) had already proven to be a reliable source for assessing land use changes at local or regional scales [15,16], they had not been tested for FLA identification until the study developed by Stryjakiewicz et al. [11]. Their results not only showed that Sentinel demonstrated itself to be an efficient tool for locating abandoned arable land in urban landscapes, but also that it provided information about the area and percentage of this type of land.

Despite the above and as expressed by the majority of the above-mentioned studies, SRS provides less accurate results than maps derived from aerial imagery and extensive field work. This is because the accuracy of the abandoned land maps derived from SRS data depends on: (i) the spatial resolution provided by the satellite platform employed and (ii) the accuracy derived from the set of processes by which a satellite image is generated. In this sense we must note that, unlike in the case of imagery provided by aerial platforms, the use of satellite imagery involves the development of several processes inherent to satellite platforms, such as pre-classification (used to analyse the behaviour of the work area with regard to the seasonal phenology of the different land uses) or classification (principal component analysis, analysis of classes by means of a dendrogram, etcetera). On the basis of the above evidence, our approach proposes a framework for automatically locating and mapping FLA in urban landscapes using the textural characterization and segmentation of aerial imagery.

2. Data and Methodology

2.1. Research Approach

Based on the methodology originally developed by Ojala and Pietikäinen [17], we developed an algorithm for identifying and extracting pixels that belong to abandoned land based on a nonparametric approach to texture characterization. The workflow of our approach is shown in Figure 2. The texture characterization was carried out by using two textural descriptors: local binary pattern (LBP) and contrast (C). LBP has become a really powerful measure of image texture, showing excellent results in terms of accuracy and computational complexity in many empirical studies [18]. This approach to texture characterization has already been successfully tested in many conflation processes between vector data and imagery in which road intersections are automatically extracted from imagery and then used as control points in order to accomplish the matching procedure [19,20]. After obtaining the map of abandoned land, we assessed the effectiveness of the proposed methodology from two different focuses: (i) the accuracy of the segmentation results (from a positional point of view) and (ii) the efficiency for locating abandoned land (as a specific type of land use) in urban areas particularly affected by rapid urbanization. For both cases, we used the same sample of test plots. Specifically for the evaluation of the segmentation results, the boundaries of these plots were extracted from cadastral databases, while for checking the land use, two inspection procedures were carried out: (i) an external validation based on a visual inspection procedure through field visits and (ii) a revision of the LBP/C parameters belonging to the sample of test plots.

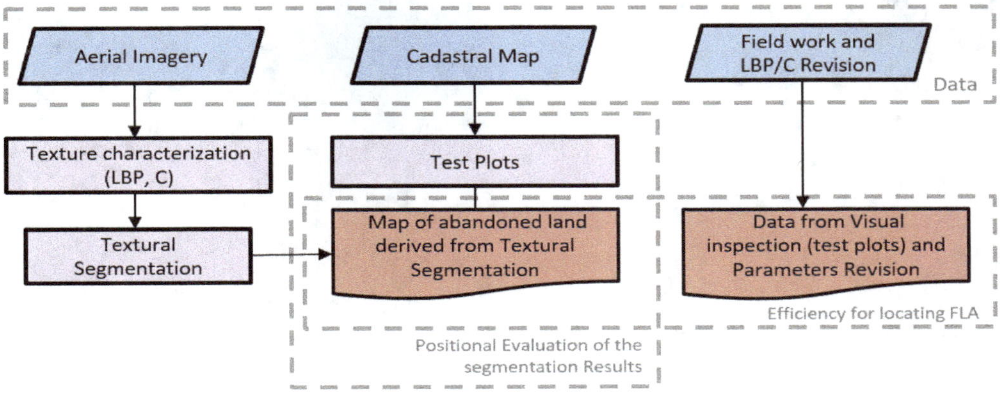

Figure 2. Workflow of the proposed methodology.

2.2. Study Area and Data

In former socialist countries, the transition from command- to market-oriented economies was followed by massive landownership transfers [21], leading to highly fragmented land use and a decrease in agriculture profitability with the consequent abandonment of the land. In this sense, Poland was not an exception.

In Poland, considerable FLA occurred during the period 1996–2002 and mainly concerned the grasslands. FLA appears to be located mainly in the east of the country, where there are small and diversified holdings, as opposed to the west, where during the communist period, many large holdings were state-owned farms [1]. With regard to urban areas and for this same period, the loss of UAA was estimated at 764,032 ha, corresponding to 36% of the total (urban and rural) decrease (30%, if we consider the net result). In municipalities with a density of over 200 inh/km^2, 35% of the UAA (633,193 ha) was lost during the period.

Specifically, Poznań was chosen as the case study area (Figure 3). According to the classification established by CORINE Land Cover, agricultural land covers 33% of the area of Poznań, and arable land clearly dominates (82%) the structure of this agricultural land. The main reason for choosing Poznań was because it has been greatly affected by landscape fragmentation and FLA over the past decade. As mentioned in Section 1, this abandonment has mainly been driven by permissive legislation or even by non-compliance with planning regulations. However, there has been another key factor in FLA: demographic evolution. Poznań has a specific demographic evolution, characterized by different flows occurring at the same time: (i) migration from some rural areas to urban areas, (ii) migration from city centers to peri-urban areas, and (iii) expansion of housing (catching up with the new lifestyle standards). Pushed by the development of the economy, these population flows create housing and infrastructure needs. The need for land is therefore becoming important in these urban areas.

Finally, and with regard to the data, we used high resolution aerial imagery provided by World_Imagery (MapServer) [22]. World_Imagery provides satellite and aerial imagery (with resolutions ranging from 0.5 m/pixel to 1 m/pixel) from many parts of the world. For further information on this data source, including the terms of use, visit online at [22]. In our case, the resolution of the image selected was 1 m/pixel.

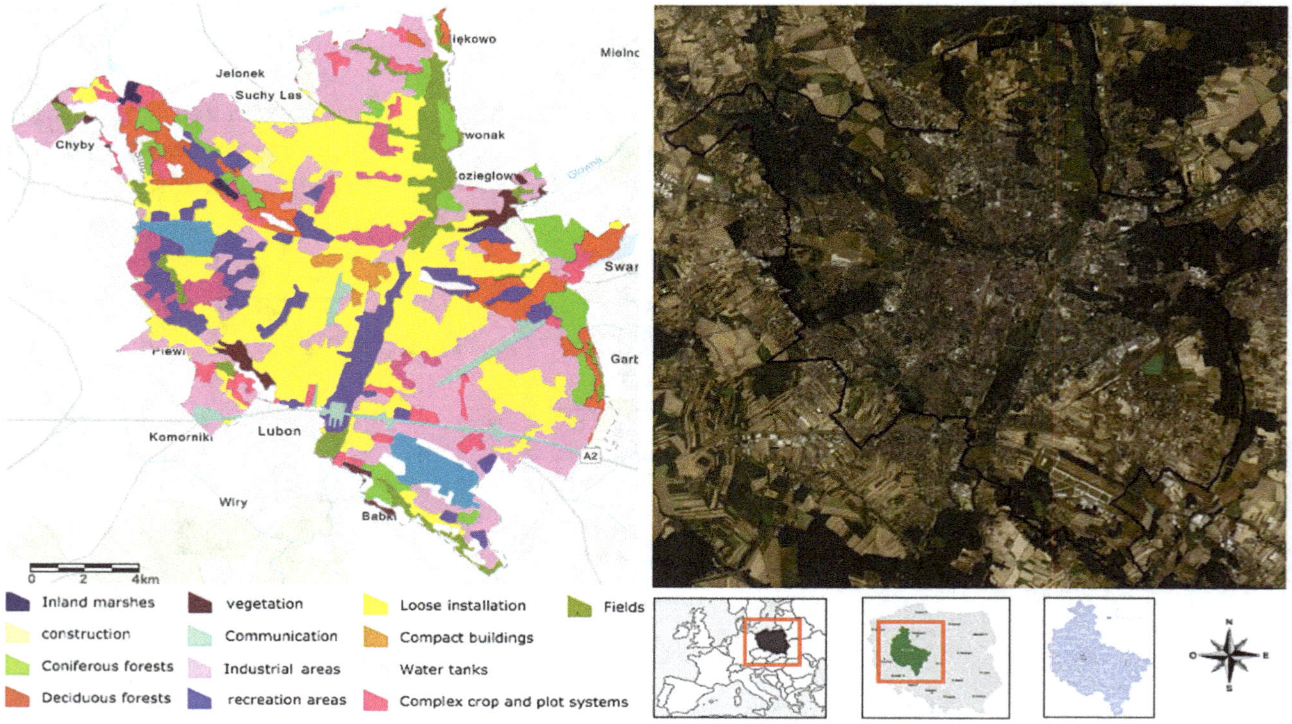

Figure 3. Location of the study area (Poznań), classification established by CORINE Land Cover and the high resolution aerial image selected.

2.3. *Abandoned Farmland Extraction Algorithm*

The main segmentation techniques based on observations of pixel tone analyze the shape of the greyscale histogram and classify the histogram clusters based on their sizes, splitting the original image. Although the resulting regions are visually different, they behave similarity with respect to certain statistical properties. However, this uniform statistical behavior is not usual in the majority of real imagery regions. Consequently, the use of pixel tone to segment imagery does not seem to be the best option [20]. To overcome this challenge, we developed an algorithm for extracting pixels from images based on a local-nonparametric approach to texture analysis. Textural segmentation can be defined as the process of partitioning an image into different parts together, which belong to the same object class. For this purpose, textural information is extracted from the image and used to develop tasks such as recognition, classification, and analysis [23]. In recent years, very discriminative and computationally efficient local texture descriptors have been developed [23]. Among all of these, our algorithm uses two main textural measures:

- Local binary pattern (LBP): LBP is a local texture descriptor capable of characterizing small texture regions [23]. LBP is a simple, yet very efficient texture operator that thresholds the neighboring pixels based on the value of the current pixel [24]. Due to its discriminative power, the LBP texture operator has become a popular approach in several applications, highlighting textural analysis procedures [25]. Figure 4 shows the procedure for computing LBP values. First, each central pixel is compared with its eight neighbors. This 3 × 3 neighborhood must be thresholded by the value of the center pixel; the neighbors having a smaller value than that of the central pixel will have bit 0, and the other neighbors having a value equal to or greater than that of the central pixel will have bit 1. Then these binary values of the pixels in the thresholded neighborhood must be multiplied by the weights given to the corresponding pixels. Finally, the values of the eight pixels will be added to obtain a number for this neighborhood. If we computed the LBP histogram over an entire region, it may be used for describing its texture. In addition, LBP achieves high levels of accuracy in textural characterization processes compared to other texture operators.
- A contrast measure (C): LBP descriptors efficiently capture the local spatial patterns. However, whereas LBP is invariant against any monotonic grey scale transformation, we must combine it with a simple contrast measure C to make it even more powerful. The computation of C is also addressed in Figure 4.

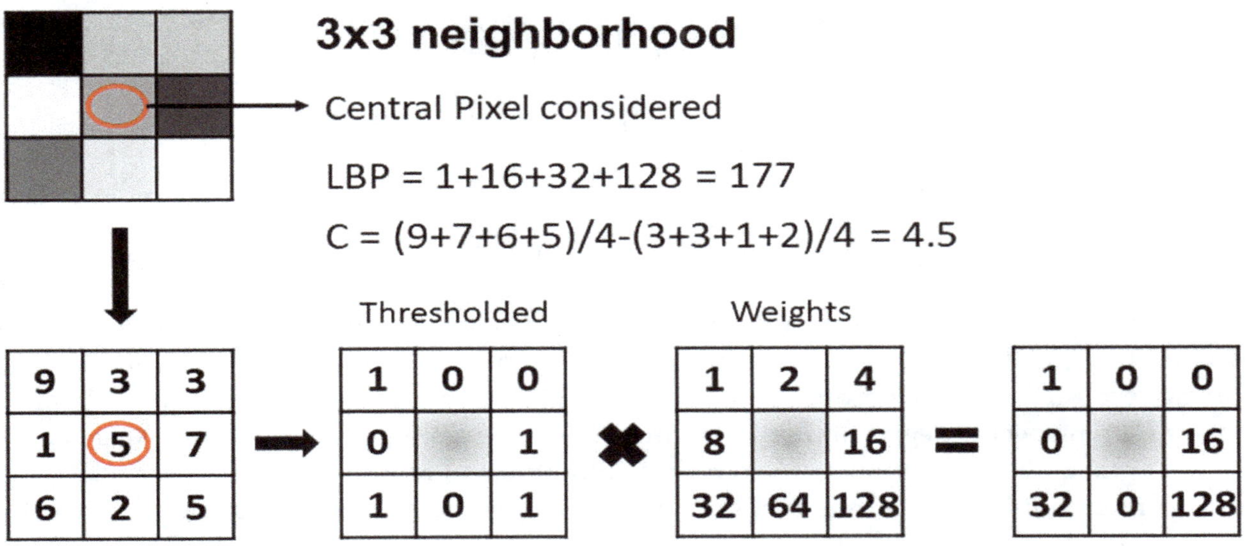

Figure 4. Computation of local binary patterns (LBP) and the contrast measure (C).

Six major steps define our FLA extraction procedure:

(1) Texture characterization

As mentioned in the previous paragraphs, texture is not an image property that may be associated with a single pixel. That is why the image was decomposed into a grid where each of its cell included a fixed number of pixels. Its LBP/C distribution was approximated by a discrete two-dimensional histogram (array) of 256 × b pixels, where b is the number of bins for contrast measure C (Figure 5).

This number of bins is chosen as a trade-off between the discriminative power and the stability of the texture description [17].

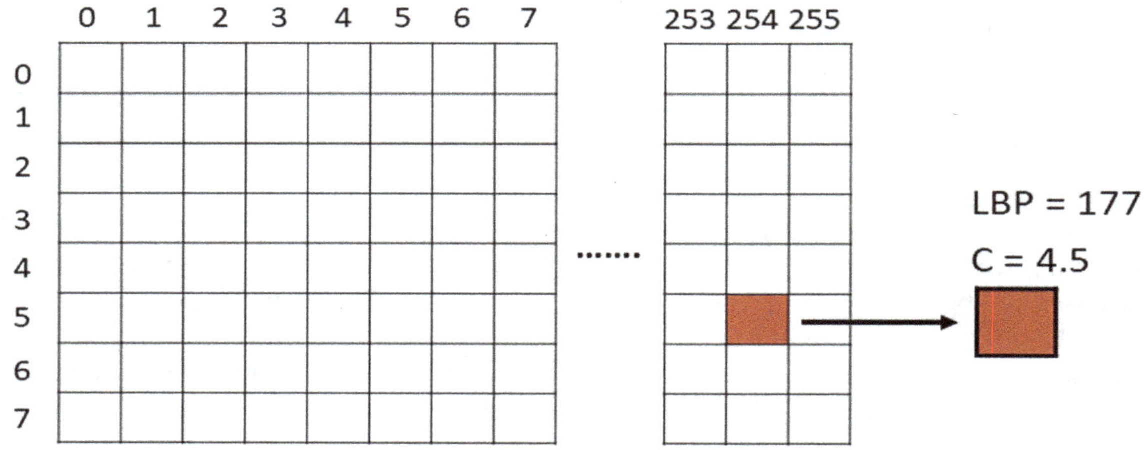

Figure 5. Bidimensional array of LBP/C.

Once these histograms were obtained, the fundamental objective was to compare *LBP/C* distributions identifying two textures as equal or different. The tool that allowed us to make this identification was the *G* statistic [26]. Compared with other similarity statistic measures, such as log-cumulative distance or Jeffrey divergence [17,23], which present low discrimination capability, the *G* value (Equation (1)) is a non-parametric log-probabilistic statistic commonly used to assess the similarity between two histograms (*A* and *B*) when a high level of discrimination is required. Thus, the higher the value, the lower the similarity between them. In addition, in Equation (1), *N* represents the number of bins and *fi* represents the frequency at bin *i*.

$$G = 2\left(\begin{array}{c} \left[\sum_{A,B}\sum_{i=0}^{N} f_i \log f_i\right] - \left[\sum_{A,B}\left(\sum_{i=0}^{N} f_i\right)\log\left(\sum_{i=0}^{N} f_i\right)\right] - \\ \left[\sum_{i=0}^{N}\left(\sum_{A,B} f_i\right)\log\left(\sum_{A,B} f_i\right)\right] + \left[\sum_{A,B}\sum_{i=0}^{N} f_i\right)\log\left(\sum_{A,B}\sum_{i=0}^{N} f_i\right)\Big] \end{array} \right) \quad (1)$$

(2) Hierarchical structure generation

With the main goal of optimizing the storage capacity and therefore improving the computation time, an approach based on a pyramidal representation was adapted to texture segmentation. Instead of performing image segmentation based on a single representation of the input image, a pyramid segmentation algorithm describes the contents of the image using multiple representations with decreasing resolution [27]. According to Marfil et al. [27], this type of algorithm exhibits interesting properties with respect to segmentation algorithms based on a single representation such as: (i) the detection of global features of interest and representing them at low resolution levels [28,29], (ii) the reduction of noise and the processing of local and global features within the same framework [30], and (iii) reduction in the complexity of the image segmentation task [28,29]. Specifically, we used a segmentation algorithm based on regular pyramids.

The general principle of our pyramidal approach was briefly described by Jolion and Montanvert [31]. According to that principle, a pyramidal architecture is completely defined if we specify how a new level is built (Figure 6a) and how a parent is linked to its children (Figure 6b). This parent-child relationship is defined by the reduction window and may be extended by transitivity down to the base level (Figure 6a). The set of children of one cell in the base level is named its receptive field [27], and it defines the embedding of this vertex in the original image. With regard to the efficiency of a pyramidal representation to solve a segmentation procedure, we must note that it is strongly influenced by the data structure used within the pyramid, limiting the information that may be encoded at each level, and the reduction scheme used, which defines its dynamics in terms of height and preservation of details [29].

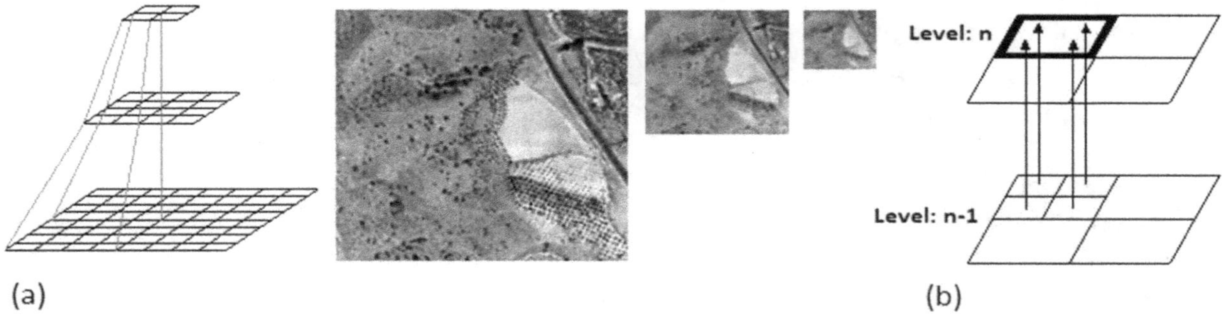

Figure 6. Hierarchical structure. (**a**) Set of pyramidal structures from texture; (**b**) parent link.

In our pyramidal structure, a cell, at level l, represents a $2^l \times 2^l$ square in the input image, and the base of the structure is composed by the *LBP/C* distributions. Each pyramid cell, denoted by *(x, y, l)*, has the following parameters associated with it:

- Homogeneity, denoted by *H(x, y, l)*. Homogeneity ranged from 1 (if the four cells immediately underneath had the same texture) to 0 (in any other case). The setting of *H* was based on a uniformity test. Thus, the four cells had the same texture if a measure of relative dissimilarity within that region was lower than a certain threshold U, ($G_{max}/G_{min} < U$). U must be set in such a way as to ensure the detection and differentiation of textures, preventing wherever possible the inclusion of two regions with different textures in the same class. For this reason, it is advisable to choose a value for U as small as possible.
- Texture, denoted by *T(x, y, l)*. The texture of one cell was calculated as the sum of the *LBP/C* distributions of the four cells immediately underneath if, and only if, the cell was homogeneous. Otherwise, the value of T(x, y, l) was set to a fixed value (T_{NH}).
- Parent link, denoted by $(X,Y)_{(x,y,l)}$. If $H_{(x,y,l)}$ was equal to 1, the values of the parent links of the four cells immediately underneath were set to (x, y). Otherwise, these four parent links were set to a null value.
- The centroid, denoted by $C_{(x,y,l)}$. $C_{(x,y,l)}$, represents the center of mass of the base region associated with (x, y, l).
- Histogram. Each parent link stored the two-dimensional histogram, which characterized the texture of the image region represented by this node. In order to optimize the storage capacity and to improve the computation time, if a node (located at the l level) represented a region of homogenous texture, all the nodes located at lower levels (until the end of the pyramid) did not store their corresponding histograms, their texture being characterized by the histogram stored in the parent link.

After completing the hieratical structure generation (Step Number 2), all the cells belonging to this structure with a homogeneity value equal to 1 and having no parent were linked to homogeneous regions at the base, defining initial image segmentation.

(3) Grow of homogeneous cells

In this step, the algorithm linked cells whose parent link values were null. Basically, a cell *(x, y, l)* was linked to the parent of its neighbors *(xp, yp, l+1)* when two cells had the same texture.

(4) Homogeneous cells fusion

The neighbor cells, $(x1, y1, l)$ and $(x2, y2, l)$, were merged if the following four conditions were true:

- $(X, Y)_{(x1,y1,l)}$ = null. Therefore, the cell had no parent.
- $(X, Y)_{(x2,y2,l)}$ = null. Therefore, the cell had no parent.
- The cells had a homogeneous texture. $H_{(x1,y1,l)} = 1$ & $H_{(x2,y2,l)} = 1$.
- The cells had the same texture.

(5) Pixel-wise

Since pixels at the base level of the pyramid were treated as blocks, the resolution of the segmented image R was represented by Equation (2), where S is the resolution of the initial image and A is the area of the pixels averaged to create a cell.

$$R = S/A \tag{2}$$

In order to improve the resolution of the output image and therefore to decrease the segmentation error, a post-processing step was developed. This post-processing was based on a type of segmentation procedure called soft segmentation [32], in which each pixel can belong to more than one region, thereby avoiding mistakes in region boundaries. When an image is segmented, pixels near region boundaries are usually intermediate in value between the regions, and they can be placed in either of them during the cell fusion [27]. Considering this, first, our algorithm recursively increased the resolution of all blocks in texture region boundaries until those boundaries were a pixel wide and then built the segmentation probability map. This probability map indicated the probability of each pixel that belonged to one of the predefined classes, and it was built on the basis of the texture parameter $(LBP, C)_{FLA}$ values learned from the training procedure explained in the next paragraph. From these data, several closed segmentation contours were generated, and the final textural segmentation was obtained.

(6) FLA zones' extraction

After resolving the segmentation issue, we had to identify and separate FLA zones from the rest. We had an image that contained multiple zones such as vegetation, constructions, etc. In this phase, we extracted and rebuilt the FLA zones from the image. In order to extract the FLA zones, we needed to remove the objects that did not possess the textural properties of FLA. Without losing generality, our main assumption about FLA on images was: unlike other vegetation zones or construction zones, which could have many different texture models, FLA zones usually had the same kind of texture model due to the representative properties of the terrain and vegetation that characterized them. Based on this assumption, we trained the system on different areas of the image to learn the FLA zones' texture parameters $(LBP, C)_{FLA}$; thereby generating a set of thresholds for these texture parameters. In this regard, it is very important to recall that although LBP achieves high levels of accuracy in textural characterization processes compared with other texture operators, these thresholds (linked to the training procedure) generate uncertainty during the classification process. In addition, and in order to increase as much as possible the robustness of the results, these training areas should not include the cadastral plots employed for evaluating the segmentation results (test plots) (described in the next subsection).

2.4. Evaluation of Segmentation Results

As several authors and studies have pointed out [17,18,23], it is not easy to assure that an algorithm provides a final "good" segmentation of the image. In general, "it is not clear what a 'good' segmentation is" [29], and it is necessary to have the opinion of a human expert who decides the final accuracy of the studied algorithm. Therefore, the measure of the segmentation quality depends on human intuition and could be different for distinct observers. In order to overcome this inconvenience, we developed a methodology that sought to avoid distortions and biases caused by human intervention in the process.

According to the classification established by Marfil et al. [27], our method could be classified as a discrepancy method. Discrepancy methods are quantitative and empirical procedures that use the ideal segmentation (reference data) for assessing the quality of the segmentation obtained with an algorithm (tested or assessed data) by counting the differences between both datasets. If these differences are expressed in terms of spatial position or location, then discrepancy methods have the same conceptual framework as the automatic procedure employed for assessing the positional accuracy of geospatial databases (using linear elements) developed by Ruiz-Lendínez et al. [33]. Specifically, these authors used the single buffer overlay method (SBOM) (Figure 7). Based on buffer generation on the line of the source of greater accuracy (Q) (reference data), this method determines the percentage of the controlled line (X) (tested or assessed data) that is within this buffer (Figure 7a). By increasing the width of the buffer, a probability distribution of inclusion of the controlled line inside the buffer of the source of greater accuracy is obtained (Figure 7c). Finally, Figure 7b shows the adaptation of this metric to the line-closed case carried out by Ruiz-Lendínez et al. [33].

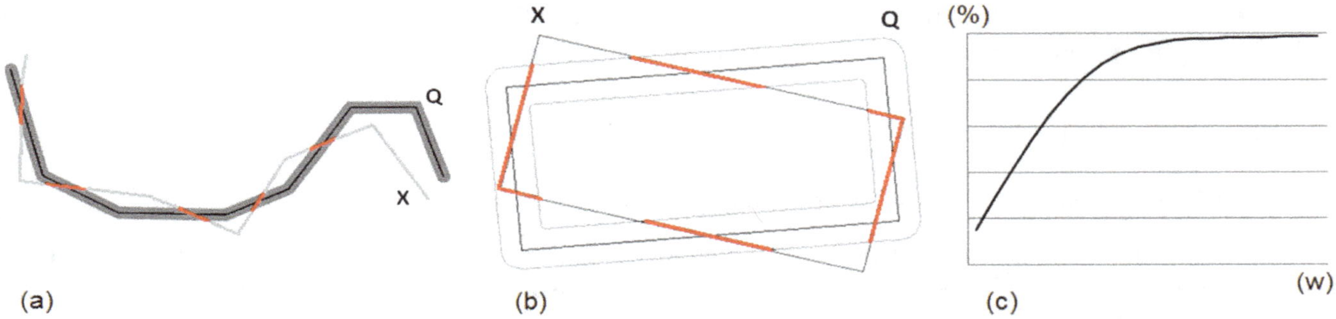

Figure 7. (**a**) Buffer generation by means of single buffer overlay method (SBOM). (**b**) Adaptation to the line-closed case (polygons or plots). (**c**) Empirical probabilistic distribution function.

In view of the above, discrepancy methods do not depend on human intuition, but in compensation, they present the problem that having a previous ideal segmentation is necessary. In many fields, research based on pattern recognition might be a problem. However, in the case of geographical data, finding a source of greater accuracy (reference data) is relatively simple. In our specific case, the ideal segmentation (reference data) is represented by a sample of cadastral plots (test plots) whose geometry is known (a priori) with high accuracy, and as test data, we used these same plots obtained by applying our abandoned farmland extraction algorithm. In this second case (test data), the boundaries of the test plots were obtained by means of a vectorization procedure.

Finally, in order to implement our methodology, we employed all the parcels present in the control sample, obtaining an aggregated distribution curve. The number and characteristics of this set of plots is specified in the next subsection.

2.5. *Evaluation of the Efficiency of Our Approach for Locating Abandoned Land*

As mentioned in the Introduction section, the effectiveness of the proposed methodology was assessed not only from a segmentation results perspective, but also taking into account the efficiency of our approach for locating the abandoned land. In this second case, the accuracy in training the algorithm to learn the FLA zone texture parameters $(LBP, C)_{FLA}$ was key to achieving an accurate land use classification, with LBP values able to appear out of the range characteristic of FLA zones, increasing uncertainty during the classification process. For this reason, and in order to check the land use, two inspection procedures were carried out: (i) an external validation based on a visual inspection procedure through field visits and (ii) a revision of the LBP/C parameters belonging to the sample of test plots. Since we had two different observations from two different inspection procedures, we applied Cohen's Kappa coefficient [34] in order to determine the land use. Cohen's Kappa coefficient *(k)* (Equation (3)) measures the concordance between two different observations in their corresponding classifications of N elements in C mutually exclusive categories. K is a more robust measure than the simple calculation of the percentage of concordance, since it takes into account the agreement that occurs by chance.

$$k = \frac{P_r(a) - P_r(e)}{1 - P_r(e)} \tag{3}$$

where $P_r(a)$ is the relative observed agreement between both observations and $P_r(e)$ is the hypothetical probability of agreement by chance using the observed data to calculate the probabilities of each observer (inspection procedure) randomly classifying each category. With regard to the k value interpretation: (i) if both observations fully agree, then $k = 1$, and (ii) if there is no agreement between the observations other than what would be expected by chance (as defined by $P_r(e)$), then $k = 0$ [34].

Finally, and with regard to the number and characteristics of the test plots, we selected 40 plots uniformly distributed over the study area. Therefore, $N = 40$ elements (sample of plots), and $C = 2$ categories (land classified as abandoned or land classified as other use). Obviously, in order to increase as much as possible the robustness of the results and avoid possible biases, these plots did not overlap with training areas employed for learning the FLA texture parameters.

3. Results

3.1. *Mapping Abandoned Farmland Derived From Textural Segmentation of the Aerial Image*

Based on a visual interpretation of the derived map, the model was able to capture the pattern of abandoned land in the areas affected by rapid urbanization. Figure 8 shows that: (i) abandoned arable land was located mainly in the northern and western parts of the city, which corresponded to the main direction of the urbanization process of Poznań; (ii) by comparing the results obtained with the data provided by CORINE Land Cover, 2275 ha (40.3%) of arable land within the city limits were abandoned; and (iii) the area of abandoned arable land was almost 9.2% of the city's area.

3.2. *Evaluation of Segmentation Results*

As mentioned in Section 2.3, we used SBOM (specifically, an adaptation of its metric to the line-closed case) to assess the quality of the segmentation obtained with our algorithm (Figure 9a). Figure 9b presents the resulting aggregated distribution function by applying SBOM on our two datasets, that is to say, plot boundary lines from the cadastral data and plot boundary lines from the results provided by our abandoned farmland extraction algorithm (segmentation boundaries) using buffers with widths from 1 to 5 m. It is important to note that in this second case, the boundaries of these test plots were obtained by means of a vectorization procedure.

The aggregated curve obtained with our method showed a distribution function of the uncertainty for several levels of confidence. Figure 9b shows a value around 2.4 m for a 95% level of confidence. Taking into account the resolution of the image and the uncertainty associated with the vectorization procedure, this value seemed to be a priori acceptable.

In any case, and in order to check the goodness of this measure, it was tested by means of the empirical evaluation method proposed by Liu and Yang [35], which was based on calculating the differences between the original image and the segmented image using the area of the segmented regions as a reference parameter. In accordance with this and taking into account the mean value of the test plot area, a variation in the width of the buffer of 2.4 m (±1.2 m) on the perimeter of the plots represented less than 0.35% of the segmentation surface.

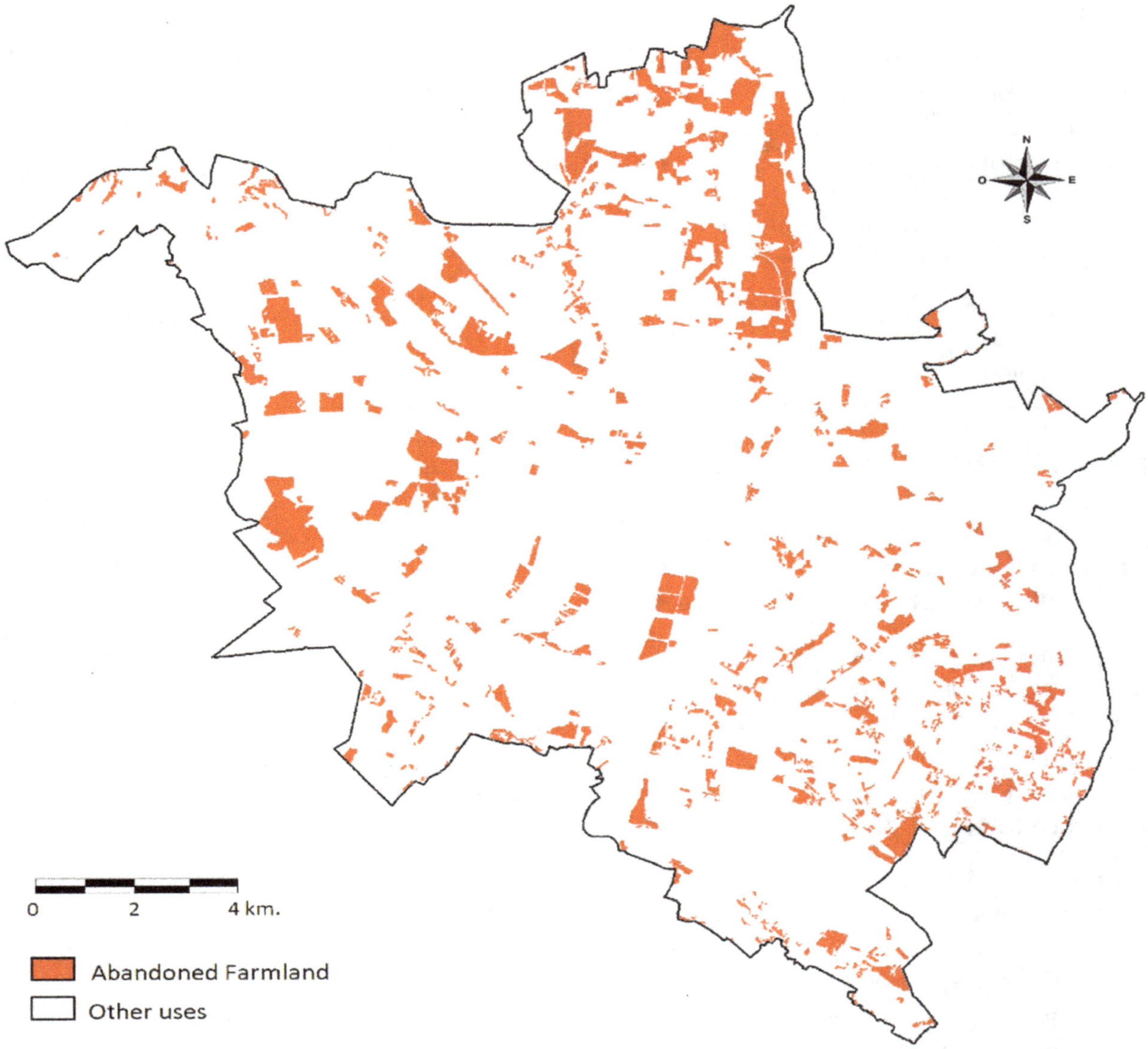

Figure 8. Map of abandoned land as derived from textural segmentation of the aerial image.

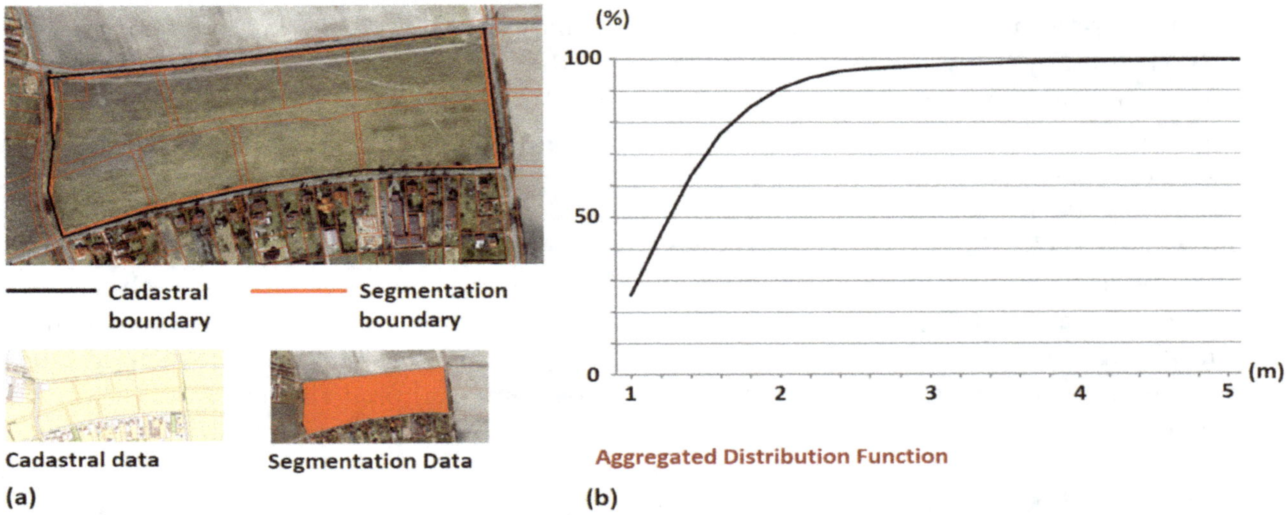

Figure 9. (**a**) Single buffer overlay method (SBOM) adapted to the line-closed case: the cadastral boundary represents the line of the source of greater accuracy (Q), and the segmentation boundary represents the controlled line (X); (**b**) aggregated distribution function.

3.3. Evaluation of the Efficiency of Our Approach for Locating Abandoned Land

As mentioned in Section 2.4, in order to determine the land use obtained from our derived map (Figure 8), we applied Cohen's Kappa coefficient in two different inspection procedures: (i) a visual inspection procedure through field visits and (ii) a revision of the LBP/C parameters belonging to the sample of test plots. Table 1 shows the results of the analysis of each inspection procedure.

Table 1. Results of the analysis of each inspection procedure.

		Field visit	
		Abandoned Land	**Other Uses**
Revision of LBP/C parameters	Abandoned Land	36	2
	Other uses	1	1

The values located on the first diagonal (36,1) represent the total number of plots in which there was agreement between both inspection procedures. The second diagonal, composed of the values of one and two, represent those plots in which there was disagreement between them. Taking into account these results, the percentage of agreement was 92.5%. In order to calculate the probability that the agreement between both inspection procedures was due to chance, we must take into account that:

- The number of plots classified as abandoned land by means of the field visit procedure was 37, and the number of plots classified as "other uses" was three. Consequently, by means of this procedure, 92.5% of plots were classified as abandoned land.
- The number of plots classified as abandoned land by means of revision of the LBP/C parameters was 38, and the number of plots classified as "other uses" was two. Consequently, by means of this procedure, 95% of plots were classified as abandoned land.

Therefore, the probability that both procedures classified the plots as abandoned land at random was 87.8%, while the probability that both procedures classified the plots as "other uses" at random was 0.37%. From this, we deduced that the probability that the agreement between both inspection procedures was due to random was 88.2%. Finally, applying Equation (3), we obtained a K value equal to 0.82. This value represented a high level of concordance between both observations (inspection procedures).

4. Discussion

This paper presented a quick and easy method for mapping abandoned land in cities affected by rapid urbanization by using textural characterization of high resolution aerial imagery. Our method, which employed free data provided by *World_Imagery (MapServer)* [22], was relevant because: (i) it produced land use maps under conditions in which time and human resources were scarce; and (ii) it had significant socioeconomic impact.

Concerning the first aspect (time), our approach was particularly efficient in comparison with other alternatives such as SRS. Mapping abandoned land using high resolution aerial imagery did not require the pre-processing phases or classification procedures inherent to satellite platforms. Such procedures were replaced by a training phase on different areas of the image to learn the FLA zones' texture parameters $(LBP, C)_{FLA}$, significantly reducing the computational time necessary to obtain the final map. With regard to the computation process itself, the computational efficiency of our algorithm depended on (i) the threshold U (see the section "Hierarchical Structure Generation") and (ii) the number of textural regions in the image. To achieve the segmentation of higher resolution images, the algorithm required more computation time because in this case, a value for U as small as possible was required.

As mentioned in the Introduction section, our approach to texture characterization was already successfully tested in other studies [19,20]. Although both their purpose and the features used differed from those addressed here, these studies were very useful when the relationship between computation time, image resolution, and threshold "U" was analyzed because the results obtained could be extrapolated to other types of segmentation processes. In our case, this was particularly useful for setting U.

Figure 10 shows the relationship obtained for the parameters mentioned above: time, U, and resolution. According to the values shown in Figure 10 and taking into account the resolution of our image (1 m), the value for the threshold U was set to 1.5. Finally, the computation time obtained for our case of study was 15.3 s. This value was higher than expected. This was due to: (i) the large size of our study area and (ii) the need to process the entire image because in our case, it was not possible to apply localized spatial filters, such as localized template matching (LTM) [36]. In this respect, we must start pointing out that our test platform was an Intel Core i5 3.8 GHz with 16 GB of memory.

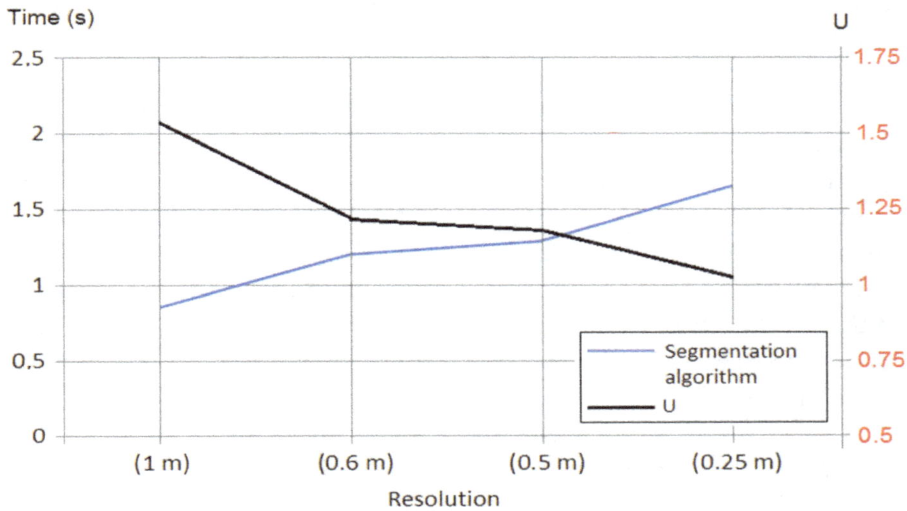

Figure 10. Computation time vs. image resolution and U.

Concerning socioeconomic impact, locating trends and patterns of FLA could help authorities in developing urban policies, providing objective information to municipalities where official statistics on land abandonment are unreliable or of low quality. Our approach could be used by local authorities that cannot allocate significant resources for land change monitoring. In addition, it provided a useful tool for spatial management, as well as for cadastral tax purposes and could be applied to any city, metropolitan area, or spatial region.

5. Conclusions

This study helped to understand how the textural characterization of high resolution aerial imagery could be used for locating and mapping FLA. We employed a textural segmentation algorithm that had already been successfully tested in other types of processes based on the automatic extraction of geographical entities from imagery. As in those processes, our algorithm proved to be an effective and valuable tool for locating and extracting abandoned farmland from imagery in areas affected by rapid urbanization. Among all local texture descriptors, we employed texture descriptors based on LBP combined with a simple contrast measure to make our method more powerful. LBP provided us with knowledge about the spatial structure of the local image texture and was very efficient from a computational point of view. In addition, LBP achieved high levels of accuracy in textural characterization processes compared with other texture operators. With regard to the structure, our algorithm was implemented on a pyramidal representation, which significantly reduced computation time and assured robustness.

The effectiveness of the proposed methodology was assessed from two different perspectives: (i) the accuracy of the segmentation results and (ii) the efficiency for locating FLA. In the first case, we developed an innovative methodology based on the procedures employed for assessing the positional accuracy of geospatial databases. This methodology provided us an uncertainty value (of the segmented plots) of around 2.4 m with regard to the source of greater accuracy employed (cadastral data). In the second case, we employed two different empirical inspection procedures, measuring (and confirming) the concordance between them.

With regard to the results, the model obtained allowed us to identify the patterns of abandoned land in the city. These patterns corresponded to the main direction of the urbanization process of Poznań, that is to say to those areas affected by rapid urbanization in the city. The area of abandoned arable land was almost 9.2% of the city's area. Finally, comparing the results obtained with the data provided by CORINE Land Cover, 2275 ha (40.3%) of arable land within the city limits were abandoned.

Finally, we must note that our study could support local authorities in adapting urban planning for a more efficient use of land.

Finally, this study represents only a further step towards the improvement in the location and management of abandoned land. Thus, although our algorithm solved the problem in an effective way, the inspection procedures carried out revealed some inconsistences during the process of classification, possibly derived from the training phase (developed to learn the FLA zones' texture parameters $(LBP, C)_{FLA}$). In this sense, the use of artificial intelligence mechanisms could help to classify FLA zones much more efficiently. This would imply the inclusion in the process of other texture descriptors that would increase the levels of discrimination. On the other hand, although our approach identified FLA zones, it was unable to determine the quality of the soil they occupied, which made it impossible to assess the real scope of the problem.

All these challenges must be addressed in future works. Thus, we plan the incorporation of genetic algorithms (GA) in the learning process mentioned above, adding new (and more powerful) texture descriptors. In this respect, some previous studies [37] already demonstrated that GA was an appropriate and efficient learning method for problems where spatial variables were implied. With regard to the quality of the soil occupied by FLA zones, we plan to compare our map of abandoned land with maps of soil quality provided by local and regional authorities. This will allow us to verify if a significant part of arable land that is of the biggest importance for agriculture is still being cultivated in the city or, conversely, that these soils should be urgently protected.

References

1. Pointereau, P.; Coulon, F.; Girard, P.; Lambotte, M.; Stuczynski, T.; Sanchez-Ortega, V.; Del Rio, A. *Analysis of Farmland Abandonment and the Extent and Location of Agricultural Areas that are Actually Abandoned or are in Risk to be Abandoned*; Anguiano, E., Bamps, C., Terres, J., Eds.; Institute for Environment and Sustainability, Joint Research Centre, European Commission: Luxembourg, 2008.
2. Grădinaru, S.; Kienast, F.; Psomas, A. Using multi-seasonal Landsat imagery for rapid identification of abandoned land in areas affected by urban sprawl. *Ecol. Indic.* **2019**, *96*, 79–86. [CrossRef]
3. Benayas, J.; Martins, A.; Nicolau, J.; Schulz, J. Abandonment of agricultural land: An overview of drivers and consequences. *CAB Rev. Perspect. Agric. Vet. Sci. Nutr. Nat. Resour.* **2007**, *2*, 1–14. [CrossRef]
4. Morán, N.; Hernández, V.; Zazo, A.; Simón, M. Multifuncionalidad, preservación y retos futuros de la agricultura peri-urbana en la Europa mediterránea. In *Agricultura Urbana Integral, Ornamental y Alimentaria*; Briz, J., De Felipe, I., Eds.; Ministerio de Agricultura, Alimentación y Medio Ambiente Press: Madrid, Spain, 2015; pp. 153–170. [CrossRef]
5. Wulder, M.; Hall, R.; Coops, N.; Franklin, S. High spatial resolution remotely sensed data for ecosystem characterization. *Bioscience* **2004**, *54*, 511–521. [CrossRef]
6. Huzui, A.; Abdelkader, A.; Pătru -Stupariu, I. Analysing urban dynamics using multi-temporal satellite images in the case of a mountain area, Sinaia (Romania). *Int. J. Digit. Earth* **2011**, *6*, 1–17.

7. Kolecka, N.; Kozak, J.; Kaim, D.; Dobosz, M.; Ginzler, C.; Psomas, A. Mapping secondary forest succession on abandoned agricultural land with LiDAR point clouds and terrestrial photography. *Remote Sens.* **2015**, *7*, 8300–8322. [CrossRef]
8. Grădinaru, S.; Lojă, C.; Onose, D.; Gavrilidis, A.; Pătru-Stupariu, I.; Kienast, F.; Hersperger, A. Land abandonment as precursor of built-up development at the sprawling periphery of former socialist cities. *Ecol. Indic.* **2015**, *57*, 305–313. [CrossRef]
9. Estel, S.; Kuemmerle, T.; Alcántara, C.; Levers, C.; Prishchepov, A.; Hostert, P. Mapping farmland abandonment and recultivation across Europe using MODIS NDVI time series. *Remote Sens. Environ.* **2015**, *163*, 312–325. [CrossRef]
10. Liu, N.; Harper, R.; Handcock, R.; Evans, B.; Sochacki, S.; Dell, B.; Walden, L.; Liu, S. Seasonal Timing for Estimating Carbon Mitigation in Revegetation of Abandoned Agricultural Land with High Spatial Resolution Remote Sensing. *Remote Sens.* **2017**, *9*, 545. [CrossRef]
11. Stryjakiewicz, T.; Królewicz, S.; Ruiz-Lendinez, J.J.; Mickiewicz, B.; Motek, P. Abandoned agricultural land quantification in urban areas using high resolution satellite imagery. In Proceedings of the RSA Central and Eastern Europe Conference: Metropolises and Peripheries of CEE Countries: New Challenges for EU, National and Regional Policies, Lublin, Poland, 11–13 September 2019.
12. Baumann, M.; Kuemmerle, T.; Elbakidze, M.; Ozdogan, M.; Radeloff, V.C.; Keuler, N.S.; Prishchepov, A.V.; Kruhlov, I.; Hostert, P. Patterns and drivers of post-socialist farmland abandonment in Western Ukraine. *Land Use Policy* **2011**, *28*, 552–562. [CrossRef]
13. Prishchepov, A.; Radeloff, V.; Baumann, M.; Kuemmerle, T.; Muller, D. Effects of institutional changes on land use: Agricultural land abandonment during the transition from state-command to market-driven economies in post-Soviet Eastern Europe. *Environ. Res. Lett.* **2012**, *7*, 024021. [CrossRef]
14. Prishchepov, A.; Radeloff, V.; Dubinin, M.; Alcantara, C. The effect of Landsat ETM/ETM + image acquisition dates on the detection of agricultural land abandonment in Eastern Europe. *Remote Sens. Environ.* **2012**, *126*, 195–209. [CrossRef]
15. Bazzi, H.; Baghdadi, N.; El Hajj, M.; Zribi, M.; Minh, D.H.T.; Ndikumana, E.; Courault, D.; Belhouchette, H. Mapping Paddy Rice Using Sentinel-1 SAR Time Series in Camargue, France. *Remote Sens.* **2019**, *11*, 887. [CrossRef]
16. Carrasco, L.; O'Neil, A.; Morton, R.; Rowland, C. Evaluating Combinations of Temporally Aggregated Sentinel-1, Sentinel-2 and Landsat 8 for Land Cover Mapping with Google Earth Engine. *Remote Sens.* **2019**, *11*, 288. [CrossRef]
17. Ojala, T.; Pietikäinen, M. Unsupervised texture segmentation using feature distributions. *Pattern Recognit.* **1999**, *32*, 477–486. [CrossRef]
18. Maenpaa, T.; Pietikäinen, M. Texture analysis with local binary patterns. In *Handbook of Pattern Recognition and Computer Vision*; Chen, C., Wang, P., Eds.; University of Massachusetts Dartmouth Press: Dartmouth, MA, USA, 2005; pp. 197–216. [CrossRef]
19. Ruiz-Lendínez, J.J.; Rubio-Campos, T.J.; Ureña-Cámara, M.A. Automatic extraction of road intersections from images based on texture characterization. *Surv. Rev.* **2011**, *43*, 212–225. [CrossRef]
20. Ruiz-Lendínez, J.J.; Maćkiewicz, B.; Motek, P.; Stryjakiewicz, T. Method for an automatic alignment of imagery and vector data applied to cadastral information in Poland. *Surv. Rev.* **2019**, *51*, 123–134. [CrossRef]
21. Kuemmerle, T.; Muller, D.; Griffiths, P.; Rusu, M. Land use change in Southern Romania after the collapse of socialism. *Reg. Environ. Change* **2009**, *9*, 1–12. [CrossRef]
22. World Imagery. Available online: http://goto.arcgisonline.com/maps/World_Imagery (accessed on 10 February 2020).
23. Pietikäinen, M.; Zhao, G. Two decades of local binary patterns: A survey. In *Advances in Independent Component Analysis and Learning Machines*; Academic Press: Oxford, UK, 2015; pp. 175–210.
24. Malhotra, A.; Sankaran, A.; Mittal, A.; Vatsa, M.; Singh, R. Fingerphoto Authentication Using Smartphone Camera Captured Under Varying Environmental Conditions. In *Using Computer Vision, Pattern Recognition and Machine Learning Methods for Biometrics*; Academic Press: Oxford, UK, 2017; pp. 119–144.
25. Dornaika, F.; Moujahid, A.; El Merabet, Y.; Ruichek, Y. A Comparative Study of Image Segmentation Algorithms and Descriptors for Building Detection. In *Handbook of Neural Computation*; Academic Press: Oxford, UK, 2017; pp. 591–606.
26. Sokal, R.; Rohlf, F. *Introduction to Biostatistics*; W.H. Freeman & Co. Ltd: Gordonsville, Virginia, USA, 1987.

27. Marfil, R.; Molina-Tanco, L.; Bandera, A.; Rodríguez, J.; Sandoval, F. Pyramid segmentation algorithms revisited. *Pattern Recognit.* **2006**, *39*, 1430–1451. [CrossRef]
28. Cho, K.; Meer, P. Image segmentation from consensus information. *Comput. Vis. Image Underst.* **1997**, *68*, 72–89. [CrossRef]
29. Kropatsch, W.; Haxhimusa, Y. Grouping and segmentation in a hierarchy of graphs. In *Computational Imaging II*; SPIE Press; Digital Library: Bellingham, WA, USA, 2004; pp. 193–204.
30. Bister, M.; Cornelis, J.; Rosenfeld, A. A critical view of pyramid segmentation algorithms. *Pattern Recognit. Lett.* **1990**, *11*, 605–617. [CrossRef]
31. Jolion, J.M.; Montanvert, A. The adaptive pyramid, a framework for 2D image analysis. *Comput. Vis. Image Underst.* **1992**, *55*, 339–348. [CrossRef]
32. Prewer, D.; Kitchen, L.J. Soft image segmentation by weighted linked pyramid. *Pattern Recognit. Lett.* **2001**, *22*, 123–132. [CrossRef]
33. Ruiz-Lendínez, J.J.; Ariza-López, F.J.; Ureña-Cámara, M.A. Automatic positional accuracy assessment of geospatial databases using line-based methods. *Surv. Rev.* **2013**, *45*, 332–342. [CrossRef]
34. Cohen, J. A coefficient of agreement for nominal scales. *Educ. Psychol. Meas.* **1960**, *20*, 37–46. [CrossRef]
35. Liu, J.; Yang, Y. Multi-resolution color image segmentation. *IEEE Trans. Pattern Anal. Mach. Intell.* **1994**, *16*, 689–700.
36. Chen, C.; Knoblock, C.; Shahabi, C. Automatically conflating road vector data with orthoimagery. *Geoinformatica* **2006**, *10*, 495–530. [CrossRef]
37. Ruiz-Lendínez, J.J.; Ariza-López, F.J.; Ureña-Cámara, M.A. A polygon and point-based approach to matching geospatial features. *ISPRS Int. J. Geo-Inf.* **2017**, *6*, 399. [CrossRef]

PERMISSIONS

All chapters in this book were first published in MDPI; hereby published with permission under the Creative Commons Attribution License or equivalent. Every chapter published in this book has been scrutinized by our experts. Their significance has been extensively debated. The topics covered herein carry significant findings which will fuel the growth of the discipline. They may even be implemented as practical applications or may be referred to as a beginning point for another development.

The contributors of this book come from diverse backgrounds, making this book a truly international effort. This book will bring forth new frontiers with its revolutionizing research information and detailed analysis of the nascent developments around the world.

We would like to thank all the contributing authors for lending their expertise to make the book truly unique. They have played a crucial role in the development of this book. Without their invaluable contributions this book wouldn't have been possible. They have made vital efforts to compile up to date information on the varied aspects of this subject to make this book a valuable addition to the collection of many professionals and students.

This book was conceptualized with the vision of imparting up-to-date information and advanced data in this field. To ensure the same, a matchless editorial board was set up. Every individual on the board went through rigorous rounds of assessment to prove their worth. After which they invested a large part of their time researching and compiling the most relevant data for our readers.

The editorial board has been involved in producing this book since its inception. They have spent rigorous hours researching and exploring the diverse topics which have resulted in the successful publishing of this book. They have passed on their knowledge of decades through this book. To expedite this challenging task, the publisher supported the team at every step. A small team of assistant editors was also appointed to further simplify the editing procedure and attain best results for the readers.

Apart from the editorial board, the designing team has also invested a significant amount of their time in understanding the subject and creating the most relevant covers. They scrutinized every image to scout for the most suitable representation of the subject and create an appropriate cover for the book.

The publishing team has been an ardent support to the editorial, designing and production team. Their endless efforts to recruit the best for this project, has resulted in the accomplishment of this book. They are a veteran in the field of academics and their pool of knowledge is as vast as their experience in printing. Their expertise and guidance has proved useful at every step. Their uncompromising quality standards have made this book an exceptional effort. Their encouragement from time to time has been an inspiration for everyone.

The publisher and the editorial board hope that this book will prove to be a valuable piece of knowledge for researchers, students, practitioners and scholars across the globe.

LIST OF CONTRIBUTORS

Gianluigi Busico and Elisabetta Giuditta
DiSTABiF—Department of Environmental, Biological and Pharmaceutical Sciences and Technologies, Campania University "Luigi Vanvitelli", Via Vivaldi 43, 81100 Caserta, Italy

Nerantzis Kazakis
Department of Geology, Laboratory of Engineering Geology & Hydrogeology, Aristotle University of Thessaloniki, 54124 Thessaloniki, Greece

Nicolò Colombani
Department of Materials, Environmental Sciences and Urban Planning, Polytechnic University of Marche, Via Brecce Bianche 12, 60131 Ancona, Italy

Griselda Vázquez-Quintero and María C. Valles-Aragón
Facultad de Ciencias Agrotecnológicas, Universidad Autónoma de Chihuahua, Ave. Pascual Orozco s/n, Campus, Chihuahua, Chihuahua 31350, Mexico

Jesús A. Prieto-Amparán, Alfredo Pinedo-Alvarez, Carlos R. Morales-Nieto and Federico Villarreal-Guerrero
Facultad de Zootecnia y Ecología, Universidad Autónoma de Chihuahua, Periférico Francisco R. Almada Km 1, Chihuahua, Chihuahua 31453, Mexico

Recep Çelik
Eng. Faculty Civil Eng. Dpt, Dicle University, 21280 Diyarbakır, Turkey

Zhoufeng Wang, Xiangqi He, Chen Zhang and Jianwei Xu
School of Geoscience and Technology, Southwest Petroleum University, Chengdu 610500, China

Yujun Wang
Hefei Surveying and mapping Institute, Hefei 230031, China

Ting Zhou, Anyi Niu, Jiaojiao Ma and Songjun Xu
School of Geography, South China Normal University, Guangzhou 510000, China

Zhanpeng Huang
International Envirotech Limited, Hong Kong 999077, China

Jianjun Chen
College of Geomatics and Geoinformation, Guilin University of Technology, 12 Jiangan Road, Guilin 541004, China

Xuning Zhao and Huizi Zhang
College of Geomatics and Geoinformation, Guilin University of Technology, 12 Jiangan Road, Guilin 541004, China

Yu Qin
State Key Laboratory of Cryospheric Sciences, Northwest Institute of Eco-Environment and Resources, Chinese Academy of Sciences, 320 Donggang West Road, Lanzhou 730000, China

Shuhua Yi
Institute of Fragile Ecosystem and Environment, Nantong University, 999 Tongjing Road, Nantong 226007, China
School of Geographic Sciences, Nantong University, 999 Tongjing Road, Nantong 226007, China

Rafael M. Navarro Cerrillo, Guillermo Palacios Rodríguez, Inmaculada Clavero Rumbao and Miguel Ángel Lara
Department of Forestry Engineering, Laboratory of Silviculture, Dendrochronology and Climate Change, DendrodatLab-ERSAF, Instituto Interuniversitario de Investigación del Sistema Tierra de Andalucía (IISTA), University of Cordoba, Campus de Rabanales, Crta.IV, km. 396, E-14071 Córdoba, Spain

Francisco Javier Bonet
Department of Ecology, University of Cordoba, Campus de Rabanales, Crta.IV, km. 396, E-14071 Córdoba, Spain

Francisco-Javier Mesas-Carrascosa
Department of Graphic Engineering and Geomatics, University of Cordoba, Campus de Rabanales, Crta. IV, km. 396, E-14071 Córdoba, Spain

Caihong Ma
State Key Laboratory of Remote Sensing Science, Institute of Remote Sensing and Digital Earth, Chinese Academy of Sciences, Beijing 100101, China
Remote Sensing and Digital Earth, Chinese Academy of Sciences, Beijing 100094, China
Sanya Institute of Remote Sensing, Sanya 572029, China

Zheng Niu
State Key Laboratory of Remote Sensing Science, Institute of Remote Sensing and Digital Earth, Chinese Academy of Sciences, Beijing 100101, China
Remote Sensing and Digital Earth, Chinese Academy of Sciences, Beijing 100094, China

Yan Ma, Fu Chen, Jin Yang and Jianbo Liu
Remote Sensing and Digital Earth, Chinese Academy of Sciences, Beijing 100094, China

Przemysław Kupidura, Katarzyna Osińska-Skotak, Katarzyna Lesisz and Anna Podkowa
Department of Photogrammetry, Remote Sensing and Spatial Information Systems, Faculty of Geodesy and Cartography, Warsaw University of Technology, 1, 00-661 Warsaw, Poland

Yijia Xiao, Yanming Chen, Xiaoqiang Liu, Zhaojin Yan, Liang Cheng and Manchun Li
Jiangsu Provincial Key Laboratory of Geographic Information Science and Technology, Key Laboratory for Land Satellite Remote Sensing Applications of Ministry of Natural Resources, School of Geography and Ocean Science, Nanjing University, Nanjing 210023, China
Collaborative Innovation Center of South China Sea Studies, Nanjing 210023, China

Ting Wang, Ziqing Xia, Yiping Peng and Zhenhua Liu
College of Natural Resources and Environment, South China Agricultural University, Guangzhou 510642, China

Yonghua Qu
State Key Laboratory of Remote Sensing Science, Jointly Sponsored by Beijing Normal University and the Institute of Remote Sensing Applications of CAS, Beijing 100875, China
Beijing Key Laboratory for Remote Sensing of Environment and Digital Cities, Beijing Normal University, Beijing 100875, China
School of Geography, Beijing Normal University, Beijing 100875, China

Ayesha Maqbool
Department of Computer Software Engineering, Military College of Signal, National University of Sciences and Technology, Islamabad 44000, Pakistan

Zain ul Abideen Usmani
Department of Information Security, Military College of Signal, National University of Sciences and Technology, Islamabad 44000, Pakistan

Farkhanda Afzal and Alia Razia
Department of Humanities & Basic Sciences, Military College of Signal, National University of Sciences and Technology, Islamabad 44000, Pakistan

José Lázaro Amaro-Mellado
Department of Graphic Engineering, University of Seville, 41092 Seville, Spain
National Geographic Institute of Spain, Andalusia Division, 41013 Seville, Spain

Dieu Tien Bui
Geographic Information System Group, Department of Business and IT, University of Southeast Norway, Gulbringvegen 36, N-3800 Bø i Telemark, Norway

Juan José Ruiz-Lendínez
Departamento de Ingeniería Cartográfica, Geodésica y Fotogrametría. Escuela Politécnica Superior de Jaén, Universidad de Jaén, 23071 Jaén, Spain

Index

www.ingramcontent.com/pod-product-compliance
Lightning Source LLC
LaVergne TN
LVHW081012210926
848485LV00004B/846